From Pole to Pole

Series Editors

Roland Kallenborn
Guido di Prisco
David Walton
Susan Barr

For further volumes:
http://www.springer.com/series/7132

Cinzia Verde · Guido di Prisco
Editors

Adaptation and Evolution in Marine Environments, Volume 2

The Impacts of Global Change on Biodiversity

Springer

Editors
Cinzia Verde
Institute of Protein Biochemistry
CNR
Napoli
Italy

Guido di Prisco
Institute of Protein Biochemistry
CNR
Napoli
Italy

ISSN 2193-7338 ISSN 2193-7346 (electronic)
ISBN 978-3-642-44888-1 ISBN 978-3-642-27349-0 (eBook)
DOI 10.1007/978-3-642-27349-0
Springer Heidelberg New York Dordrecht London

Printed on acid-free paper

Springer is part of Springer Science+Business Media (www.springer.com)

Preface

The Series "From Pole to Pole: Polar Environmental Research during the International Polar Year 2007–2009" was conceived to report achievements of environmental research during the 4th International Polar Year (IPY) 2007–2009. The major aim of this series is to provide updated science-based information on IPY research results and perspectives in all environmental disciplines. This multi disciplinary, multifaceted and international book series aims to ensure that polar science is a leading element in the new science of the twenty-first century.

The scientific value of the series will grow in the years to come, as the volumes will also be available in e-book format, and a continuous update on references and information sources is expected for several years, supported by the Series Editors and the Publisher.

Marine Biology is providing two Volumes under the general title "Adaptation and Evolution in Marine Environments—The Impacts of Global Change on Biodiversity". Volume 1, which has already been published, has provided a total of 11 contributions, assembled (besides an Introductory Overview) into two themes: Biodiversity and the Environment, and Response to Stress—Adaptations. This volume (Volume 2) contains 11 contributions, collected within three themes:

1. Part I: Biodiversity Evolution and Data Management
2. Part II: Evolution—A Molecular Perspective
3. Part III: Monitoring and Management.

The authors describe the concept, aims and first findings of the respective IPY projects, providing information on results and research perspectives feeding into the framework of IPY 2007–2009. Each contribution is equipped with exhaustive reference lists and relevant web page addresses. We convey our thanks to the authors and reviewers.

In a rapidly changing world with rising sea levels, exhausted fisheries, growing marine pollution and increasing global population demands, the basis for international management must be a realistic assessment of our scientific knowledge. The data from this IPY are making a crucial contribution to how we must approach the future management of our polar seas to avoid or at least minimise permanent damage to stocks, systems and functionality in the twenty-first century.

These two volumes constitute an extensive summary of the outstanding contribution to the scientific outcome of IPY provided by the marine biology projects in the polar regions. In addition they highlight the fact that this IPY will be remembered as an extraordinary international venture that has set a new model of global collaboration for complex multi-disciplinary and inter-disciplinary initiatives and for collaborations between national polar institutions that will resonate for many years to come.

Cinzia Verde
Guido di Prisco

Letter from the Editorial Team

The first two International Polar Years both failed to coordinate and distribute their assembled data adequately and to ensure its proper analysis, resulting in a less than satisfactory legacy from what had been considerable international efforts. Recognising this, the Third International Polar Year (International Geophysical Year) made extensive plans to ensure its contributions would be both accessible and used, establishing the World Data Centres as a major new initiative. In the early preparatory stages of the latest International Polar Year (IPY 2007–2009) the importance of providing for the legacy of this demanding international research effort was made clear, with priority being given to planning for well-organised dissemination and coordinated publication of the results, data evaluations and scientific findings. It was with this in mind that we proposed our publication project (IPY Project No. 79) in the form of the book series "From Pole to Pole: Environmental Research within the International Polar Year 2007–2009". With over 50,000 scientists involved in a myriad of projects, there was an obvious need for a guide to the principal findings and the key papers within environmental science fields.

The "From Pole to Pole" book series is intended to serve as a comprehensive publication framework for the documentation of environmental research activities performed during the IPY period. The book series is not intended to be a typical collection of original scientific project publications/chapters in the form of standard monographs. It is rather a bibliographic, science-based information source and a starting point for interested scientists and the public to access condensed information on specific environmental research topics within the IPY activities. The volumes will provide scientifically sound general information on the concepts, findings, and scientific motivation of the various relevant research activities and will direct the interested reader to more detailed scientific papers, web-based information and other publications which will provide the detailed data and their analyses. The compilation of citations and references within the book volumes will be an important component for the assessment of progress in each area, and the scientific significance and value will grow as the series develops.

The volumes will also be available in e-book format which will allow continuous updating of references and information sources (including internet pages and databases) by the editorial team on an annual basis, thus keeping the works topical as a living reference source.

Eleven volumes are currently planned for this series and will cover an extensive spectrum of environmental research including Adaptation and Evolution, Geomonitoring, Geology, Cryospheric Processes, Polar Biodiversity, Polar Climates, the Arctic and Southern Oceans, as well as Pollution Monitoring. It is expected that this documentation will provide a comprehensive picture of most of the environmental research performed within the IPY framework.

During the Oslo (2010) and Montreal (2012) IPY Science Conferences, scientific findings and implications were presented and evaluated. The outcome of these symposia made it very clear that the IPY efforts have contributed to a new and comprehensive understanding of global environmental processes, from both social and natural science perspectives. It was also clear that it would require continued efforts to make sure that the results of the IPY research would be easily available and properly documented for future research and evaluation processes.

This book series aims to make an important contribution to that documentation process. The editorial team is looking forward not only to assisting in the development of those volumes already planned, but also invites colleagues and experts to propose other topics not yet covered as potential volumes in the series "From Pole to Pole: Environmental Research within the International Polar Year 2007–2009".

With the initial published volume on the history of the International Polar Years (edited by Susan Barr and Cornelia Lüdecke), our concept has finally begun to be realised, and it is now being followed by the steady completion of other volumes. Marine Biology is providing two volumes on "Adaptation and Evolution in Marine Environments—The Impacts of Global Change on Biodiversity". Volume 1 has already been published and contains 11 chapters, assembled with an Introductory Overview under the themes Biodiversity and the Environment and Response to Stress—Adaptations. Volume 2 also contains 11 contributions assembled under three themes in which authors discuss the aims and findings of the respective IPY projects, describing results and perspectives feeding into IPY 2007-2009.

These two volumes provide an extensive summary of the outstanding contribution to IPY of marine biology at the poles. This contribution further highlights the IPY model of international and multidisciplinary collaboration that must be the future for understanding and managing the polar areas.

The editors and authors have completed an outstanding scientific compendium on IPY research in this field, demonstrating the extraordinary research conducted and coordinated in Polar Regions during the latest International Polar Year (2007–2009).

Ås, Norway, July 2012 — Roland Kallenborn
Naples, Italy — Guido di Prisco
Oslo, Norway — Susan Barr
Cambridge, UK — David Walton

Editorial Introduction

Over the past 130 years, scientists of all disciplines from around the world have joined forces in cooperative activities for explorating and investigating the polar regions in four occasions. Each was labelled as an "International Polar Year" (IPY), and produced advances in exploration and scientific knowledge, expanding the understanding of phenomena that influence the planet and paving the way to political agreements among governments.

The first book of this Series, "The History of the International Polar Years (IPYs)", edited by Susan Barr and Cornelia Lüdecke, is an excellent historical overview of the work and implications of IPY research activity in the past.

The Editorial Introduction of the second book of this Series, "Adaptation and Evolution in Marine Environments—The Impacts of Global Change on Biodiversity" Volume 1, summarised the previous IPYs, helping the reader to place IPY 2007–2009 into context and attempting to identify historical and international frames, before describing the concepts relevant to IPY of 11 contributions, assembled (besides an Introductory Overview) into two themes: Biodiversity and the Environment, and Response to Stress—Adaptations. This summary highlighted:

- the pioneering role of Karl Weyprecht, scientist, Austrian explorer and naval officer, in the 1st IPY (1882–1883). He felt that polar investigations could not be tackled by a single nation but needed coordinated international efforts. His belief set a legacy for the future IPYs, but Weyprecht had no chance to see the acceptance of his concepts, because he died before IPY.
- the 2nd IPY (1932–1933), with a network of stations in both polar regions and a world data centre, created and coordinated by the International Meteorological Organisation.
- the "International Geophysical Year" (IGY, 1957–8), the 3rd IPY, occurring 75 years after the 1st IPY and 25 years after the 2nd IPY. It envisaged the peaceful use of newly developed technologies. Continental drift was confirmed, enabling us to understand the formation of continents and oceans. The Van Allen Radiation Belt was discovered by a US satellite. The space age began with

the first satellites. The first estimates of the size of Antarctica's ice sheet were obtained by traversing the continent for the first time. World Data Centres were established to promote the concept of data sharing and re-use. The scientific, institutional and political legacies of IGY lasted for decades, providing countless science achievements, and continue to the present. A notable political result was the ratification of the Antarctic Treaty in 1961, which established that Antarctica would be dedicated to peaceful research.

During each of the three IPYs, scientists from all over the world together organised intensive scientific and exploration programmes in the polar regions, generating important advances in scientific and geographical knowledge. From laying the foundations of our understanding of nature's global systems to launching the modern space age, IPYs set the stage for many international scientific collaborations as well as a long-standing political accord.

Half a century after IGY, in 2007 the 4th IPY began (www.ipy.org), sponsored by ICSU and WMO. The Director of the IPY International Programme Office was David Carlson (see Volume 1 of the Series). IPY 2007–2009 has been the largest ever international programme of scientific research in the Arctic and Antarctic regions, building upon the long legacy, established in the previous IPYs, of international cooperation, scientific achievement and societal benefits. The importance and complexity of the 4th IPY deserved adequate opportunities to describe and discuss the outcomes flowing from such a vast international initiative. Two important fora were organised to meet this target: the "IPY Oslo Science Conference" (2010), and "IPY—From Knowledge to Action", which privileged the collective dissemination of the first scientific results and perspectives (2012, Montreal, Canada).

The Series of Springer books "From Pole to Pole: Polar Environmental Research during the International Polar Year 2007–2009" is a major contribution for libraries of world institutions. It is intended to complement the many articles on IPY research which are increasingly being published in scientific journals and provide a lasting focus and synthesis for key areas.

Marine Biology provides a strong contribution with two volumes on "Adaptation and Evolution in Marine Environments—The Impacts of Global Change on Biodiversity". This volume is the continuation of the first one. The chapters describe research that is part of IPY projects and will undergo developments in the decades to come, identifying linkages with investigations described in the chapters.

All the research reported in this book is in the framework of the international, multi- and cross-disciplinary programme "Evolution and Biodiversity in the Antarctic—The Response of Life to Change" (EBA, www.eba.aq). Launched by the Scientific Committee on Antarctic Research (SCAR, www.scar.org; the major organisation coordinating research in the Antarctic and Southern Ocean—SO-region) in 2004, it assembled almost one hundred teams and covered most of Antarctic biological research in the marine, terrestrial and freshwater realms. EBA was described in detail in the first chapter of Volume 1 by G. di Prisco and P. Convey.

It facilitated the integrated approach required for unravelling the role of the polar environments in modulating the Earth system, addressing key issues raised within the Antarctic Treaty System. SCAR provides both the opportunity and the framework to inform the non-biological disciplines of the value and breadth of the programme, especially in its contribution to understanding the impact of Climate Change on Antarctic ecosystems. Most SCAR nations participated in EBA, that acted as a major focus for capacity building in new SCAR members, especially those with reduced logistic and financial resources, and contributed to a wide variety of international programmes. EBA included sub-Antarctic islands, inland areas to remote nunataks as well as sites northward to the Magallanes Strait, stretching across the SO from the deep ocean to the continental shelves, and linking with northern polar studies. The objectives were to understand the evolution and diversity of life in the Antarctic, to determine how these have influenced the properties and dynamics of present Antarctic and SO ecosystems and to make predictions on how organisms and communities might respond to current and future environmental change.

EBA developed a major marine focus during IPY. Antarctica is conventionally described as having limited terrestrial biodiversity, in the form of isolated "islands" of terrestrial habitat surrounded by inhospitable ocean or ice. These fragmentary habitats provide an ideal "evolutionary laboratory", allowing questions to be addressed on both relatively short and long evolutionary time scales. This part will be illustrated in the volume of the Series addressing Terrestrial Biology.

An explicit aspect of EBA was to compare and integrate results from the marine, terrestrial and limnetic environments. The programme was interdisciplinary. It utilised enabling technologies in ecophysiology, microbiology, taxonomy, molecular biology and organismal biology. It liaised with the relevant physical, geological and historical disciplines to ensure regular interaction and use of the most recent data and insights in interpreting the biological results. It involved fieldwork and laboratory work in the Antarctic and home institutions. It required extensive international collaboration. Exploration of some areas required new technologies, for example benthic landers, remotely operated vehicles (ROVs) for the deep sea, autonomous underwater vehicles (AUVs) for work beneath ice shelves.

The timing of IPY overlapped with that of EBA; the EBA and IPY activities were conceived in parallel, and the IPY Initial Outline Science Plan (April 2004) indicated the ability of EBA to provide a significant contribution to IPY. The research and projects were all under the umbrella of EBA, and cross-linkages will continue into the future. By undertaking a focussed initiative on the spatial distribution of marine and terrestrial diversity, EBA is leaving a legacy of biodiversity information and the tools with which to explore it, a substantial contribution to IPY.

EBA will lapse in 2012–2013. To develop this field in the light of our new knowledge, the Antarctic biology community is proposing two programmes, focussed on distinct but complementary aspects of polar biology and working across marine, freshwater and terrestrial environments: "State of the Antarctic Ecosystem (AntEco)", and "Antarctic Thresholds—Ecosystem Resilience and Adaptation (AnT-ERA)". These programmes are the legacy of EBA, and they are key to both improving our understanding and protecting Antarctic biodiversity.

Eleven contributions are assembled in this Volume, in three parts: Part I, Biodiversity Evolution and Data Management; Part II, Evolution: A Molecular Perspective; Part III, Monitoring and Management. The authors describe the concept, aim and first findings of the respective IPY projects, providing information, equipped with exhaustive reference lists and relevant web pages, on results and research perspectives within the framework of IPY 2007–2009.

The ideas and concepts of each chapter are briefly outlined below.

A comment on Part I seems pertinent. Some SCAR projects are integral parts of EBA, and have been described or mentioned in many of the Chapters of Volume 1. One of the most important ones is the "Census of Antarctic Marine Life" (CAML; www.caml.aq), performed in 2004–2010 under the auspices of the "Census of Marine Life" (CoML, www.coml.org). Polar regions experience greater rates of climate change than elsewhere on the planet. The faunas are uniquely adapted to their extreme environments, and may be vulnerable to shifts in climate. There is an urgent need to establish the state of these communities, and in particular their diversity, if we are to understand the impact of climate change. CAML was a 5-year project that during IPY 2007–2009 focussed on the ice-bound oceans of Antarctica. The coincidence with IPY made CAML a once-in-a-lifetime opportunity to conduct a comprehensive study of the evolution and biology of a vast region of the Earth, comprising the part of CoML that deals with the SO. Its objective was to study the evolution of life in Antarctic waters to determine how this had influenced the diversity of the present biota, and to use these observations to predict how it may respond to future change. The project integrated knowledge across all regions, biomes, habitats and fields of study to strengthen our knowledge of ecosystem dynamics in this high-latitude ocean system. Only through a multi-scale level of investigation will a better understanding of the diversity and status of Antarctica's marine life be obtained. CAML's aims were described in detail by A Brandt in Chap. 2 of Volume 1. She illustrated the project Antarctic Benthic Deep-Sea Biodiversity—System Coupling (ANDEEP-SYSTCO; www.andeep-systco.com). The importance of CAML clearly appears in eight chapters of Volume 1. It employed genomic techniques and contributed to the project Barcode of Life. It interacted with the Arctic Ocean Diversity project (ArcOD), drawing comparisons between the Arctic Ocean and the SO. CAML discovered many new species and established a comprehensive Antarctic marine biology database. The essential feature of CAML was its international structure, involving ships of many nations. Young researchers had the opportunity to participate, both at sea and in subsequent data analysis.

Further essential information on CAML accomplishments is summarised in Chaps. 1 and 2 of this Volume (history, organisation, targets, main expeditions, main results, workshops, coordination with SCAR-MarBIN, DNA barcoding, legacy). IPY was an unprecedented effort, which involved thousands of participants from many nations and generated massive amounts of extremely diverse data, which need to be interrelated to understand environmental change and its impact on Antarctic biodiversity. SCAR Marine Biodiversity Information Network (SCAR-MarBIN, www.scarmarbin.be), and the new Antarctic Biodiversity Information Facility (ANTABIF, www.biodiversity.aq) provide a case study on the determined way in which IPY data management was approached, ensuring that the outcomes from the data collection would be freely available. The different facets of this data management and the design used for ANTABIF are discussed to ensure that ANTABIF and its successors are successful in maintaining open access to biodiversity data and grow with the research. The heterogeneity and the distribution of data will require simple, well-standardised and agile technologies, to follow the constantly changing needs of the community it is serving. IPY has shown that the polar regions are changing rapidly and the need for rapid technical and cultural changes becomes urgent. Short- and long-term strategies to facilitate the evolution of the SCAR data system towards universal access to scientific data, which could otherwise be lost, are suggested.

In Part II (Evolution: A Molecular Perspective), the effect and importance of modern techniques is strikingly evident. Polar science has taken advantage of the explosive development of molecular methods, driven by research on the human genome, which revolutionised biology and provided the tools to explore the function of individual genes. Additional tools are:

(i) the increasing use of DNA barcode sequences for as many polar marine species and research expeditions as possible, in order to maximise the taxonomical and geographical coverage of obtained sequences. There is a potentially great extent of environmental change in polar region under climate-driven fluctuations, which might lead to extinctions; a reference baseline of barcode sequences is needed, and is under way, also thanks to CAML. It will be possible to study the material in a coordinated fashion, focussing on gaps, setting priorities for the most important taxa and avoiding duplicated collections in similar regions by different research groups;
(ii) the increasing importance of molecular phylogeny, using protein and RNA sequencing in drawing evolutionary trees.

Thus, molecular biology now has the potential to revolutionise the fields of evolutionary biology and ecology.

In Chap. 3, evolution in the SO has been considered in a global context. Molecular data have been largely obtained from vertebrates (penguins, fish, seals), with some exceptions, e.g. krill. There have been only a few studies on benthic invertebrates, about speciation and connectivity, endemic radiation, cryptic speciation and historical connectivity between the Antarctic and other oceans. The recent surge of molecular data has allowed the use of genetic methods to investigate the SO biota, in some cases in parallel with the Arctic, as well as the biota with a prevalence of eurybathy, with circumpolar distributions and non-dispersive life histories. Molecular studies allow the strength of these concepts to be tested across a range of taxa with diverse life-history strategies, providing a richer understanding of speciation and connectivity within the SO ecosystem. The Polar Front and the Antarctic Circumpolar Current (ACC) are not an absolute barrier to dispersal, and organisms may actually move in and out of the SO, but the wide environmental variations experienced either side of the Polar Front is likely to hamper the survival of many species. Many cosmopolitan species are in the deep sea; at greater depths variations will be lesser and deep-sea species may cross more easily. The review addresses the difficulties in estimating evolutionary times based on a reliable molecular clock. Connectivity within the SO and with other oceans are among the themes considered. Molecular ecology and phylogenetics are in their infancy, yet they are transforming our understanding of connectivity. A large part of the findings comes from DNA barcode data.

Species bipolarity has raised the intriguing question as to whether co-specific Antarctic and Arctic populations evolved independently since separation between the Arctic and Antarctic cold-water provinces, or genetic continuity has been ensured by a trans-tropical gene flow. Several drawbacks limit classical approaches (morphology; analysis of genetic variation in nuclear-gene sequences). Ciliates are ideal organisms for the analysis of the breeding structure of microbial populations and for obtaining data which satisfy the interbreeding criterion on which the biological (Darwinian) concept of species is founded. Chapter 4 describes how ciliates govern their gene exchanges through unique sexual conjugation (or mating). More than any other group of polar microorganisms, ciliates can readily provide living laboratory material in unlimited amounts. Evidence from breeding analyses of Antarctic, Fuegian and Arctic populations of a species shows that they are genetically interconnected by gene flow and form a unique interbreeding species. These strains are mating compatible and breeding interfertile with one another and therefore share the same gene pool, indicating that bipolar populations may maintain genetic continuity in spite of ecological discontinuity. Individuals of these populations may be able to swarm and ensure a pole-to-pole gene flow by dwelling (and multiplying) in the cold currents that cross the equatorial ocean depths. This hypothesis can hardly be verified directly, but molecular biology steps in to identify nuclear (bi-parentally inherited) and mitochondrial (uni-parentally inherited) ribosomal gene sequences characterised by single nucleotide polymorphisms, powerful genetic markers of the evolutionary history of natural populations. Work with two Arctic strains shows that polymorphisms may reflect natural hybridisation between Arctic and Antarctic populations, thus covering a driving role in speciation and evolution. The authors suggest an additional synergistic force, namely the capacity of these populations to communicate and interact *via* diffusible signalling pheromones, synthesised to promote mating and growth. The pheromones are cross-reactive and their structures secure long-lasting activity and wide range of dispersal in any environment.

The next two Chapters deal with the physiological and physico-chemical role of temperature. The two questions: (i) which were the drivers of polar evolution and what was gained and lost with respect to the role and effect of these drivers, and (ii) which animal groups might have gained or lost in fitness during environmental changes, are the focus of Chap. 5. Directly or indirectly, temperature has been a fundamental driver. Other abiotic factors are shaped by temperature, e.g. gas concentrations (oxygen and CO_2), water density and viscosity (salinity is instead not influenced). Compared with the tropics, oxygen and CO_2 concentrations are almost twice as high in Antarctic waters. Gas solubility is enhanced in the cold, but diffusibility is constrained, thereby hampering gas transport, and especially oxygen uptake. Convective transport is also constrained in the cold, due to enhanced water viscosity. For a more comprehensive understanding of the evolutionary process operating in many species, the concept of oxygen and capacity limited thermal tolerance (OCLTT) was developed. OCLTT was suggested to be a suitable matrix for the integration of other stressor effects which may interfere with the levels of thermal tolerance and acclimation. If it appears rewarding to interpret functional adjustments in key groups of the Antarctic (e.g. fish, see Volume 1), which display unique features, it is also rewarding to look at benthic brachyuran and anomuran crabs. These seem to have taken these adaptations to a certain limit and live at the doorsteps of the polar oceans or have succeeded in settling in "warmer" water bodies, but are still excluded from life at the coldest temperatures.

Temperature is crucial for survival since it affects the reaction rates of the chemical reactions that occur in any living organism. Chapter 6 stresses that thermodynamic analysis is of primary importance, since activation entropy and energy make reactions possible. For organisms that do not have the capacity to keep temperature constant, the problem of that of the environment is acute. Indeed, low temperatures may render reaction rates too slow to sustain life, whereas high temperatures can accelerate them to an extent that would lead to production of unwanted or excess metabolites, due to the differential action of temperature on rates *via* activation energy. Three types of organisms are defined as a function of the temperature of their environment: *psychrophiles* that thrive in environments characterised by temperatures close to or below the freezing point of water, *mesophiles* demanding moderate temperatures and *thermophiles* that are able to withstand temperatures which in some cases exceed that of the boiling point of water. The rate of growth of a microorganism is related to the rate of the metabolic reactions, catalysed by enzymes that require proper folding and stability. The temperature of maximum growth rate is not the best possible, since it induces partial unfolding or overproduction of unwanted metabolites. In thermophiles, although high temperatures favour reaction rates, they also induce structural unfolding or misfolding, due to uncontrolled hydrophobic forces. The folding of a high number of proteins is assisted by chaperones. Maximum stability is not suitable, since structural plasticity and flexibility are required to secure interactions in a very crowded cell. In psychrophiles, the main problem is to secure metabolic fluxes by acting on the activity or relative abundance of enzymes. In evolution, production of high amounts of catalysts is not cheap, and the reaction-rate problem has been solved through enzymes displaying lower activation energy and lower thermal dependence of the activity that allows the organisms to be exposed to unusually low or high temperatures. Therefore a continuum in the adaptation of cold-adapted microorganisms, depending on their evolutionary history, is needed.

In Part III (Monitoring and Management) Chap. 7 describes satellite technology to investigate seals thriving in both polar environments. Many species spend their life in close proximity to the coast or the ice edge, where they can be observed throughout the year. However, except for the short breeding and moulting periods, several species are pelagic, and it was not until the invention of satellite-linked dive recorders that it became possible to learn about the whereabouts and behaviour of ecologically and economically important species outside the breeding season, as well as gaining knowledge of the seasonal distribution and diet composition, beyond incidental sightings and analysis of a limited number of stomach contents from animals captured in the pack ice. There was a danger of erroneous conclusions by extrapolation from such studies, if large amounts of stocks spend a considerable amount of time in open water, where their diet may be based on different preys. Satellite-linked dive recorders have been employed to determine location and diving behaviour throughout the year in the Arctic and Antarctic; it is now possible to determine haul-out patterns, useful in converting aerial sightings into population numbers. Such investigations have socio-economical importance, because two Arctic species, by numbers and habit of congregating in specific locations on the pack ice for breeding and moulting, have had significant impact on the economy of coastal communities for the last 100 years. Moreover, since the North Atlantic and the adjacent Barents Sea–White Sea hold some of the world's most important fish stocks, the ecological and economical implications of the millions of these seals are matters of concern.

Complementing each other, Chaps. 8 and 9 deal with monitoring and management issues related to environmental processes, biodiversity and global changes, and environmental assessment in Admiralty Bay, King George Island (KGI). These contributions are welcome, considering the need to improve coordination of science activities and management in KGI. A coordinated and standard approach to data gathering, observations, logistics and utilisation of infrastructure would benefit all parties operating in KGI. The authors' large effort in providing a huge amount of information will certainly be invaluable to reach this aim.

There are scientific, economic and environmental-protection imperatives for national Antarctic programmes to work together in partnership. In KGI, the complex of infrastructure, stations and logistics provide a unique opportunity to explore how cooperation across Antarctica and the SO may benefit from close coordination. There are already good examples of scientific cooperation in KGI, for example in conservation and environmental monitoring in Admiralty Bay.

The information below is taken from Kennicutt (2009 COMNAP meeting, Punta Arenas, Chile).

KGI is one of the South Shetland Islands. Cape Horn is about 900 km to the north. More than 90 % of the island is glaciated. The ice-free areas and coastal zones carry a diverse plant and animal life, including penguins, seals, petrels and rich tundra vegetation. Admiralty Bay is an Antarctic Specially Managed Area (ASMA No.1). The area is representative of the terrestrial, limnetic, coastal, near-shore, pelagic, and fjord bottom ecosystems of KGI. The ecosystem reflects the general environmental conditions prevailing in the South Shetland Islands, with rocky shores covered by subtidal macroalgal communities. Napier Rock, at the entrance of the bay, boasts a rich and diverse benthic invertebrate fauna. Fish are represented by 15 species of Nototheniidae.

KGI has the greatest concentration of national research activities in Antarctica. There are nine permanent stations and a rock airstrip. The following nations have a presence: Chile, Argentina, Poland, Germany, Uruguay, Brazil, Peru, Ecuador, China, Korea, and the Russian Federation. The Netherlands, UK and USA may also carry out research. Human activities date to the early nineteenth century, when the first sealers arrived. Within a few years the fur and elephant seals were on the verge of extinction. Whale bones can still be found on many KGI beaches. Scientific activities on KGI include meteorological, hydrological, geophysical, biological, geological, sea ice, greenhouse gases and glaciological research.

Research there in marine and terrestrial biology include physiology and adaptation of fish and krill; taxonomy and ecology of marine benthic fauna and flora (micro- and macro-algae), vascular plants, mosses and lichens; terrestrial and marine ecology; migration and dispersion of birds. There is a CliCOPEN project on the response of marine and terrestrial ecosystems to deglaciation related to regional warming. A long-term project on biology and dynamics of bird populations (mainly penguins) has been carried out since 1976. There has also been routine collection of data on distribution and reproduction of the southern giant petrel and on skuas.

These studies are particularly relevant to the science of the SCAR Standing Scientific Group on Life Sciences, including its Expert Group on Birds and Marine Mammals and its Scientific Research Programme, EBA. These studies support understanding evolution and biodiversity in the Antarctic and identifying science outcomes relevant to the conservation policies of the Antarctic Treaty System (ATS). They will help EBA to understand how evolution and diversity have led to the present ecosystems in the Antarctic and how these ecosystems will respond to future change. EBA benefits from KGI in the following objectives:

- determining the variations in diversity at different spatial scales within the Antarctic and within defined time frames.
- understanding the ability of Antarctic organisms to cope with daily, seasonal and longer-term environmental changes.
- studying ecological responses to latitudinal and environmental gradients on local, regional and global scales.
- discerning the role of natural and anthropogenic dispersal processes in gene flow and population structure.
- understanding interactions between introduced and indigenous species, effects of abiotic change on biota, and how environmental change and organism responses are linked to climate.

KGI marine biological data are deposited with SCAR-MarBIN and ANTABIF (see Chap. 2), and are of relevance to the SCAR ATS Committee that deals with protection of Antarctic species. Offshore studies have also contributed to CAML (see Chaps. 1 and 2). A study of the environment is under way, including analysis of biotic and abiotic variables. The results will serve as a baseline for future monitoring and environmental management of the ASMA, and will inform the

ATS Committee on Environmental Protection (CEP) about the design of monitoring programmes.

As a location of intense human activity and known disturbance, data and observations of contaminants at KGI are relevant to the SCAR Expert Group on Environmental Contamination in Antarctica. Visitors are included into the monitoring carried out in the framework of the IPY-ALIENS project on invasive species.

Chapter 10 discusses anthropogenic impacts on sub-Antarctic and Antarctic islands (including KGI) and the adjacent marine environments. Marine pollution, diversity changes caused by introduction of non-indigenous species, and global environmental changes (e.g. warming and thinning of the ozone layer) are the main consequences from human activities in the sub-Antarctic and Antarctic regions. Debris in seawater pollute the SO, degrade beaches, kill and injure seabirds and mammals. Floating plastics provide substrata to cosmopolitan pelagic species, while other elements sink to the deep sea (e.g. glass and metal bottles and containers). The impacts of oil spills often appear years later. In South Georgia, wastes from old whaling stations may still be detected in the deepest sediment. Human pollution over several decades is apparent at McMurdo Station and at several stations in the Maritime Antarctic. Some isolated islands remain in relatively pristine state, but risks increase with human visitation. Biodiversity may also be affected by natural processes. Sustainable management of marine ecosystems must distinguish the effects of human impacts from those related to climate variability and change. Marine reserves are needed, as establishing target areas and ecosystem reserves can greatly benefit the biota. Such areas can also be used to monitor the effects of global changes. Multidisciplinary studies, identification of impact origins, and long-term monitoring are required in order to assess the effects of human activities on Antarctic environments and biodiversity. Such studies are appropriate tools for environmental management, especially when taking biodiversity hotspots into account, and need national Antarctic programme managers to work with scientists to ensure the management is soundly based on good science.

Chapter 11 conjugates behaviour, ecology, technology, in the framework of climate change. It highlights seabirds as monitoring sentinels for polar marine ecosystems. Direct monitoring would require huge investments in equipment, personnel and logistics. Indicators, such as animals sensitive to changes, are needed to provide information on the 'health of the ecosystem'. For this, seabirds are excellent, since we know which stocks of marine organisms they feed on. Some of the best data series exist for albatrosses, penguins, puffins, making them ideal models for Polar Life Observatories (PLOs), as well as icons to call attention to human-induced changes and make governments aware of the need to respond to a major global threat. PLOs meet priorities regarding environmental protection defined by the Convention on Biological Diversity (CBD), the Intergovernmental Panel on Climate Change (IPCC), and the International Council for Science (ICSU). Bio-logging devices monitor prey stocks by evaluating distribution and availability of mesopelagic fish, squid, krill, etc. However, seabirds with attached devices do not behave like unequipped conspecifics; for example, flipper bands have a major impact on survival and breeding success of penguins. As the short reading range of Radio Frequency Identification (RFID) constrains its use for identifying individuals within crowded colonies, fixed passages are used, although for emperor penguins that breed on sea ice such passages do not exist. As RFID tags are tiny and implanted under the skin and birds cannot be visually localised inside the colony, mobile RFID antennae are used, carried by remotely controlled robots that can circulate among penguins.

In conclusion, EBA also acts as an umbrella to IPY research in Volume 2, for evolutionary and biodiversity information, molecular perspectives, and management. The programme has direct relevance to Global Change, because it addresses the impacts of the latter on biodiversity, adaptations and community dynamics, and provides information that can be extrapolated also to temperate latitudes. EBA will lapse in 2013, and steps have been taken to ensure agreement on exciting new programmes for the future.

Cinzia Verde
Guido di Prisco

Contents

Contributors

Claudio Alimenti Dipartimento di Scienze Ambientali e Naturali, University of Camerino, via Gentile III da Varano, 63032 Camerino, MC, Italy, e-mail: claudio.alimenti@unicam.it

A. Louise Allcock Department of Zoology and Ryan Institute, National University of Ireland, University Road, Galway, Ireland, e-mail: louise.allcock@gmail.com

Cristina E. Alvarez Universidade Federal do Espírito Santo, Av. Fernando Ferrari 514, Vitória–ES 29073-910, Brazil, e-mail: cristina.engel@ufes.br

Carlos A. M. Barboza Instituto de Biologia, Universidade Federal do Rio de Janeiro, Av Carlos Chagas Filho 373, Rio de Janeiro–RJ 21941-902, Brazil; Universidade Federal Fluminense, Outeiro de São João Batista s/no, Niterói–RJ 24020-141, Brazil, e-mail: carlosambarboza@gmail.com

Manuela Bassoi Instituto de Biologia, Universidade Federal do Rio de Janeiro, Av Carlos Chagas Filho 373, Rio de Janeiro–RJ 21941-902, Brazil, e-mail: manu.bassoi@gmail.com

Marcelo Bernardes Universidade Federal Fluminense, Outeiro de São João Batista s/no, Niterói–RJ 24020-141, Brazil, e-mail: uff.bernardes@gmail.com

Márcia C. Bícego Instituto Oceanográfico, Universidade de São Paulo, Praça do Oceanográfico 191, São Paulo–SP 05508-120, Brazil, e-mail: marciabicego@usp.br

Arnoldus Schytte Blix Department of Arctic Biology, University of Tromsø, Tromsø 9037, Norway, e-mail: asb000@uit.no

Elisabete S. Braga Instituto Oceanográfico, Universidade de São Paulo, Praça do Oceanográfico 191, São Paulo–SP 05508-120, Brazil, e-mail: edsbraga@usp.br

Tania A. S. Brito Fundação UNESCO-HidroEX, Av Mário Palmério 1000, Frutal–MG 38200-000, Brazil, e-mail: tania.brito@hidroex.mg.gov.br

Sandra Bromberg Instituto Oceanográfico, Universidade de São Paulo, Praça do Oceanográfico 191, São Paulo–SP 05508-120, Brazil, e-mail: bromberg@usp.br

Lúcia S. Campos Instituto de Biologia, Universidade Federal do Rio de Janeiro, Av Carlos Chagas Filho 373, Rio de Janeiro–RJ 21941-902, Brazil, e-mail: campos-lucia@biologia.ufrj.br

Belmiro M. Castro Instituto Oceanográfico, Universidade de São Paulo, Praça do Oceanográfico 191, São Paulo–SP 05508-120, Brazil, e-mail: bmcastro@usp.br

Peter Convey British Antarctic Survey, High Cross, Madingley Road, Cambridge CB3 0ET, UK, e-mail: pcon@bas.ac.uk

Thaïs N. Corbisier Instituto Oceanográfico, Universidade de São Paulo, Praça do Oceanográfico 191, São Paulo–SP 05508-120, Brazil, e-mail: tncorbis@usp.br

Bruno Danis Royal Belgian Institute of Natural Sciences, Rue Vautier 29, 1000 Brussels, Belgium; Antarctic Biodiversity Information Facility (ANTABIF), Rue Vautier 29, 1000 Brussels, Belgium, e-mail: bruno.danis@gmail.com

Graziano Di Giuseppe Dipartimento di Biologia, University of Pisa, via Volta 4, 56126 Pisa, Italy, e-mail: gdigiuseppe@biologia.unipi.it

Fernando Dini Dipartimento di Biologia, University of Pisa, via Volta 4, 56126 Pisa, Italy, e-mail: fdini@biologia.unipi.it

Heitor Evangelista Universidade do Estado do Rio de Janeiro, Rua São Francisco Xavier 524, Rio de Janeiro–RJ 20550-013, Brazil, e-mail: heitor@uerj.br

Lars P. Folkow Department of Arctic Biology, University of Tromsø, Tromsø 9037, Norway, e-mail: lars.folkow@uit.no

Roberto F. C. Fontes Campus Experimental do Litoral Paulista, Universidade Estadual Paulista, Praça Infante Dom Henrique s/no, São Vicente–São Paulo 11330-900, Brazil, e-mail: rcfontes@clp.unesp.br

Marcio Francelino Universidade Federal Rural do Rio de Janeiro, Br 465 Km 7, Seropédica–RJ 23890-000, Brazil, e-mail: marciorocha@ufrrj.br

Paula F. Gheller Instituto Oceanográfico, Universidade de São Paulo, Praça do Oceanográfico 191, São Paulo–SP 05508-120, Brazil, e-mail: paulafgheller@gmail.com

Charles Gerday Laboratory of Biochemistry, Institute of Chemistry, University of Liège, 4000 Liège, Belgium, e-mail: ch.gerday@ulg.ac.be

Vicente Gomes Instituto Oceanográfico, Universidade de São Paulo, Praça do Oceanográfico 191, São Paulo–SP 05508-120, Brazil, e-mail: vicgomes@usp.br

Huw Griffiths British Antarctic Survey, High Cross, Madingley Road, Cambridge CB3 0ET, UK, e-mail: hjg@bas.ac.uk

Eduardo Hajdu Museu Nacional, Universidade Federal do Rio de Janeiro, Quinta da Boa Vista, s/no, Rio de Janeiro–RJ 20940-040, Brazil, e-mail: eduardo.hajdu@gmail.com

Helena G. Kawall Instituto Oceanográfico, Universidade de São Paulo, Praça do Oceanográfico 191, São Paulo–SP 05508-120, Brazil, e-mail: helena.kawall@gmail.com

Rosane G. Ito Instituto Oceanográfico, Universidade de São Paulo, Praça do Oceanográfico 191, São Paulo–SP 05508-120, Brazil, e-mail: rgito@usp.br

Priscila K. Lange Instituto de Biologia, Universidade Federal do Rio de Janeiro, Av Carlos Chagas Filho 373, Rio de Janeiro–RJ 21941-902, Brazil, e-mail: prilange@gmail.com

André M. Lanna Instituto de Biologia, Universidade Federal do Rio de Janeiro, Av Carlos Chagas Filho 373, Rio de Janeiro–RJ 21941-902, Brazil, e-mail: andrebioufrj@gmail.com

Helena P. Lavrado Instituto de Biologia, Universidade Federal do Rio de Janeiro, Av Carlos Chagas Filho 373, Rio de Janeiro–RJ 21941-902, Brazil, e-mail: hpasseri@biologia.ufrj.br

Céline Le Bohec European Associated Laboratory 647 'BioSensib', Centre Scientifique de Monaco & Centre National de la Recherche Scientifique, 8 quai Antoine 1er, Monte-Carlo 98000; Principality of Monaco, and Institut Pluridisciplinaire Hubert Curien, Unité Mixte de Recherche 7178 Centre National de la Recherche Scientifique & Université de Strasbourg, 23 rue Becquerel, 67087 Strasbourg Cedex 02, France, e-mail: clebohec@centrescientifique.mc

Yvon Le Maho European Associated Laboratory 647 'BioSensib', Centre Scientifique de Monaco & Centre National de la Recherche Scientifique, 8 quai Antoine 1er, Monte-Carlo 98000; Principality of Monaco, and Institut Pluridisciplinaire Hubert Curien, Unité Mixte de Recherche 7178 Centre National de la Recherche Scientifique & Université de Strasbourg, 23 rue Becquerel, 67087 Strasbourg Cedex 02, France, e-mail: yvon.lemaho@iphc.cnrs.fr

Neusa Paes Leme Instituto Nacional de Pesquisas Espaciais, Av dos Astronautas 1.758, São José dos Campos–SP 12227-010, Brazil, e-mail: nleme@crn.inpe.br

Pierangelo Luporini Dipartimento di Scienze Ambientali e Naturali, University of Camerino, via Gentile III da Varano, 63032 Camerino, MC, Italy, e-mail: piero.luporini@unicam.it

Michel M. Mahiques Instituto Oceanográfico, Universidade de São Paulo, Praça do Oceanográfico 191, São Paulo–SP 05508-120, Brazil, e-mail: mahiques@usp.br

César C. Martins Instituto Oceanográfico, Universidade de São Paulo, Praça do Oceanográfico 191, São Paulo–SP 05508-120, Brazil, e-mail: ccmart@ufpr.br

Gabriel C. S. Monteiro Instituto Oceanográfico, Universidade de São Paulo, Praça do Oceanográfico 191, São Paulo–SP 05508-120, Brazil, e-mail: gabrielsousaconzomonteiro@gmail.com

Rosalinda C. Montone Instituto Oceanográfico, Universidade de São Paulo, Praça do Oceanográfico 191, São Paulo–SP 05508-120, Brazil, e-mail: rmontone@usp.br

Tatiana Morales Instituto de Biologia, Universidade Federal do Rio de Janeiro, Av Carlos Chagas Filho 373, Rio de Janeiro–RJ 21941-902, Brazil, e-mail: thmorales@hotmail.com

Rafael B. Moura Instituto de Biologia, Universidade Federal do Rio de Janeiro, Av Carlos Chagas Filho 373, Rio de Janeiro–RJ 21941-902, Brazil, e-mail: lytechinusvariegatus@gmail.com

Cristina R. Nakayama Campus Diadema, Universidade Federal de São Paulo, Rua Artur Riedel 275, Diadema–SP 09972-270, Brazil, e-mail: crnakayama@unifesp.br

Phan V. Ngan Instituto Oceanográfico, Universidade de São Paulo, Praça do Oceanográfico 191, São Paulo–SP 05508-120, Brazil, e-mail: phanvn@usp.br

Erling S. Nordøy Department of Arctic Biology, University of Tromsø, Tromsø 9037, Norway, e-mail: erling.nordoy@uit.no

Thayane Oackes Instituto de Biologia, Universidade Federal do Rio de Janeiro, Av Carlos Chagas Filho 373, Rio de Janeiro–RJ 21941-902, Brazil, e-mail: t.oackes@gmail.com

Rodolfo Paranhos Instituto de Biologia, Universidade Federal do Rio de Janeiro, Av Carlos Chagas Filho 373, Rio de Janeiro–RJ 21941-902, Brazil, e-mail: rodolfo@biologia.ufrj.br

Flávio D. Passos Instituto de Biologia, Universidade Estadual de Campinas, Rua Monteiro Lobato 255, Campinas–SP 13083-970, Brazil, e-mail: flaviodp@unicamp.br

Vivian H. Pellizari Instituto Oceanográfico, Universidade de São Paulo, Praça do Oceanográfico 191, São Paulo–SP 05508-120, Brazil, e-mail: vivianp@usp.br

Antonio B. Pereira Universidade Federal do Pampa, Av Antônio Trilha, São Gabriel–RS 97300-000, Brazil, e-mail: antoniopereira@unipampa.edu.br

Monica A. V. Petti Instituto Oceanográfico, Universidade de São Paulo, Praça do Oceanográfico 191, São Paulo–SP 05508-120, Brazil, e-mail: mapetti@usp.br

Hans-Otto Pörtner Integrative Ecophysiology, Alfred-Wegener-Institute, Am Handelshafen 12, 27570 Bremerhaven, Germany, e-mail: Hans.Poertner@awi.de

Sylvain Renaudier Antarctic Biodiversity Information Facility (ANTABIF), Rue Vautier 29, 1000 Brussels, Belgium, e-mail: s.renaudier@gmail.com

Carlos E. Rezende Centro de Biociências e Biotecnologia, Universidade Estadual do Norte Fluminense Darcy Ribeiro, Av. Alberto Lamego 2000, Campos dos Goytacazes-RJ 28013-602, Brazil, e-mail: crezendeuenf@yahoo.com.br

Mariane Rodrigues Instituto de Biologia, Universidade Federal do Rio de Janeiro, Av Carlos Chagas Filho 373, Rio de Janeiro–RJ 21941-902, Brazil, e-mail: marianeers@gmail.com

Luiz Henrique Rosa Instituto de Ciências Biológicas, Universidade Federal de Minas Gerais, Av Antônio Carlos, 6627, Belo Horizonte–MG 31270-901, Brazil, e-mail: lhrosa@ufop.br

Martin Sander Universidade do Vale do Rio dos Sinos, Av Unisinos 950, São Leopoldo–RS 93022-000, Brazil, e-mail: sander@bios.unisinos.br

Carlos E. G. R. Schaefer Universidade Federal de Viçosa, Av Peter Henry Rolfs s/n, Viçosa–MG 36570-000, Brazil, e-mail: carlos.schaefer@ufv.br

Stefano Schiaparelli University of Genova, Corso Europa 26, 16132 Genova, Italy, e-mail: stefano.schiaparelli@unige.it

Eduardo Secchi Instituto de Oceanografia, Fundação Universidade do Rio Grande, Av Italia, Km 8, Rio Grande–RS 96201-900, Brazil, e-mail: edu.secchi@furg.br

D. Michael Stoddart Institute for Marine and Antarctic Studies, University of Tasmania, Bag 129, Hobart 7001, Australia, e-mail: michael.stoddart@utas.edu.au

Jan M. Strugnell Department of Genetics, La Trobe Institute for Molecular Science, La Trobe University, Bundoora, Vic 3086, Australia, e-mail: j.strugnell@latrobe.edu.au

Denise R. Tenenbaum Instituto de Biologia, Universidade Federal do Rio de Janeiro, Av Carlos Chagas Filho 373, Rio de Janeiro–RJ 21941-902, Brazil, e-mail: deniser@biologia.ufrj.br

Adriana Vallesi Dipartimento di Scienze Ambientali e Naturali, University of Camerino, via Gentile III da Varano, 63032 Camerino, MC, Italy, e-mail: adriana.vallesi@unicam.it

Anton Van de Putte Royal Belgian Institute of Natural Sciences, Rue Vautier 29, 1000 Brussels, Belgium; Antarctic Biodiversity Information Facility (ANTABIF), Rue Vautier 29, 1000 Brussels, Belgium, e-mail: antonarctica@gmail.com

Victoria Wadley Australian Antarctic Division, Channel Highway, Kingston 7050, Australia, e-mail: victoria.wadley@aad.gov.au

Kathleen Walther Integrative Ecophysiology, Alfred-Wegener-Institute, Am Handelshafen 12, 27570 Bremerhaven, Germany, e-mail: kathleen.walther@awi.de

Rolf R. Weber Instituto Oceanográfico, Universidade de São Paulo, Praça do Oceanográfico 191, São Paulo–SP 05508-120, Brazil, e-mail: rweber@usp.br

Jason D. Whittington Department of Biology, Nordic Centre for Research on Marine Ecosystems and Resources under Climate Change (NorMER), Centre for Ecological and Evolutionary Synthesis (CEES), University of Oslo, P.O. Box 1066, 0316 Blindern, Norway, e-mail: jason.whittington@bio.uio.no

Astrid Wittmann Integrative Ecophysiology, Alfred-Wegener-Institute, Am Handelshafen 12, 27570 Bremerhaven, Germany, e-mail: astrid.wittmann@awi.de

Yocie Yoneshigue-Valentin Instituto de Biologia, Universidade Federal do Rio de Janeiro, Av Carlos Chagas Filho 373, Rio de Janeiro–RJ 21941-902, Brazil, e-mail: yocie@biologia.ufrj.br

Part I
Biodiversity Evolution and Data Management

Chapter 1
The Census of Antarctic Marine Life: The First Available Baseline for Antarctic Marine Biodiversity

Stefano Schiaparelli, Bruno Danis, Victoria Wadley and D. Michael Stoddart

1.1 History of the Project

The Census of Antarctic Marine Life (CAML, www.caml.aq) was a 5-year long international project that focused the attention on the ice-bound oceans of Antarctica during the International Polar Year (IPY) in 2007–08, bringing together researchers from 30 different countries and more than 50 institutions. It was one of the fifteen IPY-endorsed biological projects devoted to Antarctica (Project #83) and coordinated field operations of 18 research voyages in Antarctica during IPY and/or within the CAML life-span.

CAML's main objectives were to study the evolution of life in Antarctic waters to determine how this had influenced the diversity of the present biota and to use these observations to predict how it might respond to future change.

CAML was also one of the fourteen projects of the international Census of Marine Life (CoML, www.coml.org) (Gutt et al. 2010), each focusing on specific geographic environments or subject areas, with the aim to understand marine

S. Schiaparelli (✉)
Di.S.T.A.V.: University of Genova, Corso Europa 26, 16132 Genoa, Italy
e-mail: stefano.schiaparelli@unige.it

B. Danis
Antarctic Biodiversity Information Facility (ANTABIF),
Rue Vautier 29, 1000 Brussels, Belgium
e-mail: bruno.danis@gmail.com

V. Wadley
Australian Antarctic Division, Channel Highway, Kingston, TAS 7050 Australia
e-mail: victoria.wadley@aad.gov.au

D. Michael Stoddart
Institute for Marine and Antarctic Studies, University of Tasmania,
Bag 129, Hobart, TAS 7001 Australia
e-mail: Michael.Stoddart@utas.edu.au

C. Verde and G. di Prisco (eds.), *Adaptation and Evolution in Marine Environments, Volume 2*, From Pole to Pole, DOI: 10.1007/978-3-642-27349-0_1,

biodiversity and set reference baselines to allow measuring change. CAML's sister project was the Arctic Ocean Diversity (ArcOD), devoted to the census of Arctic marine biodiversity. Thanks to strong interaction with this project, it is now possible to draw comparisons between differences in ecological structure and dynamics of the Arctic and Southern Oceans (Gradinger et al. 2010; Gutt et al. 2010).

1.1.1 The IPY Proposal

In considering its contribution as part of the "Evolution and Biodiversity in the Antarctic (EBA)" programme, the Scientific Committee on Antarctic Research (SCAR) accepted a proposal that a Census of Antarctic Marine Life be undertaken as a major activity during IPY and in July 2004 appointed an international Steering Committee to guide the project. The Committee held its first meeting following the SCAR meeting in Bremerhaven in October 2004 with the aim to shape the project and write an application through CoML to the Alfred P. Sloan Foundation, New York, for financial support to science coordination. The application was signed by Prof. M. Stoddart, CAML Chair, and Dr. C. Summerhayes, SCAR Executive Secretary. A grant was awarded and in early 2005 V. Wadley was appointed Project Manager with her first task being the coordination of an opening workshop (Brussels, June 2005). CAML was approved as IPY project in 2005.

While CAML received its core funding from CoML and the NAPs (in-kind support) researchers had to find institutional and/or national agency funds, to enable entire teams or individuals to participate in field and laboratory work.

CAML has been a very expensive project. Funds for coordination, including planning meetings, technical data analysis, taxonomic workshops, and salary for a Project Manager, were covered by the Sloan Foundation, under the CoML programme, for a total of ca. one and a half million USD between January 2005 and December 2009. SCAR provided additional funds during the duration of CAML. Specific research projects and travel were funded by national agencies.

1.1.2 CAML Organization

CAML has been chaired by Prof. M. Stoddart and led by V. Wadley (Australian Antarctic Division, Hobart/Institute for Marine and Antarctic Studies, University of Tasmania). The CAML Steering Committee appointed by SCAR comprised: C. de Broyer (Royal Belgian Institute of Natural Sciences, Brussels, Belgium), A. Murray (University of Nevada, Desert Research Institute, Reno, USA), Lúcia de Siqueira Campos (University of de Janeiro, Brazil), P. Rodhouse (British Antarctic Survey, Cambridge, UK), D. Rodriguez (Universidad Nacional de Mar del Plata, Argentina), A. Brandt (University of Hamburg, Germany). At its first meeting in 2004 the SC co-opted several experts who had shown interest in the CAML, and all subsequent meetings included S. Schiaparelli (University of Genova, Italy), B. Danis (Royal Belgian Institute of Natural Sciences, Brussels, Belgium), H. Griffiths (British Antarctic Survey, Cambridge, UK), G. Hosie (Australian Antarctic Division, Hobart, Australia), R. Hopcroft (University of Alaska, Fairbanks, USA), J. Gutt (AWI, Bremerhaven, Germany) and many others.

1.1.3 CAML Scientific Targets

The CAML SC compiled a Science Statement (www.caml.aq/education-outreach/documents/20061114_CAMLSciStatement_000.pdf) at the Brussels workshop in 2005. It was not intended as a work plan but, it established the main questions that needed to be asked about the Antarctic ocean in order to gain knowledge about its role in provision of ecosystem services to humankind, and how these are likely to be affected by climate and global change.

CAML's philosophy was to integrate knowledge of the *Known*, *Unknown* and *Knowable* diversity, and the abundance and distribution of marine life across regions, biomes and habitats. In particular, a specific set of scientific questions was proposed for each of the benthos, pelagos and top predators, virtually considering all size classes, from microbes to mammals.

By promoting collaboration between disciplines and international programmes, CAML intended to provide a legacy of knowledge for future generations and a baseline against which to measure future change in Antarctic marine diversity.

National plans for specific studies were presented and later coordinated by the SC, based on the available ship time provided by National Antarctic Programmes.

Beside new sampling activities and purpose-designed field studies, CAML utilised unemployed taxonomic collections and promoted data mining of historical data sets in collaboration with its sister programme, Marine Biodiversity Information Network (SCAR-MarBIN, www.scarmarbin.be), another IPY project.

Within this context, CAML's five main scientific goals were summarised as follows:

1. Undertake a species inventory of the Antarctic slopes and abyssal plains
2. Undertake an inventory of benthic fauna under disintegrating ice shelves
3. Undertake an inventory of plankton, nekton and sea-ice associated biota at all levels of biological organisation from viruses to vertebrates
4. Assess critical habitats for Antarctic top predators
5. Develop a coordinated network of inter-operable databases for all Antarctic biodiversity data

In addition to the above, CAML also promoted techniques of modern molecular biology to solve evolutionary and ecological questions, thanks to the agreement with the Barcode of Life Data System (BOLD http://www.boldsystems.org/) which provided facilities for the sequencing of samples from CoML projects.

All outcomes and cruise updates were constantly maintained by the Project Manager and made available to the wider public though the CAML website.

1.2 CAML Coordination Effort

One of the main goals was the coordination of international teams and programmes in 18 research voyages (Table 1.1) with sampling activities.

Coordination occurred at several levels during phases of voyages' organisation, ranging from endorsement letters for national funding agencies to organisation of symposia and workshops during which data sets were assembled or prepared for publication. For some activities, e.g. barcoding (see Sect. 1.3.3), it was necessary to select a coordinator (R. Grant, British Antarctic Survey) who would manage the high flow of data/sequences progressively fed from voyage activities.

Sampling guidelines were produced and protocols made available to the scientific community after the first CAML workshop (Brussels 2005) such as the protocol for benthic samples (www.caml.aq/benthos/documents/Benthos-Sampling.pdf) and for methodologies for sampling benthos of deep-sea basins and abyssal plains (www.caml.aq/benthos/documents/Benthos-Deep-Sampling.pdf).

1.2.1 CAML Main Expeditions

Some 18 expeditions had direct links and/or established expressions of interest with CAML activities. Although other voyages did not establish formal collaborations with CAML, they assessed geographical targets, or produced data for benefit of the CAML community. An example of the first case is associated to the Italian RV "OGS-Explora" (PI L. de Santis, OGS, Trieste). Although not part of CAML, it swath-mapped previously unknown areas and made the acquired high-resolution bottom topography available to CAML (Post et al. 2010); the latter was used to set sampling positions for the CAML-IPY voyage CEAMARC. Portions of the surveyed areas were declared Vulnerable Marine Ecosystems (VMEs) by the Commission for the Conservation of Antarctic Marine Living Resources (CCAMLR, www.ccamlr.org) and closed to bottom fishing since 2008 (CCAMLR 2009a, b). Examples of the second type the "ANtarctic benthic DEEP-sea biodiversity: colonisation history and recent community patterns (ANDEEP)" voyages I–III in 2002 and 2005) in the Scotia and Weddell Seas (Brandt et al. 2004, 2007) which, although accomplished before, or in coincidence with the CAML beginning, brought a conspicuous series of scientific papers published in collaboration with CAML members and helped in setting the guidelines and protocols adopted during the main research IPY voyages.

Table 1.1 List of research voyages with research targets specifically linked to IPY #53 (CAML) and #83 (SCAR-MarBIN)

Country	Vessel	Date	Area	Project or voyage name (IPY proposal)
Au, F, J	*Aurora Australis*	11/07–01/08	TA, GLC	CEAMARC, CCAMLR, CASO
Br	*Ary Rongel*	02/07	KGI	MABIREH (IPY#156)
Br	*Ary Rongel*	11/07–12/07; 11/08–03/09	SA, WS, BeS, AB	SOS-CLIMATE; SASSI; CRAC-ICE; ICED-IPY; CASO (IPY# 911)
Dk	*Vaederren*	08/06–03/07	BS; AP	GALATHEA 3
F, Au, J, Be	*L'Astrolabe*	12/07–01/08	TA, GLC	CEAMARC
G	*Polarstern*	11/06–01/07	WS	ANT XXIII/8
G	*Polarstern*	11/08–02/09	SA	ANT-XXIV/2; ANDEEP-SYSTCO (IPY#16, SCACE; IPY#111)
J	*Umitaka Maru*	12/07–01/08	SR; TA; GLC	STAGE, NIDR
J, Au, F	*Umitaka Maru*	01/08–02/08	TA, GLC	CEAMARC
NZ	*Tangaroa*	01/08–03/08	RS	IPY-CAML (TAN0802) (CPM IPY#343)
Pe	*Humboldt*	01/06–02/06; 12/06–01/07; 01/07–02-07; 01/08–02/08	AB	ABBED; MABIREH (IPY#156)
Pe	*Humboldt*	01/08–02/08, 01/09–02/09	SubA; DP; BS; AB; EI	
Po	*Polar Pioineer*	01/06; 03/07–04/07	SS, KGI, AB	
RF	*Akademik Fedorov*		Haakon VII Sea	SASIE (IPY#818)
UK	*James Clark Ross*	02/06–04/06; 02/08–04/08	ScS; BeS; AS; WS	BIOPEARL I (JR147) & II (JR179) (IPY#417, ICCED; IPY#949)
USA	*Yuzhmorgeologia*	02/08–03/08; 02/09–03/09	SOI	CCAMLR + CAML
USA	*Gould*	01/06–02/06; 01/07–02/07; 01/08–02/08; 04/07–05/07	CSI; PS	NSF polar programs (IPY#77)
Ve	*Oyarvide*	02/08–03/08	KGI	First Venezuelan Antarctic expedition

Whenever possible, information about voyage data report or web sites is reported. Countries: *Au* Australia; *Be* Belgium; *Br* Brazil; *Dk* Denmark; *F* France; *G* Germany; *J* Japan; *NZ* New Zealand; *Pe* Peru; *Po* Poland; *RF* Russian Federation; *UK* United Kingdom; *USA* United States; *Ve* Venezuela. Sites: *AB* Admiralty Bay; *AP* Antarctic Peninsula; *AS* Amundsen Sea; *BeS* Bellingshausen Sea; *BS* Bransfield Strait; *CSL* Cape Shirreff Livingston; *DP* Drake Passage; *EI* Elephant Island; *GLC* George V Land Coast; *KGI* King George Island; *PS* Palmer Station; *RS* Ross Sea; *SA* South Atlantic; *SR* Syowa Region; *ScS* Scotia Sea; *SS* South Shetlands; *SubA* Sub-Antarctic; *TA* Terre Adélie; *WS* Weddell Sea

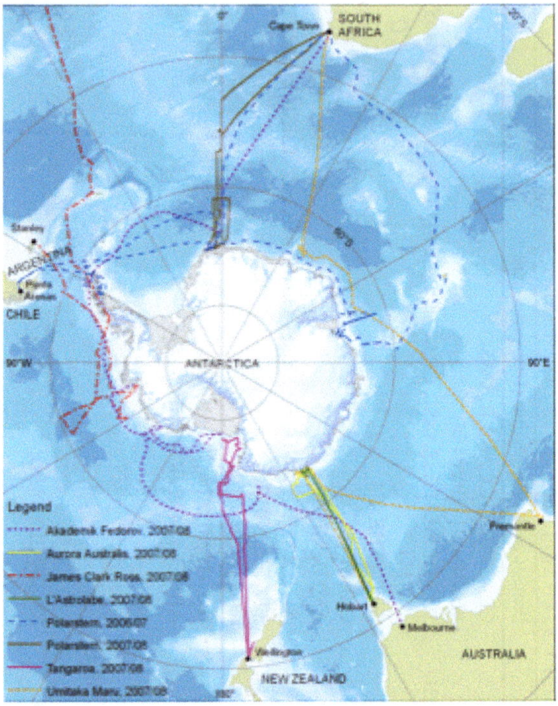

Fig. 1.1 Tracks of main research voyages under the CAML umbrella in the 2006–2008

This report only lists the voyages undertaken in the CAML framework (Table 1.1) and show the routes of the major ones (Fig. 1.1). Other information about smaller research projects can be found in the CAML web site.

1.3 CAML Main Results

1.3.1 Distributional Records

In 2010, immediately after the end of CoML, over 1 million distribution records, representing over 5,200 species (validated by taxonomic experts) were present in the SCAR-MarBIN information systems (Griffiths 2010, Fig. 4). More data sets are continuously added with the completion of specific studies and will increase our spatial coverage of sampling stations around Antarctica giving to end users updated information about species distributions and state of the art.

This impressive number of records is the largest repository of Antarctic marine georeferenced biodiversity information ever compiled and is freely available through Internet (De Broyer et al. 2012c; www.scarmarbin.be).

1.3.2 Coordination with SCAR-MarBIN

In parallel with CAML, SCAR-MarBIN was born in the framework of the International Polar Year (IPY) (De Broyer et al. 2012c). The project's objective is clear-cut: to offer free and open access to Antarctic marine biodiversity data, and provide a means to manage data arising from CAML's exceptional sampling effort.

SCAR-MarBIN was inspired by the Antarctic Treaty (Art. 3.1.c): "In order to promote international cooperation in scientific investigation in Antarctica [...] Scientific observations and results from Antarctica shall be exchanged and made freely available."

SCAR-MarBIN was quickly adopted by SCAR as its marine biodiversity information network. Based in Brussels, the project has been mainly funded by the Belgian Science Policy Office since 2005. SCAR-MarBIN's mandate was to build a web portal offering a single access point to Antarctic marine biodiversity data, providing a thematic information Node for global networks, such as the Ocean Biogeographic Information System (OBIS, www.iobis.org), or the Global Biodiversity Information Facility (GBIF, www.gbif.org).

Thanks to a shared vision, and an extremely tight collaboration, SCAR-MarBIN grew as a successful tool, serving the SCAR community and beyond. As a tangible result, SCAR-MarBIN has built the first dynamic Register of Antarctic Marine Species (RAMS) (De Broyer et al. 2012a), which builds the taxonomic backbone to the information system and is maintained by an international board of specialists, as well as a biogeographic data system, giving access to hundreds of datasets which used to be scattered all over the world. For the first time, scientists have direct access to raw data on biodiversity, which, before SCAR-MarBIN, was only available from the researchers who gathered it; the data were stored on their computers and the only public mention of it was in papers they wrote.

Based on this, many projects are being developed, ranging from the new Antarctic Biodiversity Information Facility (ANTABIF, www.biodiversity.aq), harmonising the access to marine, terrestrial and limnetic biodiversity data (Danis et al. 2012c), to data-derived products (such as the Antarctic Field Guides, http://afg.biodiversity.aq, Danis et al. 2012a) or the Biogeographic Atlas of the Southern Ocean (http://atlas.biodiversity.aq, De Broyer et al. 2012b).

These commons embody the continuation of the Census of Antarctic Marine Life, in the form of an electronic ecosystem involving researchers, institutes, data, computers, networks and databases. At the moment of writing these lines, CAML-generated data is still flowing in this ecosystem.

Further details about the Antarctic Biodiversity Information Networks can be found in a dedicated chapter "Connecting biodiversity data during the IPY: the path towards Polar Science" in this book series (Danis et al. 2012b).

A full list of CAML-related datasets can be consulted on the ANTABIF data portal (http://data.biodiversity.aq).

1.3.3 DNA Barcoding

CAML aimed to collate DNA barcode sequences for as many Antarctic marine species and Antarctic research expeditions as possible, in order to maximise the taxonomical and geographical coverage of obtained sequences.

This large-scale effort was a CAML priority, given the potentially great extent of environmental change in polar region under climate-driven fluctuations, which might lead to severe extinctions, and to the fact that, before CAML, a reference baseline of barcode sequences was substantially lacking for Antarctica (Grant and Linse 2009).

Thanks to the collaborations with the Marine Barcode of Life project and the Canadian Centre for DNA Barcoding, it was possible to start large-scale DNA barcoding projects for many of the major CAML voyages.

Under the CAML umbrella, material obtained in the framework of different expeditions was studied in a coordinated fashion, by focusing on gaps, setting priorities for most important taxa and avoiding any loss of resources deriving by potentially duplicated collecting effort in similar regions by different research groups (Grant et al. 2011).

In this way, large collections of organisms and sequences have been achieved and several researchers were able to collate data sets with thousands of sequences (e.g. Hemery et al. 2012).

1.3.4 Published Results and Journal Special Issues

Several journal special issues have appeared in the last years entirely or only partly dedicated to specific CAML voyages or to national Projects linked to CAML. Here we report the main references concerning special issues related (entirely or partially) to CAML-related data, in order to address the reader to the information. Tens of other CAML-related papers have been and are being published in many journals but it is not possible here to give an account about all these.

An Antarctic Science special issue (Antarctic Sci 20, Special Issue 3) dedicated to the project "Biodiversity, Phylogeny, Evolution and Adaptive Radiation of life in Antarctica (BIOPEARL)", managed by the British Antarctic Survey, appeared in 2008 (Linse 2008b). This project had two voyages, one in the Scotia Sea in 2006 (BIOPEARL I) and one in the Amundsen Sea in 2008 (BIOPEARL II). The special issue includes eleven papers: Linse (2008a); Griffiths et al. (2008); Linse et al. (2008); Strugnell et al. (2008); Rock et al. (2008); Barnes (2008); Barnes et al. (2008); Kaiser et al. (2008); Pearce (2008); Allen and Smellie (2008).

In 2010, a special issue of Polar Science (Volume 4, Issue 2), "Antarctic Biology in the 21st Century—Advances in and beyond IPY", assembled papers presented at the 10th SCAR International Biology Symposium, Hokkaido

University, Sapporo 2009 (Fukuchi and Conlan 2010; Stoddart 2010). Of the 22 contribution, 13 are on marine themes: Coppola et al. (2010); de Pascale et al. (2010); Jadwiszczak (2010); Koubbi et al. (2010); Lautredou et al. (2010); Massom and Stammerjohn (2010); McLeod et al. (2010); Naito et al. (2010); Russo et al. (2010); Sato-Okoshi et al. (2010); Takahashi et al. (2010); Toda et al. (2010); Vallesi et al. (2010).

The international project "Cooperative East Antarctic Marine Census project (CEAMARC)", was conducted by a consortium of 3 ships from Australia, France and Japan, and scientists/students from several nations. Eighteen contributions from this project were published in a special issue of Polar Science (Vol 5, Issue 2) (Hosie et al. 2011): Amakasu et al. (2011a, b); Causse et al. (2011); Cherel et al. (2011); Dettaï et al. (2011a); Giraldo et al. (2011); Koubbi et al. (2011a, c); Lacarra et al. (2011); Mayzaud et al. (2011); Moteki et al. (2011); Ono et al. (2011); Smith et al. (2011); Swadling et al. (2011); Takahashi et al. (2011); Vallet et al. (2011a, b).

ANtarctic benthic DEEP-sea biodiversity: colonisation history and recent community patterns—SYSTem COupling (ANDEEP-SYSCO), was the prosecution of the ANDEEP I-III expeditions, aimed at defining responses of the abyssal benthos to seasonal and longer-term changes in primary productivity (Brandt et al. 2011b). Ten papers were published in a special issue of Deep Sea Research II (Vol 58, Issues 19–20): Brandt et al. (2011a); Brenke et al. (2011); Flores et al. (2011); Göcke and Janussen (2011); Janussen and Rapp (2011); Pawlowski et al. (2011); Veit-Köhler et al. (2011); Wilmsen and Schüller (2011); Würzberg et al. (2011a, b).

In 2009, the South American Consortium on Antarctic Marine Biodiversity, (LA CAML), organised a LA CAML/BioMAntar/COMARGE Integrated Workshop on Antarctic-South America interactions. Ten papers were published in a special issue of Oecologia Australis (Campos et al. 2011; Vol 15, Issue 1), "Antarctic—South American Interactions in the Marine Environment (ASAI)" [www.oecologiaaustralis.org/ojs/index.php/oa/issue/view/29 (Aguayo–Lobo et al. 2011; Costa et al. 2011; Da Rocha Fortes and Absalão 2011; de Moura Barboza et al. 2011; di Prisco and Verde 2011; Gutt and Arntz, 2011; Ivar do Sul et al. 2011; Krüger and Petry 2011; Rodrigues et al. 2011; Verde et al. 2011)].

Major achievements were presented in the CAML Final Symposium "Diversity and Change in Southern Ocean Ecosystems", hosted by the Italian National Antarctic Museum, Genova 2009. From the 42 presentations, 25 were assembled in a special issue (Schiaparelli and Hopcroft 2011) of Deep Sea Research II, Vol 58, Issues 1–4: Allcock et al. 2011; Arango et al. 2011; Bowden et al. 2011; De Broyer et al. 2011; Dettaï et al. 2011b; Díaz et al. 2011; Eléaume et al. 2011; González-Wevar et al. 2011; Griffiths et al. 2011; Gutt et al. 2011; Hardy et al. 2011; Havermans et al. 2011; Kaiser et al. 2011; Koubbi et al. 2011b; Mühlenhardt-Siegel 2011; Nakayama et al. 2011; O'Driscoll et al. 2011; O'Loughlin et al. 2011; Olguín and Alder 2011; Post et al. 2011; Rapp et al. 2011; Schrödl et al. 2011; Siciński et al. 2011; Strugnell et al. 2011; Würzberg et al. 2011c.

Table 1.2 CAML main workshops and events

Location	Date	Topic
London (UK)	1–4 Oct 10	CoML Final meeting Presentations by all field programs
Rio de Janeiro (Brazil)	3–4 Nov 09	Antarctic-South American Interactions in the Marine Environment (ASAI)
Genova (Italy)	**17–20 May 09**	**CAML Final Symposium**
Villefranche sur Mer (France)	13–16 May	Polar synthesis macroscope
Long Beach, California (USA)	17–21 Feb 09	CoML all program meeting
Valencia (Spain)	9–10 Nov 08	Meeting of CAML Steering Committee
Durham, North Carolina (USA)	26–28 Oct 08	Polar Synthesis Macroscope
St Petersburg (Russia)	08–11 Jul 08	SCAR-IASC Open Science Conference
Bonn (Germany)	19–30 May 08	European Conference of Parties, Biodiversity Conference
Bialowieza, Poland	**4–8 Jun 07**	**CAML Scientific Steering Committee and Invited Experts meeting**
Tokyo (Japan)	11–12 Oct 06	CAML Top Predators Working Group
Concepcion (Chile)	Aug 06	Oficina Latinoamericana para el Census of Antarctic Mar Life (OLA-CAML)
Hobart (Australia)	15 Jul 06	Southern Ocean Observation System (SOOS) Workshop
Hobart (Australia)	11 Jul 06	Workshop: Processes of Colonisation and Dispersal—how they shape the Biodiversity of Antarctic Marine Ecosystems and CAML Scientific Steering Committee meeting
Brussels (Belgium)	**27–30 May 05**	**First CAML workshop**

In *bold* are the major events, where guidelines (Brussel, 2005), coordination between cruises and targets (Bialowieza, 2007) and final results (Genova, 2009) were discussed

1.3.5 Workshop Organisation

The milestone workshop for CAML was that held in Brussels in May 2005. The draft Science Plan was produced and finalised in September, after consultation with the SC. This Plan was circulated within nations and marine Antarctic science organisations to obtain encourage involvement in CAML and to help national projects to get funded by national institutions/funding agencies. In other workshops, partial results were discussed and/or data set organised (Table 1.2). The CoML Final Meeting with presentations by all field programmes was the last formal meeting of CAML, which ended in December 31st, 2010.

1.4 The CAML Legacy

The CAML scientific legacy can be summarised in four main points:

- Legacy sampling sites (identifiable by GPS) from which biodiversity data have been obtained and that can be re-sampled at intervals in the future in order to track changes in marine biodiversity.
- Collections of marine specimens that may be used for biological research for many years to come.
- Publication of data sets that can be freely accessed (so far, 66 data sets relating to CAML are available at http://data.biodiversity.aq (last search in April 10th, 2012).
- Baseline of Antarctic DNA barcodes (COI) that enabled and will enable to answer important questions regarding marine genetic diversity and distribution of species and their links with areas outside the Polar Front.

Moreover, CAML facilitated the establishment of regional networks of Institutions that worked at the national level in a similarly coordinated way. This is the case of the Latin America CAML (LA CAML), where seven Latin-American countries (Chile, Peru, Ecuador, Brazil, Argentina, Venezuela, Uruguay) operated together in order to maximise gaining Antarctic biodiversity data. Twenty-five projects were run by LA CAML researchers (www.caml.aq/news/la-caml-projects.html).

After a meeting in Hobart (July 2006), following the Open Science Conference, SCAR began establishing the Southern Ocean Observing System (SOOS, http://www.scar.org/treaty/atcmxxxiii/ATCM33_ip050_e.pdf). SOOS was established in 2010, with a SC reporting to SCAR and to the Scientific Committee on Ocean Research (SCOR), to coordinate multi-disciplinary pan-Antarctic long-term monitoring systems which will help to assess the impacts of global change on Southern Ocean ecosystems. A Secretariat has been established at the Institute for Marine and Antarctic Studies at the University of Tasmania, for 2010–2015.

1.5 Concluding Remarks

During its five-year time span CAML generated considerable interest in Antarctic marine biodiversity, with press releases, television and radio interviews given by CAML scientists. It appeared in CoML press releases and publicity. Based on the biological complexity reported by CEAMARC, CCAMLR declared two areas off Terre Adélie as "Vulnerable Marine Ecosystems" that, while not formally protected, should be avoided by bottom fishers. As an important catalyst for the launch of SOOS, and as the major data-provider for SCAR MarBIN, CAML's greatest legacy has been the renewed interest in Antarctic marine biodiversity. It is to be hoped that SCAR and other influential bodies are able to build on its achievements and further understand the biological complexity of the vast circum-Antarctic Southern Ocean, particularly as climate change is expected to bring about significant changes in the decades ahead.

Acknowledgments CAML gratefully acknowledges the financial assistance received from CoML for coordination, without which international workshops and coordination meetings would not have been possible. It also acknowledges the support from National Antarctic Programmes which provided ship time. IPY and CoML together provided a unique set of conditions for this important research to be undertaken. This is CAML contribution #79.

References

Aguayo–Lobo A, Acevedo JR, Brito JL, Acuña PG, Bassoi M, Secchi ER, Dalla Rosa L (2011) Presence of the leopard seal, *Hydrurga leptonyx* (de Blainville, 1820), on the coast of Chile: an example of the Antarctica—South America connection in the marine environment. Oecol Aust 15:69–85

Allcock AL, Barratt I, Eléaume M, Linse K, Norman MD, Smith PJ, Steinke D, Stevens DW, Strugnell JM (2011) Cryptic speciation and the circumpolarity debate: a case study on endemic Southern Ocean octopuses using the COI barcode of life. Deep-Sea Res II 58:242–249

Allen CS, Smellie JL (2008) Volcanic features and the hydrological setting of Southern Thule, South Sandwich Islands. Antarct Sci 20:301–308

Amakasu K, Ono A, Hirano D, Moteki M, Ishimaru T (2011a) Distribution and density of Antarctic krill (*Euphausia superba*) and ice krill (*E. crystallorophias*) off Adélie Land in austral summer 2008 estimated by acoustical methods. Polar Sci 5:187–194

Amakasu K, Ono A, Moteki M, Ishimaru T (2011b) Sexual dimorphism in body shape of Antarctic krill (*Euphausia superba*) and its influence on target strength. Polar Sci 5:179–186

Arango CP, Soler-Membrives A, Miller KJ (2011) Genetic differentiation in the circum—Antarctic sea spider *Nymphon australe* (Pycnogonida; Nymphonidae). Deep-Sea Res II 58:212–219

Barnes DKA (2008) A benthic richness hotspot in the Southern Ocean: slope and shelf cryptic benthos of Shag Rocks. Antarct Sci 20:263–270

Barnes DKA, Linse K, Enderlein P, Smale D, Fraser KPP, Brown M (2008) Marine richness and gradients at Deception Island, Antarctica. Antarct Sci 20:271–280

Bowden DA, Schiaparelli S, Clark MR, Rickard GJ (2011) A lost world? Archaic crinoid-dominated assemblages on an Antarctic seamount. Deep-Sea Res II 58:119–127

Brandt A, De Broyer C, Gooday AJ, Hilbig B, Thomson MRA (2004) Introduction to ANDEEP (ANtarctic benthic DEEP-sea biodiversity: colonization history and recent community patterns)—a tribute to Howard L. Sanders Deep-Sea Res II 51:1457–1465

Brandt A, Ebbe B, Gooday AJ (2007) Introduction to ANDEEP, summary and outlook. Deep-Sea Res II 54:1645–1651

Brandt A, Bathmann U, Brix S, Cisewski B, Flores H, Göcke C, Janussen D, Krägefsky S, Kruse S, Leach H, Linse K, Pakhomov E, Peeken I, Riehl T, Sauter E, Sachs O, Schüller M, Schrödl M, Schwabe E, Strass V, van Franeker JA, Wilmsen E (2011a) Maud Rise—a snapshot through the water column. Deep-Sea Res II 58:1962–1982

Brandt A, Ebbe B, Bathmann U (2011b) Southern Ocean biodiversity—from pelagic processes to deep-sea response. Deep-Sea Res II 58:1945–1947

Brenke N, Guilini K, Ebbe B (2011) Characterization of the seafloor at the SYSTCO stations based on video observations and ground truthing. Deep-Sea Res II(58):2043–2050

Campos SL, Bassoi M, Nakayama C, Valentin YY, Lavrado HP, Menot L, Sibuet M (2011) A (2011) Antarctic—South American interactions in the marine environment: a COMARGE and CAML effort through the South American consortium on Antarctic marine biodiversity. Oecol Aust 15:5–22
Causse R, Ozouf-Costaz C, Koubbi P, Lamy D, Eléaume M, Dettaï A, Duhamel G, Busson F, Pruvost P, Post A, Beaman RJ, Riddle MJ (2011) Demersal ichthyofaunal shelf communities from the Dumont d'Urville Sea (East Antarctica). Polar Sci 5:272–285
CCAMLR (2009a) Conservation measure 22-06. Bottom fishing in the Convention Area. http://www.ccamlr.org/pu/e/e_pubs/cm/09-10/22-06.pdf. Accessed 10 April 2012
CCAMLR (2009b) Conservation Measure 22-07. Interim measure for bottom fishing activities subject to Conservation Measure 22-06 encountering potential vulnerable marine ecosystems in the Convention Area. http://www.ccamlr.org/pu/e/e_pubs/cm/09-10/22-07.pdf. Accessed 10 April 2012
Cherel Y, Koubbi P, Giraldo C, Penot F, Tavernier E, Moteki M, Ozouf-Costaz C, Causse R, Chartier A, Hosie G (2011) Isotopic niches of fishes in coastal, neritic and oceanic waters off Adélie land, Antarctica. Polar Sci 5:286–297
Coppola D, Giordano D, Vergara A, Mazzarella L, di Prisco G, Verde C, Russo R (2010) The hemoglobins of sub-Antarctic fishes of the suborder Notothenioidei. Polar Sci 4:295–308
Costa ES, Ayala L, Ivar do Sul JA, Coria NR, Sanchez-Scaglioni RE, dos Santos Alves MA, Petry MV, Piedrahita P (2011) Antarctic and sub-Antarctic seabirds in South America: a review. Oecol Aust 15:59–68
Da Rocha Fortes R, Absalão RS (2011) Biogeography and connectivity between western South American and Antarctic marine molluscs. Oecol Aust 15:111–123
Danis B, De Broyer C, Clarke A, Schiaparelli S (eds) (2012a) The SCAR Antarctic field guides. http://afg.biodiversity.aq and http://afg.scarmarbin.be. Accessed April 10 2012
Danis B, Van de Putte A, Renaudier S, Griffiths A (2012b) Connecting biodiversity data during the IPY: the path towards polar science. In: Verde C, di Prisco G (eds) Adaptation and evolution in marine environments—The impacts of global change on biodiversity, vol 2. Springer, Berlin
Danis B, Van de Putte A, Youdjou N, Segers S (eds) (2012c) ANTABIF: the Antarctic biodiversity information facility. http://www.biodiversity.aq. Accessed 10 April 2012
De Broyer C, Danis B, with 64 SCAR-MarBIN Taxonomic Editors (2011) How many species in the Southern Ocean? Towards a dynamic inventory of the Antarctic marine species. Deep-Sea Res II 58:5–17
De Broyer C, Clarke A, Koubbi P, Pakhomov E, Scott F, Vanden Berghe W, Danis B (eds) (2012a) The SCAR-MarBIN register of Antarctic marine species (RAMS). http://www.scarmarbin.be/scarramsabout.php. Accessed 10 April 2012
De Broyer C, Koubbi Ph (eds) (2012b) The SCAR biogeographic atlas of the Southern Ocean. http://www.atlas.biodiversity.aq. Accessed 10 April 2012
De Broyer C, Renaudier S, Danis B (eds) (2012c) SCAR-MarBIN: the Antarctic marine biodiversity information network. http://www.scarmarbin.be. Accessed 10 April 2012
de Moura Barboza CA, Bendayan de Moura R, Monnerat Lanna A, Oackes T, Campos LS (2011) Echinoderms as clues to Antarctic ~ South American connectivity. Oecol Aust 15:86–110
de Pascale D, Giuliani M, De Santi C, Bergamasco N, Amoresano A, Carpentieri A, Parrilli E, Tutino ML (2010) PhAP protease from *Pseudoalteromonas haloplanktis* TAC125: Gene cloning, recombinant production in *E. coli* and enzyme characterization. Polar Sci 4:285–294
Dettaï A, Adamowizc SJ, Allcock L, Arango CP, Barnes DKA, Barratt I, Chenuil A, Couloux A, Cruaud C, David B, Denis F, Denys G, Díaz A, Eléaume M, Féral J-P, Froger A, Gallut C, Grant R, Griffiths HJ, Held C, Hemery LG, Hosie G, Kuklinski P, Lecointre G, Linse K, Lozouet P, Mah C, Monniot F, Norman MD, O'Hara T, Ozouf-Costaz C, Piedallu C, Pierrat B, Poulin E, Puillandre N, Riddle M, Samadi S, Saucède T, Schubart C, Smith PJ, Stevens DW, Steinke D, Strugnell JM, Tarnowska K, Wadley V, Ameziane N (2011a) DNA barcoding and molecular systematics of the benthic and demersal organisms of the CEAMARC survey. Polar Sci 5:298–312

Dettaï A, Lautredou A-C, Bonillo C, Goimbault E, Busson F, Causse R, Couloux A, Cruaud C, Duhamel G, Denys G, Hautecoeur M, Iglesias S, Koubbi P, Lecointre G, Moteki M, Pruvost P, Tercerie S, Ozouf C (2011b) The actinopterygian diversity of the CEAMARC cruises: barcoding and molecular taxonomy as a multi-level tool for new findings. Deep-Sea Res II 58:250–263

di Prisco G, Verde C (2011) Evolution and biodiversity in the Antarctic—the response of life to change (EBA): role and legacy. Oecol Aust 15:23–31

Díaz A, Féral JP, David B, Saucède T, Poulin E (2011) Evolutionary pathways among shallow and deep-sea echinoids of the genus *Sterechinus* in the Southern Ocean. Deep-Sea Res II 58:205–211

Eléaume M, Beaman RJ, Griffiths HJ, Best B, Riddle MJ, Wadley V, Rintoul SR, Hemery LG, Améziane N (2011) Near-bottom current direction inferred from comatulid crinoid feeding postures on the Terre Adélie and George V shelf, East Antarctica. Deep-Sea Res II 58: \163–169

Flores H, van Franeker J-A, Cisewski B, Leach H, Van de Putte AP, Meesters EHWG, Bathmann U, Wolff WJ (2011) Macrofauna under sea ice and in the open surface layer of the Lazarev Sea, Southern Ocean. Deep-Sea Res II 58:1948–1961

Fukuchi M, Conlan KE (2010) Xth SCAR international biology symposium on antarctic biology in the 21st century—advances in and beyond IPY: a brief overview. Polar Sci 4:93–96

Giraldo C, Cherel Y, Vallet C, Mayzaud P, Tavernier E, Moteki M, Hosie G, Koubbi P (2011) Ontogenic changes in the feeding ecology of the early life stages of the Antarctic silverfish (*Pleuragramma antarcticum*) documented by stable isotopes and diet analysis in the Dumont d'Urville Sea (East Antarctica). Polar Sci 5:252–263

Göcke C, Janussen D (2011) ANT XXIV/2 (SYSTCO) Hexactinellida (Porifera) and bathymetric traits of Antarctic glass sponges (incorporating ANDEEP-material); including an emendation of the rediscovered genus *Lonchiphora*. Deep-Sea Res II 58:2013–2021

González-Wevar CA, David B, Poulin E (2011) Phylogeography and demographic inference in *Nacella* (*Patinigera*) *concinna* (Strebel, 1908) in the western Antarctic Peninsula. Deep-Sea Res II 58:220–229

Gradinger R, Bluhm BA, Hopcroft RR, Gebruk AV, Kosobokova K, Sirenko B, Węsławski (2010) Marine life in the Arctic. In: McIntyre AD (ed) Life in the world's oceans: diversity, distribution, and abundance, 1st edn. Blackwell Publishing Ltd, Oxford, pp 183–202

Grant RA, Linse K (2009) Barcoding antarctic biodiversity: current status and the CAML initiative, a case study of marine invertebrates. Polar Biol 32:1629–1637

Grant RA, Griffiths HJ, Steinke D, Wadley V, Linse K (2011) Antarctic DNA barcoding; a drop in the ocean? Polar Biol 34:775–780

Griffiths HJ (2010) Antarctic marine biodiversity—what do we know about the distribution of life in the Southern Ocean? PLoS ONE 5(8):e11683

Griffiths HJ, Linse K, Barnes DKA (2008) Distribution of macrobenthic taxa across the Scotia Arc, Southern Ocean. Antarct Sci 20:213–226

Griffiths HJ, Danis B, Clarke A (2011) Quantifying Antarctic marine biodiversity: the SCAR-MarBIN data portal. Deep-Sea Res II 58:18–29

Gutt J, Arntz WE (2011) South American and Antarctic interactions in the marine environment—biological relationships, general relevance and research perspective. Oecol Aust 15:171–174

Gutt J, Hosie G, Stoddart M (2010) Marine life in the Antarctic. In: McIntyre AD (ed) Life in the world's oceans: diversity, distribution, and abundance, 1st edn. Blackwell Publishing Ltd, Oxford, pp 203–220

Gutt J, Barratt I, Domack E, d'Udekem d'Acoz C, Dimmler W, Grémare A, Heilmayer O, Isla E, Janussen D, Jorgensen E, Kock K-H, Lehnert LS, López-Gonzáles P, Langner S, Linse K, Manjón-Cabeza ME, Meißner M, Montiel A, Raes M, Robert H, Rose A, Sañé Schepisi E, Saucède T, Scheidat M, Schenke H-W, Seiler J, Smith C (2011) Biodiversity change after climate-induced ice-shelf collapse in the Antarctic. Deep-Sea Res II 58:74–83

Hardy C, David B, Rigaud C, De Ridder C, Saucède T (2011) Ectosymbiosis associated with cidaroids (Echinodermata: Echinoidea) promotes benthic colonization of the seafloor in the Larsen Embayments, Western Antarctica. Deep-Sea Res II 58:84–90

Havermans C, Nagy GT, Sonet G, De Broyer C, Martin P (2011) DNA barcoding reveals new insights into the diversity of Antarctic species of *Orchomene* sensu lato (Crustacea: Amphipoda: Lysianassoidea). Deep-Sea Res II 58:230–241

Hemery LG, Eléaume M, Roussel V, Améziane N, Gallut C, Steinke D, Cruaud C, Couloux A, Wilson NG (2012) Comprehensive sampling reveals circumpolarity and sympatry in seven mitochondrial lineages of the Southern Ocean crinoid species *Promachocrinus kerguelensis* (Echinodermata). Mol Ecol 21:2502–2518

Hosie GW, Koubbi P, Riddle MJ, Ozouf-Costaz C, Moteki M, Fukuchi M, Ameziane N, Ishimaru T, Goffart A (2011) CEAMARC, the collaborative east Antarctic marine census for the census of antarctic marine life (IPY # 53): an overview. Polar Sci 5:75–87

Ivar do Sul JA, Barnes DKA, Costa MF, Convey P, Costa ES, Campos LS (2011) Plastics in the Antarctic environment: are we looking only at the tip of the iceberg? Oecol Aust 15:150–170

Jadwiszczak P (2010) Penguin response to the Eocene climate and ecosystem change in the northern Antarctic Peninsula region. Polar Sci 4:229–235

Janussen D, Rapp HT (2011) Redescription of *Jenkina articulata* Brøndsted from the deep Eckström Shelf, E-Weddell Sea, Antarctica and a comment on the possible mass occurrence of this species. Deep-Sea Res II 58:2022–2026

Kaiser S, Barnes DKA, Linse K, Brandt A (2008) Epibenthic macrofauna associated with the shelf and slope of a young and isolated Southern Ocean island. Antarct Sci 20:281–290

Kaiser S, Griffiths HJ, Barnes DKA, Brandão SN, Brandt A, O'Brien PE (2011) Is there a distinct continental slope fauna in the Antarctic? Deep-Sea Res II 58:91–104

Koubbi P, Ozouf-Costaz C, Goarant A, Moteki M, Hulley P-A, Causse R, Dettaï A, Duhamel G, Pruvost P, Tavernier E, Post AL, Beaman RJ, Rintoul SR, Hirawake T, Hirano D, Ishimaru T, Riddle M, Hosie G (2010) Estimating the biodiversity of the East Antarctic shelf and oceanic zone for ecoregionalisation: example of the ichthyofauna of the CEAMARC (Collaborative East Antarctic Marine Census) CAML surveys. Polar Sci 4:115–133

Koubbi P, Hulley P-A, Pruvost P, Henri P, Labat J-P, Wadley V, Hirano D, Moteki M (2011a) Size distribution of meso- and bathypelagic fish in the Dumont d'Urville Sea (East Antarctica) during the CEAMARC surveys. Polar Sci 5:195–210

Koubbi P, Moteki M, Duhamel G, Goarant A, Hulley P-A, O'Driscoll R, Ishimaru T, Pruvost P, Tavernier E, Hosie G (2011b) Ecoregionalization of myctophid fish in the Indian sector of the Southern Ocean: results from generalized dissimilarity models. Deep-Sea Res II 58:170–180

Koubbi P, O'Brien C, Loots C, Giraldo C, Smith M, Tavernier E, Vacchi M, Vallet C, Chevallier J, Moteki M (2011c) Spatial distribution and inter-annual variations in the size frequency distribution and abundances of *Pleuragramma antarcticum* larvae in the Dumont d'Urville Sea from 2004 to 2010. Polar Sci 5:225–238

Krüger L, Petry MV (2011) On the relation of Antarctic and sub-Antarctic seabirds with abiotic variables of South and Southeast Brazil. Oecol Aust 15:51–58

Lacarra M, Houssais M-N, Sultan E, Rintoul SR, Herbaut C (2011) Summer hydrography on the shelf off Terre Adélie/George V Land based on the ALBION and CEAMARC observations during the IPY. Polar Sci 5:88–103

Lautredou AC, Bonillo C, Denys G, Cruaud C, Ozouf-Costaz C, Lecointre G, Dettaï A (2010) Molecular taxonomy and identification within the Antarctic genus *Trematomus* (Notothenioidei, Teleostei): How valuable is barcoding with COI? Polar Sci 4:333–352

Linse K (2008a) Antarctic marine biodiversity—a taxonomic crisis? Antarct Sci 20:209

Linse K (2008b) The BIOPEARL expedition to the Scotia Sea in 2006. Ant Sci 20:211–212

Linse K, Walker LJ, Barnes DKA (2008) Biodiversity of echinoids and their epibionts around the Scotia Arc, Antarctica. Antarct Sci 20:227–244

Massom RA, Stammerjohn SE (2010) Antarctic sea ice change and variability—physical and ecological implications. Polar Sci 4:149–186

Mayzaud P, Chevallier J, Tavernier E, Moteki M, Koubbi P (2011) Lipid composition of the Antarctic fish *Pleuragramma antarcticum.* Influence of age class. Polar Sci 5:264–271
McLeod DJ, Hosie GW, Kitchener JA, Takahashi KT, Hunt BPV (2010) Zooplankton atlas of the Southern Ocean: the SCAR SO-CPR survey (1991–2008). Polar Sci 4:353–385
Moteki M, Koubbi P, Pruvost P, Tavernier E, Hulley P-A (2011) Spatial distribution of pelagic fish off Adélie and George V Land, East Antarctica in the austral summer 2008. Polar Sci 5:211–224
Mühlenhardt-Siegel U (2011) Cumacean (Peracarida, Crustacea) endemism and faunal overlap in Antarctic deep-sea basins. Deep-Sea Res II 58:68–73
Naito Y, Bornemann H, Takahashi A, McIntyre T, Plötz J (2010) Fine-scale feeding behavior of Weddell seals revealed by a mandible accelerometer. Polar Sci 4:309–316
Nakayama CR, Kuhn E, Araújo ACV, Alvalá PC, Ferreira WJ, Vazoller RF, Pellizari VH (2011) Revealing archaeal diversity patterns and methane fluxes in Admiralty Bay, King George Island, and their association to Brazilian Antarctic Station activities. Deep-Sea Res II 58:128–138
O'Loughlin PM, Paulay G, Davey N, Michonneau F (2011) The Antarctic region as a marine biodiversity hotspot for echinoderms: Diversity and diversification of sea cucumbers. Deep-Sea Res II 58:264–275
O'Driscoll RL, Macaulay GJ, Gauthier S, Pinkerton M, Hanchet S (2011) Distribution, abundance and acoustic properties of Antarctic silverfish (*Pleuragramma antarcticum*) in the Ross Sea. Deep-Sea Res II 58:181–195
Olguín HF, Alder VA (2011) Species composition and biogeography of diatoms in Antarctic and Subantarctic (Argentine shelf) waters (37–76°S). Deep-Sea Res II 58:139–152
Ono A, Moteki M, Amakasu K, Toda R, Horimoto N, Hirano D, Ishimaru T, Hosie GW (2011) Euphausiid community structure and population structure of *Euphausia superba* off Adélie Land in the Southern Ocean during austral summer 2003, 2005 and 2008. Polar Sci 5:146–165
Pawlowski J, Fontaine D, Aranda da Silva A, Guiard J (2011) Novel lineages of Southern Ocean deep-sea foraminifera revealed by environmental DNA sequencing. Deep-Sea Res II 58:1996–2003
Pearce DA (2008) Biodiversity of the bacterioplankton in the surface waters around Southern Thule in the Southern Ocean. Antarct Sci 20:291–300
Post AL, O'Brien PE, Beaman RJ, Riddle MJ, De Santis L (2010) Physical controls on deep-water coral communities on the George V Land slope, East Antarctica. Antarct Sci 22:371–378
Post AL, Beaman RJ, O'Brien PE, Eléaume M, Riddle MJ (2011) Community structure and benthic habitats across the George V Shelf, East Antarctica: trends through space and time. Deep-Sea Res II 58:105–118
Rapp HT, Janussen D, Tendal OS (2011) Calcareous sponges from abyssal and bathyal depths in the Weddell Sea, Antarctica original research article. Deep-Sea Res II 58:58–67
Rock J, Costa FO, Walker DI, North AW, Hutchinson WF, Carvalho GR (2008) DNA barcodes of fish of the Scotia Sea, Antarctica indicate priority groups for taxonomic and systematics focus. Antarct Sci 20:253–262
Rodrigues E, Suda CNK, Rodrigues Junior E, Feijó de Oliveira M, dos Santos Carvalho C, Sree Vani G (2011) Antarctic fish metabolic responses as potential biomarkers of environmental impact. Oecol Aust 15:124–149
Russo R, Riccio A, di Prisco G, Verde C, Giordano D (2010) Molecular adaptations in Antarctic fish and bacteria. Polar Sci 4:245–256
Sato-Okoshi W, Okoshi K, Sasaki H, Akiha F (2010) Shell structure characteristics of pelagic and benthic molluscs from Antarctic waters. Polar Sci 4:257–261
Schiaparelli S, Hopcroft RR (2011) The census of Antarctic marine life: diversity and change in southern ocean ecosystems. Deep-Sea Res II 58:1–4
Schrödl M, Bohn JM, Brenke N, Rolán E, Schwabe E (2011) Abundance, diversity, and latitudinal gradients of southeastern Atlantic and Antarctic abyssal gastropods. Deep-Sea Res II 58:49–57

Siciński J, Jażdżewski K, De Broyer C, Presler P, Ligowski R, Nonato EF, Corbisier TN, Petti MAV, Brito TAS, Lavrado HP, BŁażewicz-Paszkowycz M, Pabis K, Jażdżewska A, Campos LS (2011) Admiralty Bay benthos diversity—a census of a complex polar ecosystem. Deep-Sea Res II 58:30–48

Smith MB, Labat J-P, Fraser AD, Massom RA, Koubbi P (2011) A GIS approach to estimating interannual variability of sea ice concentration in the Dumont d'Urville Sea near Terre Adélie from 2003 to 2009. Polar Sci 5:104–117

Stoddart M (2010) Antarctic biology in the 21st century—advances in, and beyond the international polar year 2007–2008. Polar Sci 4:97–101

Strugnell JM, Collins MA, Allcock AL (2008) Molecular evolutionary relationships of the octopodid genus *Thaumeledone* (Cephalopoda: Octopodidae) from the Southern Ocean. Antarct Sci 20:245–251

Strugnell JM, Cherel Y, Cooke IR, Gleadall IG, Hochberg FG, Ibáñez CM, Jorgensen E, Laptikhovsky VV, Linse K, Norman M, Vecchione M, Voight JR, Allcock AL (2011) The Southern Ocean: source and sink? Deep-Sea Res II 58:196–204

Swadling KM, Penot F, Vallet C, Rouyer A, Gasparini S, Mousseau L, Smith M, Goffart A, Koubbi P (2011) Interannual variability of zooplankton in the Dumont d'Urville sea (139°E – 146°E), east Antarctica, 2004–2008. Polar Sci 5:118–133

Takahashi KT, Hosie GW, Kitchener JA, McLeod DJ, Odate T, Fukuchi M (2010) Comparison of zooplankton distribution patterns between four seasons in the Indian Ocean sector of the Southern Ocean. Polar Sci 4:317–331

Takahashi KT, Hosie GW, McLeod DJ, Kitchener JA (2011) Surface zooplankton distribution patterns during austral summer in the Indian sector of the Southern Ocean, south of Australia. Polar Sci 5:134–145

Toda R, Moteki M, Ono A, Horimoto N, Tanaka Y, Ishimaru T (2010) Structure of the pelagic cnidarian community in Lützow–Holm Bay in the Indian sector of the Southern Ocean. Polar Sci 4:387–404

Vallesi A, Alimenti C, Di Giuseppe G, Dini F, Pedrini B, Wüthrich K, Luporini P (2010) The water-born protein pheromones of the polar protozoan ciliate, *Euplotes nobilii*: coding genes and molecular structures. Polar Sci 4:237–244

Vallet C, Beans C, Koubbi P, Courcot L, Hecq J-H, Goffart A (2011a) Food preferences of larvae of Antarctic silverfish *Pleuragramma antarcticum* Boulenger, 1902 from Terre Adélie coastal waters during summer 2004. Polar Sci 5:239–251

Vallet C, Labat J-P, Smith M, Koubbi P (2011b) Interannual variations in euphausiid life stage distribution in the Dumont d'Urville Sea from 2004 to 2008. Polar Sci 5:166–178

Veit-Köhler G, Guilini K, Peeken I, Sachs O, Sauter EJ, Würzberg L (2011) Antarctic deep-sea meiofauna and bacteria react to the deposition of particulate organic matter after a phytoplankton bloom. Deep-Sea Res II 58:1983–1995

Verde C, Giordano D, Russo R, Riccio A, Coppola D, di Prisco G (2011) Evolutionary adaptations in Antarctic fish: the oxygen-transport system. Oecol Aust 15:40–50

Wilmsen E, Schüller M (2011) Diversity and distribution of Polychaeta in deep Antarctic and Subantarctic waters along the Greenwich meridian. Deep-Sea Res II 58:2004–2012

Würzberg L, Peters J, Brandt A (2011a) Fatty acid patterns of Southern Ocean shelf and deep sea peracarid crustaceans and a possible food source, foraminiferans. Deep-Sea Res II 58: 2027–2035

Würzberg L, Peters J, Flores H, Brandt A (2011b) Demersal fishes from the Antarctic shelf and deep sea: A diet study based on fatty acid patterns and gut content analyses. Deep-Sea Res II 58:2036–2042

Würzberg L, Peters J, Schüller M, Brandt A (2011c) Diet insights of deep-sea polychaetes derived from fatty acid analyses. Deep-Sea Res II 58:153–162

[illegible] De Broyer C, [illegible] P, [illegible] R, [illegible] (eds) [illegible] *Biogeographic Atlas of the Southern Ocean* [illegible]

[illegible]

[illegible] Sci [illegible] 2 [illegible] 2009 [illegible]

[illegible] (2012) [illegible] origin of Antarctic [illegible]

Chapter 2
Connecting Biodiversity Data During the IPY: The Path Towards e-Polar Science

Bruno Danis, Anton Van de Putte, Sylvain Renaudier and Huw Griffiths

2.1 IPY and the Need for Data Sharing

The International Polar Year (IPY) was a unprecedented effort in polar research committing thousands of participants from many nations (Carlson 2010), with a common objective: to describe and understand complex systems using multiple approaches, in an integrated fashion. IPY scientists have collected enormous amounts of data in many formats, ranging from multimedia, hourly to millennial time series, isotope ratios and fractions, energy and material fluxes, species identification and distribution patterns, disease types and rates, genetic sequences, samples identifiers, singular events and gradual processes, to sociological studies (Carlson 2011).

As a result of this approach, IPY scientists now have to deal with a "data deluge" which, if well utilized, has potential to considerably change our vision on

B. Danis (✉) · A. Van de Putte
Royal Belgian Institute of Natural Sciences, Rue Vautier 29, 1000 Brussels, Belgium
e-mail: bruno.danis@gmail.com

A. Van de Putte
e-mail: antonarctica@gmail.com

B. Danis · A. Van de Putte · S. Renaudier
Antarctic Biodiversity Information Facility (ANTABIF), Rue Vautier 29, 1000 Brussels, Belgium
e-mail: s.renaudier@gmail.com

B. Danis
Laboratoire de Biologie Marine, CP160/15, Université Libre de Bruxelles, Avenue FD Roosevelt, 50, 1050 Brussels, Belgium

H. Griffiths
British Antarctic Survey, High Cross, Madingley Road, Cambridge, CB3 0ET, UK
e-mail: hjg@bas.ac.uk

C. Verde and G. di Prisco (eds.), *Adaptation and Evolution in Marine Environments, Volume 2*, From Pole to Pole, DOI: 10.1007/978-3-642-27349-0_2,

many polar systems. More and more, simulations are based on these data, and in turn produce very large amounts of newly generated data. Even if data-intensive science emphasis has been on dealing with large quantities of data (Hey and Hey 2006; Hey et al. 2009; Newman and Ellisman 2003), their heterogeneity, together with the arising need to interconnect them to understand complex problems, is an even more overwhelming challenge.

The cultural diversity of the data collectors involved in the IPY presents yet another challenge for data managers. Convincing them of the usefulness of sharing their data requires a lot of imagination, and a well-balanced usage of sticks and carrots (the latter being far more efficient). For example, depending on the cultural background of the scientists, the tools to reward their efforts towards data publication, will not be the same. This approach requires rapid adaptation efforts from both parties.

IPY data managers have established a working vision to guide their efforts for a useful vision for science in general (Parsons et al. 2011). Data should be discoverable, open, linked, useful, and safe.

More specifically, data should be:

1. capable of being located, identified, and generally accessible using simple tools
2. openly accessible. We find the current Wikipedia definition describes our intentions well: "Open data is a philosophy and practice requiring that certain data be freely available to everyone, without restrictions from copyright, patents or other mechanisms of control"
3. citable. So scientists can be acknowledged for their effort to make these data available and to provide a mechanism of validation of these efforts
4. interrelated and connected. There are many ways to interconnect data and we use the term "linked data" rather generally
5. usable for a practical, advantageous purpose or in several ways by defined but possibly very different users
6. protected from risk, corruption, and loss; now and over the long term.

These are simple terms or concepts, but they sum up the primary objectives that should be sought in managing diverse scientific data.

As biologists, the perspective on data management displays certain similarities to that described by Star and Griesemer (1989): the huge amounts of data generated during the IPY can be referred to as a "data ecosystem", including the people, technical infrastructures and technologies as well as the multiple interactions between them. These systems undergo selection processes and evolve by adapting to constantly changing needs from the communities they are serving. In this respect, there are many exciting aspects to be explored by IPY data managers. For example, IPY data ecosystems present large numbers of niches that emerge, co-evolve, and survive (or do not) depending on environmental context. If this context changes rapidly, components of the ecosystems need to cope with these variations, or they will go extinct. Ecosystems are considered most healthy when diverse, and this is why it is important that IPY data are discoverable and available in a large number of contexts. The system also includes new and legacy technologies and ever-growing data along with their countless potential mutations, provoked or undesired. The system must be understood holistically in order to optimize not just the individual components, for the benefit of society.

To help understand the IPY data ecosystem and its components and interactions, we describe an example of information network, SCAR-MarBIN (Scientific Committee on Antarctic Research—Marine Biodiversity Information Network, www.scarmarbin.be) and its successor (ANTABIF, the Antarctic Biodiversity Information Facility, www.biodiversity.aq), which both focus on biodiversity data and which took up the challenge of managing data from an extremely ambitious IPY program, the Census of Antarctic Marine Life (CAML).

2.2 The Antarctic Biodiversity Data Ecosystem

Initially, the SCAR-MarBIN initiative was launched during the CAML and IPY timeframe to provide the community with the tools to establish a web-based inventory of the Antarctic marine biodiversity, from micro-organisms to whales, that would be openly accessible to all. In the course of the IPY, CAML coordinated international expeditions that collected an unprecedented mass of information on the diversity, occurrence and abundance of Antarctic marine species. During this effort, thousands of species new to science have been discovered, in particular in the Antarctic deep sea (Brandt et al. 2007).

At the beginning, the intention of SCAR-MarBIN was to bring in as many data types as possible, that made sense for biodiversity research. But given the extreme variability of data that could potentially be considered as biodiversity-relevant (ranging from molecular to ecosystems realms) our endeavour had to adjust its objectives to a more pragmatic approach, with a philosophy to focus on robust standards that would allow the system to expand in the future, jointly with new projects, and with other disciplines. Hence, SCAR-MarBIN focused on three main data types: taxonomy (the names and classification of living things), biogeographic data (where these living things are found) and metadata (descriptions of datasets).

Establishing a complete and accurate inventory of the presently described Antarctic marine fauna and flora had never been attempted so far, and required a strong commitment of the Antarctic research community, as well as a robust data management system. Thanks to a collective effort of a board of 64 taxonomic editors, each of them being a renown specialist, and the support of the World Register of Marine Species, the first Register of Antarctic Marine Species (RAMS) was compiled and published (De Broyer and Danis 2010), building upon a first block contributed by Clarke and Johnston (2003). RAMS includes an accurate list of more than 8,200 valid species with an up-to-date systematic classification comprising more than 17,500 taxon names. This taxonomic information is constantly completed by the experts. RAMS contributes to the "taxonomic backbone" of the SCAR-MarBIN, ANTABIF, OBIS and GBIF networks, to establish a dynamic, knowledge-based information system on Antarctic marine biodiversity for the future.

The biogeographic information component of SCAR-MarBIN (ANTOBIS) was progressively built by linking databases, distributed across the planet. A single access is now available through the SCAR-MarBIN data portal, which totals 198 datasets and 1.3 million records. This information can be visualized, mapped and downloaded from the portal, for free.

Although it is focused primarily on biogeographic data, SCAR-MarBIN also develops tools to improve its usefulness for conservation and sustainable management. Data can be processed and presented in a variety of ways, allowing the integration of biogeographic data into the development of policy for Antarctic conservation. Furthermore, current data can be used to construct a panel of indicators to detect changes in biodiversity and hence adapt conservation strategies.

The new ANTABIF project builds upon these grounds and has been building a completely new architecture, strongly relying on Open Source standards and technologies provided by the Global Biodiversity Information Facility (GBIF). This architecture includes a series of websites, all focusing on a specific data type or data-derived product. The philosophy is to continue providing free and open access to Antarctic biodiversity data, while at the same time building new projects to showcase the power of a linked, community-driven tool.

2.3 Findings and Motivations

CAML and its sister project, SCAR-MarBIN, have been instrumental in facilitating, supporting and highlighting the work of students and early-career Antarctic scientists. CAML enabled this new generation of polar scientists to attend meetings, workshops and even scientific voyages to Antarctica during IPY through direct funding and in-kind support. SCAR-MarBIN made new datasets and resources freely available to all, enabling students and scientists around the world to access information and data previously only available in a handful of large institutes. Perhaps most significantly, CAML gave this new generation of scientists a voice and presence in the international scientific arena and the global media. Many of the CAML highlights, online diaries and press releases focused on the work of young scientists at the cutting edge of polar science and helped to bring this work to the attention of the world.

The legacies of CAML and the ongoing development of SCAR-MarBIN have begun to include practical tools and resources for use in conservation, planning and management in the Antarctic and Southern Ocean. As a result of a single CAML cruise, the Commission for the Conservation of Antarctic Living Resources (CCAMLR) declared two Vulnerable Marine Ecosystems in late 2008, based on images and samples collected by the R/V Aurora Australis. The areas, each comprising around 400 km^2, have a high diversity of marine life, in particular habitat-forming hydrocorals and sponges. They are also home to species previously unknown to science. This declaration ensures that these unique areas are not damaged by indiscriminate fishing practices. The wealth of biodiversity, distribution and biogeographic data collected under the CAML umbrella form a unique insight into the distribution of marine life in the Southern Ocean. Such a resource, whilst incredibly valuable for scientific research, becomes the basis of a range of data-derived products including scientific synthesis papers, the Antarctic Field Guides and Biogeographic Atlas. These products feed into international efforts to create and classify new Marine Protected Areas and Vulnerable Marine Ecosystems within the Southern Ocean.

To attract more users and showcase the usefulness of the SCAR information networks, a series of data derived products are developed, also using a community-based approach. As examples, ANTABIF and SCAR-MarBIN are developing the Antarctic Field Guides (AFG) and the Biogeographic Atlas of the Southern Ocean (BASO) projects.

The objective of the AFG project is to develop a community-driven, online identification aid, which takes the form of an interactive, customizable field guide (afg.biodiversity.aq). The field guide offers access to dynamically generated information (built on the fly from the content of RAMS, SCAR-MarBIN, ANT-ABIF, and AFG custom databases), and allows the user to create his/her own field guide to the Antarctic wildlife, in function of the region he/she is interested in, or the taxa he/she wants to focus on. The user can download and share his custom field guide as a pdf file, and allow others to build upon it. As part of a community effort, a substantial amount of information has already been made available by A Clarke and D Barnes (British Antarctic Survey) and S Schiaparelli (University of Genova). The content of the AFG is progressively built and focuses on high-quality pictures, which are useful for the identification of Antarctic organisms. In the long run, we hope to couple the AFG with an online Identification Keys system to further help Antarctic scientists to identify what they find in the field using modern, interoperable tools.

The Antarctic Field Guides is an Open Source project.

The multi-authored Biogeographic Atlas of the Southern Ocean (BASO) will provide an up-to-date synthesis of Antarctic and sub-Antarctic biogeographic knowledge and a new comprehensive online resource for visualisation, analysis and modelling of species distribution. The major objectives of the Biogeographic Atlas of the Southern Ocean are to establish a new synthesis of the biogeography of the Southern Ocean (patterns and processes), covering benthos, zooplankton, nekton, birds and seals, to provide a benchmark of current biogeographic knowledge, to model predictive distribution of species and assemblages according to abiotic factors and to develop dynamic editions of the Atlas, including a dedicated webportal (atlas.biodiversity.aq) and electronic books, connected to our network of interoperable databases, allowing easy access to the new biogeographic syntheses. It is envisioned to offer access to updated occurrence and environmental data, customized mapping and data visualisation and analysis, thereby facilitating the ecoregionalization process and the modelling of predictive distribution of species and assemblages in the context of environmental changes. So far a total of over 60 specialists (taxonomists, ecologists, modellers, developers) are contributing data and expertise to this project. In the long term, it is hoped that this approach can be expanded to other realms.

The Biogeographic Atlas of the Southern Ocean is an Open Source project.

2.4 The Biodiversity Data Paper Concept

The new Data Paper concept is a new incentive for biodiversity researchers to publish their data (Chavan and Penev 2011). As discussed previously, free and open access to biodiversity data is essential to feed emerging data-intensive science, and also to develop informed, adaptative conservation of Antarctic biodiversity, currently challenged by rapid environmental change. Even if SCAR-MarBIN gives access to an unprecedented of expert-validated data, we suspect that a significant amount of primary data are not yet easily accessible nor discoverable. Among potential impediments, one is a lack of incentives to data publishers for publishing of their data resources. This activity is often considered as time-consuming and unrewarded in the short term. The Data Paper concept brings up conventional scholarly publication of enriched metadata (detailed description of datasets).

Integrated in the new ANTABIF workflow, the GBIF Integrated Publishing Toolkit (ipt.biodiversity.aq) is an online informatics suite which allow users (individual scientists, institutions,…) to standardize, manage, document and publish their primary biodiversity data. Once the user decides, new datasets are exposed on the web, and become available to various data aggregators, which expose the data using dedicated data portals. The user can opt for the option of generating a manuscript (Data Paper), from the IPT metadata, and submit it for peer-reviewing and eventual publication in a journal accepting this type of contribution.

Using the IPT, the mechanism is in place for incentivizing data publishers efforts towards easy, efficient and enhanced publishing, dissemination, sharing and re-use of biodiversity data. The Data Paper is one possible mechanism to offer scholarly reward for efforts and investment by scientists in enriching metadata and publishing them as citable academic papers.

The pioneering Data Paper concept arises from a joint effort led by GBIF in collaboration with Pensoft Publishers through its journals *Zookeys*, *PhytoKeys*, *MycoKeys*, *BioRisk*, *NeoBiota*, *Nature Conservation* and the forthcoming *Biodiversity Data Journal* which all accept IPT-generated manuscripts.

It is hoped that in addition to rewarding those publishing their biodiversity data, the Data Papers concept will also speed up the publishing of fit-for-use, well documented biodiversity data resources. However, to be proven, the concept requires a high degree of commitment by the SCAR communities, starting by the biodiversity researchers and expanding towards other scientific fields.

2.5 The Future: Towards True Integration

ANTABIF (Antarctic Biodiversity Information Facility), funded by the Belgian Science Policy Office, takes a step further and is building a dedicated Antarctic biodiversity data portal providing access to a distributed network of contributing database, according to the principles of the Global Biodiversity Information Facility. Building upon the SCAR-MarBIN experience, ANTABIF gives access to biodiversity data from marine, terrestrial and limnetic realms.

The design of ANTABIF architecture has been the object of special attention in order to offer an added value to the user and partners, by integrating the best available resources and standards, such as those developed in the frame of SCAR-MarBIN V2 (data.scarmarbin.be), building upon webservice-oriented, dynamic technologies. Careful attention is also taken to make the best possible use of GBIF informatics tools. Using this approach, the Antarctic biodiversity networks will be interoperable with many entities, allowing the publication of the data in many different contexts, leveraging the power of Web2.0 and collective intelligence.

The ANTABIF technological ecosystem is largely based on the progress made in the framework of the SCAR-MarBIN V2 data portal. It is based on 100 % Open Source solutions, and integrates the latest GBIF informatics components (Harvesting and Indexing Toolkit (HIT) and Integrated Publishing Toolkit (IPT)). The philosophy is to try and see as far as possible by adopting robust standards, which have been widely accepted by their respective community. The main principles are details in Fig. 2.1.

Another important aspect in the design of the ANTABIF infrastructure is the use of advanced, highly customizable search engines. The search engines are schema-free and document based, which means that the user can enter a complex query in the search box. As a community-based initiative, we also thought that the data portal should be able to learn from the behaviour of its users, a feature that is embedded in the current design. As users query the data systems, the search engine is capable of refining the hits it will feedback to the user. ANTABIF uses customized GBIF's IT infrastructure tools, which are foreseen to become widely adopted by the biodiversity science community. For example, the Integrated Publishing Toolkit (IPT) allows ANTABIF to push newly published data to global scale networks such as GBIF and OBIS. GBIF's Harvesting and Indexing Toolkit (HIT) allows ANTABIF to harvest data from various sources, for example from institutions with relatively advanced IT infrastructure. To ensure no dataset is harvested twice, GBIF maintains a registry, which keeps a trace of all existing and new biodiversity datasets available for publication.

A similar approach is adopted for metadata, to maximize the probabilities to discover available datasets through search engines such as Google. The metadata is stored in a central repository, the Global Change Master Directory (GCMD), which also serves as a repository for other entities, such as the Australian Antarctic Data Center (AADC), OBIS, SCAR-MarBIN and SCAR (through the Antarctic Master Directory, AMD). Using this integrated approach, ANTABIF is in line with

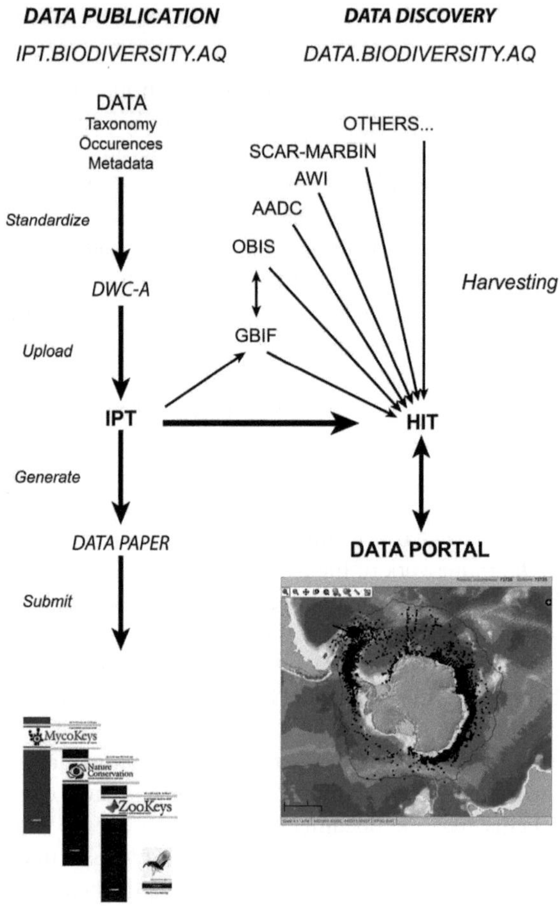

Fig. 2.1 How ANTABIF builds geospatial data content, and interacts with other components of the data ecosystem

the vision developed by IPY data managers, allowing primary biodiversity data to be discoverable, open, linked, useful and safe.

We have high hopes that certain elements of the architectural design adopted for ANTABIF is transposable to other disciplines than biodiversity: adopting open standards, and promoting them, developing a few open source tools which can be used in a wide variety of contexts, offering a simple, flexible and responsive platform to discover, visualize and access data openly, is a rather generic method. Our approach, once checked against reality (and extensively tested by its community in an iterative way), needs to be promoted and communicated to a wider audience, so our efforts can be rewarded by being adopted in other contexts, or networks. Again, this demands strong commitment from the community of users.

At the basis of SCAR-MarBIN's success, we believe that personal communication is a key factor. In the context of ANTABIF, once the online products are launched, constant efforts should be sought between data providers and data managers, to ensure the latter do not get disconnected from the community they serve. This need for communication is not new, and was already stressed by Vickery (1999), finding that "Scientists and technologists need to interact and communicate". Data managers are not just technologists, and having them actually take part in field activities together with scientists, who should reversely also play an active role in data curation and tool developments, is an elegant way of maintaining mutual trust and understanding of each other's reality. Only in this way will scientists feel they are part of their own data ecosystem, and take more responsibility in its management and get involved in their developments. This facet of scientific data management was already stated by Star and Ruhleder (1996): "We must ask users to meet designers halfway by learning their language and developing an understanding of the design domain. If designers are at fault for assuming that all user requirements can be formally captured and codified, users are often equally at fault for expecting "magic bullets"—technical systems that will solve social or organizational problems".

Another idea is to ensure technologies adapt to the evolving behaviour of the users, by becoming more sophisticated and flexible. Giving access to new types of information including actual samples, digital data, documentation, images, multimedia, publications, etc., is a good example. These objects could be linked together across disciplines as in a data network. Collaborative research efforts are highly needed to explore how to capture and enrich this kind of contextual information. Another crucial (and complex) aspect is the capture of data uncertainties related to the newly available knowledge and assumptions of any data collector (Parsons et al. 2010).

The need for research and development in polar biodiversity informatics highlights another important issue in what we are facing in sustaining data systems: sponsorship of the data ecosystem. During the IPY, the final report of the Data Policy and Management Subcommittee includes the following note: "In the period leading up to the start of IPY, data stewardship was undervalued, despite robust data management plans within the IPY Framework Document, the strong recommendations of the ICSU Program Area Assessment, and telling examples from earlier international projects" (Parsons et al. 2011).

At the beginning of the IPY it was extremely difficult to obtain data management funding, especially for broad international or interdisciplinary support. SCAR-MarBIN was an exception to this rule, as the Alfred P. Sloan Foundation through seed money, and the Belgian Science Policy Office supporting the setup of the network in congruence with the Census of Antarctic Marine Life initiative. By the end of the IPY, more governments began to endorse information systems, which was in fact too late. It is time for funding agencies to realize that if they fund data collection, which in the case of polar research can be extremely costly, they also need to fund data stewardship, acknowledging their responsibility in enhancing the data availability, which is a key to improving Polar Science. In some cases, progressively shifting towards an internationally funded model makes a lot of sense, as information networks benefit the wide community, taking the forms of a common.

Ideally, sponsors should be actively engaged in the planning and execution of data systems for ambitious research programs, offering the possibility to have a very long planning period. The ideal situation would be one where the scientists personally know the data managers, know which data centre they can submit their data to, and are familiar with standards. This was not the case during the IPY during which data systems were thought of, when expeditions were ongoing, or were over.

We are facing many challenges, and the fact that the data are still highly distributed and stored in many different types of institution in many different formats is probably one of the most fundamental ones. The diversity of the needs of the data users are also extremely diverse and often unpredictable. Some short-term technical solutions to these challenges include the use of more open, cloud-based approaches of data casting, using simpler approaches in the architectural design of information systems, based on proven concepts and robust standards, and iterate to progressively interoperate with other systems and communities.

To be consistent with some of the norms guiding scientific research (such as rigor, transparency, integrity, and reproducibility), open access to data should actually be a central rule: scientific data should be viewed as a common good that is open (except for legitimate ethical reasons). Also, data should be used in a framework where data providers are given fair attribution for their contribution to these commons. The example of the Data Paper concept is one step in this direction. Data managers should ensure the integrity of the data that they are entrusted to and provide the necessary documentation and context to support robust science. This is of course a long-term vision, but a simple element, such as requiring research proposals to include detailed data management and archiving plans, would probably help.

In the course of the IPY, the rapid and dramatic changes in the polar regions environment has been extensively documented, and we can expect that the demand for rigorous, transparent and scrupulously documented data will increase dramatically in the near future. Many challenges remain, but it does not help to deny our collective responsibility to keep the data safe, to avoid vanishing of common knowledge before it even reaches future generations.

2.6 Summary

The International Polar Year was an unprecedented research effort, which involved thousands of participants from many nations. This effort generated massive amounts of extremely diverse data, which need to be interrelated to understand complex problems such as environmental change and its impact on Antarctic biodiversity. We use the experience of SCAR Marine Biodiversity Information Network (SCAR-MarBIN), and the new Antarctic Biodiversity Information Facility (ANTABIF) as a case study to examine an example of IPY data management approaches seeking to address issues around such complex scientific questions. In the present chapter, we discuss the different facets of this data management experience, and the design used for ANTABIF to be transposed to other contexts.

To ensure that ANTABIF and its successors are successful in maintaining open access to Antarctic biodiversity data, the heterogeneity and the distributed nature of data will require technologies to be much simpler, well standardized and agile to follow the constantly changing needs of the community it is serving. The IPY has shown that the Polar Regions are changing rapidly and the need for rapid technical and cultural changes, towards true openness becomes urgent. We suggest several short and long-term strategies to facilitate the evolution of the overall SCAR data ecosystem towards universal access to high quality scientific data, which would otherwise be lost.

Acknowledgments This book chapter is contribution #78 to the Census of Antarctic Marine Life initiative. The authors wish to thank the vast community involved in the biodiversity information networks, and in particular the Scientific and Follow-up Committees of SCAR-MarBIN and ANTABIF. The two initiatives are funded by the Belgian Science Policy Office and are implemented within the Belgian Biodiversity Platform. We would also like to thank the different sponsors, including: the Alfred P. Sloan Foundation, The Total Foundation for biodiversity, the Census of Antarctic Marine Life, The Alfred Wegener Institute, the German Research Foundation, the Scientific Committee on Antarctic Research, the Australian Antarctic Division, the Netherlands Organization for Scientific Research.

References

Brandt A, Gooday AJ, Brandão SN, Brix S, Brökeland W, Cedhagen T, Choudhury M et al (2007) First insights into the biodiversity and biogeography of the Southern Ocean deep sea. Nature 447:307–311 doi:10.1038/nature05827

Carlson DJ (2010) Why do we have a 4th IPY? In: Barr S, Lüdecke C (eds) The history of the international polar years (IPYS). Springer, Berlin, pp 299–312

Carlson DJ (2011) IPY 2007–2008: where the threads of the double helix and Sputnik intertwine. In: Huettmann F (ed) Protection of the three poles. Springer, Japan

Chavan V, Penev L (2011) The data paper: a mechanism to incentivize data publishing in biodiversity science. BMC Bioinformatics 12 (Suppl 15):S2. doi:10.1186/1471-2105-12-S15-S2, http://www.biomedcentral.com/1471-2105/12/S15/S2

Clarke A, Johnston NM (2003) Antarctic marine benthic diversity. Oceanogr Mar Biol Ann Rev 41:47–114

De Broyer C, Danis B (2010) How many species in the Southern Ocean? Towards a dynamic inventory of the Antarctic marine species. Deep-Sea Research II: Topical Studies in Oceanography. doi:10.1016/j.dsr2.2010.10.007 http://dx.doi.org/10.1016/

Hey T, Hey J (2006) e-Science and its implications for the library community. Libr Hi Tech 24:515–528

Hey T, Tansley S, Tolle K (eds) (2009) The fourth paradigm: data-intensive scientific discovery. Microsoft Research, USA

Newman HB, Ellisman MH (2003) Orcutt JA (2003) Data-intensive e-science frontier research. Commun ACM 46:68–77

Parsons MA, Duerr R, Minster JB (2010) Data citation and peer-review. Eos, Transactions of the American Geophys Union 91:297–298

Parsons MA, Godoy O, LeDrew E, de Bruin TF, Danis B, Tomlinson S, Carlson D (2011) A conceptual framework for managing very diverse data for complex, interdisciplinary science. J Inform Sci (Oct). doi:10.1177/0165551511412705, http://jis.sagepub.com/cgi

Star SL, Griesemer JR (1989) Institutional ecology, 'translations' and boundary objects: amateurs and professionals in Berkeley's Museum of Vertebrate Zoology, 1907–39. Soc Stud Sci 19:387–420

Star SL, Ruhleder K (1996) Steps toward an ecology of infrastructure: Design and access for large information spaces. Inform Syst Res 7:111

Vickery B (1999) A century of scientific and technical information. J Documentation 55:476–527

Part II
Evolution: A Molecular Perspective

Chapter 3
Southern Ocean Evolution in a Global Context: A Molecular Viewpoint

Jan M. Strugnell and A. Louise Allcock

Molecular data can be used to pinpoint both contemporary and historical forces acting on biota but until recently such data have been largely obtained only from vertebrates, such as penguins (Baker et al. 2006), fish (Kuhn and Gaffney 2006; Rogers et al. 2006), and seals (Curtis et al. 2009), whose mobile adult stages are less affected by the barriers imposed by abiotic forces than are invertebrates. Exceptions include research focused on commercially important pelagic taxa, primarily krill (Goodall-Copestake et al. 2010; Batta-Lona et al. 2011). The few early studies on benthic invertebrates indicated the potential use of molecular data in interpretation of Antarctic speciation and connectivity by providing evidence of limited gene flow (Allcock et al. 1997), endemic radiation (Held 2000), cryptic speciation (Held 2003) and historical connectivity between the Antarctic and other oceans (Lörz and Held 2004). Progress was intermittent for several years but a recent surge of molecular data has allowed concepts generally associated with Antarctic biology, but often unproven, to be investigated using genetic methods. Such concepts include the Southern Ocean biota being commonly described as highly endemic (e.g. Brandt et al. 2007a), but in some cases common with the Arctic (Stepanjants 2006); and the biota showing a prevalence of eurybathy (Brey et al. 1996), possessing circumpolar distributions (Clarke and Johnston 2003), and non-dispersive life histories (Pearse et al. 2009). Molecular studies are allowing the strength of these concepts to be tested across a range of taxa with diverse life

J. M. Strugnell (✉)
Department of Genetics, La Trobe Institute for Molecular Science,
La Trobe University, Bundoora, Vic 3086, Australia
e-mail: j.strugnell@latrobe.edu.au

A. L. Allcock
Department of Zoology, Ryan Institute, National University of Ireland Galway,
University Road, Galway, Ireland
e-mail: louise.allcock@gmail.com

C. Verde and G. di Prisco (eds.), *Adaptation and Evolution in Marine Environments, Volume 2*, From Pole to Pole, DOI: 10.1007/978-3-642-27349-0_3,

history strategies. Patterns of congruence and conflict are emerging and together these studies are providing a richer understanding of speciation and connectivity within the Southern Ocean ecosystem.

3.1 A Brief Climatic, Oceanographic and Tectonic History of the Southern Ocean

The origins of the Southern Ocean fauna and flora are complex: an endemic biota which has evolved in situ is enriched by organisms that have migrated or dispersed into the Southern Ocean from South America via the Scotia Arc and/or from adjacent deep-water basins (Clarke 2008; Clarke and Crame 2010). Marked shifts in the climatic, oceanographic and tectonic histories of the Antarctic region throughout the Cenozoic (65 Ma-present) must necessarily have played major roles in shaping the extant biodiversity. These shifts are likely to have directly influenced the evolution of the Southern Ocean biota by impacting levels of isolation from, or connectivity to, the world's other ocean basins.

In the early Cenozoic the climate in the Antarctic region was relatively warm and ice free. A deep seaway between the South Tasman Rise and East Antarctica developed about 32 Ma while the Drake Passage probably opened to deep-water flow prior to 29 Ma (Lawver and Gahagan 2003) but possibly not until 24 Ma (Pfuhl and McCave 2007) (Fig. 3.1). Development of the Antarctic Circumpolar Current followed and at around the same time the first continent-wide ice sheet formed. Since the opening of these gateways, the most extensive tectonic activity in the Antarctic has been associated with the formation of the Scotia Sea.

The islands of the Scotia Arc comprise relatively ancient rocks as well as volcanically young rocks (e.g. South Sandwich Islands ~5 Ma). The floor of the Scotia Sea is relatively young and results from sea-floor spreading over the last 30 million years (Thomson 2004). South Georgia reached its present position by eastward and then northward movement from a position between the Antarctic Peninsula and South America (Lawver and Gahagan 2003).

Almost continual cooling followed the opening of the Tasman Gateway and Drake Passage but a marked decline is detected in the mid-Miocene (c. 14 Ma). This appears to be associated with the formation of the East Antarctic Ice Sheet and is concurrent with increased Antarctic Bottom Water formation and intensification of the global thermohaline circulation (Diekman et al. 2004).

Antarctic ice-sheet size has fluctuated with Milankovitch cycles: ice-core data (EPICA 2004) provide detailed temperature records for the last 800,000 years showing the 100,000 year periodicity of glaciation cycles during this time. Sediment-core data suggest at least 38 cycles, of varying periodicity, in the past 5 million years (Naish et al. 2009). Evidence suggests that the West Antarctic Ice Sheet has collapsed on multiple occasions to allow a trans-Antarctic seaway

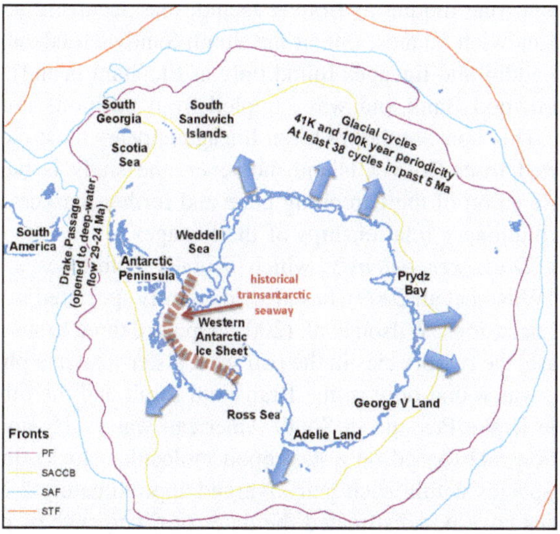

Fig. 3.1 The Antarctic, indicating key locations mentioned in the text, the proposed historical seaway between the Ross Sea and the Weddell Sea (Vaughan et al. 2011), and the important fronts of the Antarctic Circumpolar Current (ACC). The ACC flows clockwise around Antarctica. It is bounded by the Sub Tropical Front (STF) to the north and the Southern Boundary Current (SACCB) to the south, with most of the transport between the sub-Antarctic front (SAF) and the Polar Front (PF). The arrows indicate the movement of ice sheets during glacial cycles

between the Ross and Weddell Seas (Pollard and DeConto 2009). Models suggest this occurred most frequently prior to 3 Ma but has also occurred more recently.

Most evidence for the effects of glacial cycles comes from the last glacial period. Glaciological models, supported by marine geological and geophysical data [see summary in Thatje et al. (2005)] suggest a massive advance of Antarctic ice-sheets during the last glacial maximum (LGM), a fall in global-sea levels and a grounding of ice-sheets out across much of the continental shelf of the Southern Ocean, with a limited number of shelf-edge locations (e.g. western Ross Sea, Prydz Bay, George Vth land) free from grounded ice.

3.2 The Antarctic Circumpolar Current as a Barrier

The Antarctic Circumpolar Current (ACC) marks the boundary between the Southern Ocean and the Atlantic, Pacific and Indian Oceans. Its role in dispersal, and as a barrier to dispersal, has been widely speculated upon. Genetic studies have given more precise information as to how various species interact with the ACC.

A study of the pycnogonid *Colossendeis megalonyx* (Krabbe et al. 2010) showed that the ACC has apparently acted as a vicariant barrier in this species. Six distinct clades or lineages were detected in the study. Five were found only south of the Polar Front (one unique to Bouvet Island, one occurring at Bouvet Island and the South Sandwich Islands, one at the South Sandwich Islands and Elephant Island, and two additional lineages found only at Elephant Island). The sixth was confined to Burdwood Bank and was sampled from positions north of the sub-Antarctic Front. This non-Southern Ocean lineage appears most closely related to a lineage reported from Bouvet Island, however, the study is based on a short section (561 base pairs) of the barcoding gene and further markers are required to establish the evolutionary relationships of the lineages.

The sea-slug *Doris kerguelensis,* which actually comprises a suite of many cryptic species (Wilson et al. 2009), has also apparently speciated in response to the ACC on several occasions. Wilson et al. (2009) reported three 'trans-Drake' species pairs. In each case, the two species in the pair were sister-taxa in a phylogenetic tree and one lineage was found only in the Bransfield Strait and the other only on the other side of the Drake Passage in South American waters. By applying the best available mutation rate (based on a gastropod molecular clock), the authors suggested that the species within each pair diverged approximately 2 Ma.

Several studies have now estimated the dates of divergence of pairs of species that exist on either side of the ACC. Studies on echinoderms using echinoderm-calibrated molecular clocks indicate that the South American seastar *Odontaster penicillatus* diverged from its Southern Ocean sister species, *O. pearsei,* about 1 Ma (Janosik et al. 2011), the South American sea urchin *Sterechinus agassizi* diverged from its Southern Ocean sister species, *S. neumayeri*, about 5 Ma (Diaz et al. 2011) and that cryptic lineages of the brittle star *Astrotoma agassizi* diverged on either side of the ACC about 1.5 Ma (Hunter and Halanych 2008). González-Wevar et al. (2010) used a relaxed clock method that allows different evolutionary rates on different tree branches to estimate the date at which the Southern Ocean limpet *Nacella concinna* diverged from non-Antarctic *Nacella* species. Using the earliest known dates for various fossils to constrain other nodes on the tree, they estimated divergence to have occurred between about 5 and 9 Ma. Research on limatulids illustrates the very real problems with estimating divergence times (see later). Page and Linse (2000) also used fossil constraints (but with a strict clock) to estimate the date of divergence of the Southern Ocean bivalve *Limatula ovalis* from its sub-Antarctic congener *L. pygmaea.* Considering each of three genes separately, they found widely differing divergence time estimates that ranged from 0.24-2.87 Ma to 6.81 to 19.12 Ma. Nonetheless, these dates are still all after the onset of the ACC, suggesting that the common ancestors of these species were able to cross the ACC, even if the extant species do not.

Evidence that species currently cross the ACC is provided by nemerteans. Thornhill et al. (2008) reported distinct lineages of *Parbolasia corrugata* on either side of the Drake Passage, but a subsequent study by Mahon et al. (2010) detected two lineages that cross the Polar Front in the Drake Passage region.

Similarly, a study of lysianassoid amphipods suggests some lineages do cross the ACC (Havermans et al. 2011). Sampling was concentrated in the Southern Ocean but a single sample of *Orchomenopsis cavimanus* was obtained from the Magellan region (Burdwood Bank). Cryptic speciation was present and one lineage appeared to be confined to the Scotia Sea. However, the other was found throughout the entire sampling region which extended from the eastern Weddell Sea to the Burdwood Bank in the area of the Sub-Antarctic Front providing some evidence that this species disperses across the ACC close to South America.

Some species may actually use the ACC for dispersal. A molecular study of the notothenioid fish *Gobionothen gibberifrons* noted gene flow was unidirectional and concluded that larval dispersal in the ACC accounted for this (Matschiner et al. 2009).

The sub-Antarctic lineage of the cryptic species complex *Durvillea antarctica*, the buoyant bull kelp, likely disperses in its adult stage throughout the sub-Antarctic islands as algal rafts in the ACC (Fraser et al. 2010). Furthermore, Nikula et al. (2010) found molecular support for a rafting hypothesis in two sub-Antarctic amphipod species commonly found in *D. antarctica* holdfasts. There is also biological evidence in support of rafting in the brooding isopod *Septemserolis septemcarinata*. This species has enlarged pereiopods for climbing onto substrata and only rare dispersal events (the majority from west to east in accordance with flow in the ACC) were required to explain the apparent gene flow between widely spaced populations (Leese et al. 2010) situated within the ACC but on both sides of the Polar Front.

These studies demonstrate that the Polar Front and the ACC do not form an absolute barrier to dispersal and that eddies which spin off north and south are likely to play a role in carrying organisms in and out of the Southern Ocean (Barnes and Clarke 2011). Nonetheless, the wide variation in environmental parameters experienced either side of the Polar Front is likely to hamper the survival of many organisms that cross this potential barrier. In the deep sea, this variation will be less and there is probably a greater likelihood of deep-sea species crossing this barrier. Indeed many truly cosmopolitan species (see later) are found in the deep sea.

3.3 Connectivity with Other Oceans

The exchange of marine species from South American waters with the Southern Ocean has been widely noted (e.g. Dell 1972) and in recent years it has been recognised that fewer Southern Ocean species are endemic than had previously been thought. For example, Griffiths et al. (2009) estimate that the number of Southern Ocean endemics is close to 50 %, meaning that 50 % of species occur either side of this potential barrier. Furthermore they note the increasing presence of endemics with increasing distance from South America, suggesting that the Scotia Arc may be a route of species migration in and out of the Southern Ocean. Certainly this region provides the shortest geographic distance between Southern Ocean and non-Southern Ocean shallow waters, and our knowledge is biased towards shallow waters.

Fig. 3.2 **a** Phylogenetic relationships of Southern Ocean endemic and deep-sea octopuses. ▶ Bayesian phylogenetic tree based on the results of the relaxed phylogenetic analysis utilizing the seven genes: *rhodopsin, pax-6, octopine dehydrogenase* (ODH), 12S rDNA, 16S rDNA, cytochrome oxidase subunit I (COI) and cytochrome oxidase subunit III (COIII) of 12 Antarctic octopus species, seven deep-sea octopus species and 15 outgroup taxa. The topology is that from the posterior sample which has the maximum sum of posterior probabilities on its internal nodes. Each node in the tree is labelled with its posterior probability, * indicates a posterior probability of 1.0. The divergence times correspond to the mean posterior estimate of their age in millions of years. The genera *Adelieledone* (*dark blue*), *Pareledone* (*blue*), and *Megaleledone* (*light blue*) are endemic to Antarctic waters. The deep-sea genera, *Graneledone* (*red*), *Velodona* (*yellow*), and *Thaumeledone* (*orange*) are a monophyletic group and are nested within the Antarctic clade. The deep-sea clade was estimated to have originated around 33 million years ago (Ma; 95 % HPD interval 5–64 Ma). The three deep-sea genera were estimated to have diverged from one another around 15 Ma (95 % HPD interval 1–36 Ma). This figure is an abridged version of that published in Strugnell et al. (2008). **b** Bayesian phylogenetic tree based on the results of the relaxed phylogenetic analysis using six genes: *rhodopsin*, 12SrDNA, 16SrDNA, cytochrome oxidase subunit I (COI), cytochrome oxidase subunit III (COIII) and cytochrome *b* (cyt*b*) of 25 *Benthoctopus* individuals, 2 *Vulcanoctopus* individuals and 7 outgroup taxa. The topology is that from the posterior sample which has the maximum product of posterior clade probabilities. Each node in the tree is labelled with its posterior probability, '*' indicates posterior probability of 1.0. The divergence times correspond to the mean posterior estimate of their age in millions of years. The *Benthoctopus* species collected from the Southern Ocean (south of the Polar Front) are shown in dark blue and represent a paraphyletic group. Other *Benthoctopus* species collected from the Southern Hemisphere are shown in light blue. *Benthoctopus* and *Vulcanoctopus* species collected from the Northern Hemisphere are shown in *red*. Outgroup taxa are shown in black. '+' indicates species that possess an ink sac. Clade1 was estimated to have had a common ancestor around 16 Ma (95 % HPD interval 4–32 Ma), Clade 2 was estimated to have had a common ancestor around 5 Ma (95 % HPD interval 1–11 Ma) and Clade 3 was estimated to have had a common ancestor around 6 Ma (95 % HPD interval 1–13 Ma). Clade 2 was estimated to have diverged from its sister clade around 14 Ma (95 % HPD interval 4-27 Ma). This figure is an abridged version of that published in Strugnell et al. (2011)

3.3.1 Southern Ocean: Source and Sink?

Recent phylogenetic studies have demonstrated historical connectivity between the Southern Ocean and other ocean basins on deep evolutionary timescales. Both polar emergence and submergence has occurred on Antarctic margins, with taxa that originated in the deep sea moving into the Southern Ocean and vice versa.

Strong evidence for polar submergence comes from phylogenetic studies of Asellota isopods (Raupach et al. 2009) based on 18S rRNA sequences, and octopods (Strugnell et al. 2008) employing mitochondrial (12S rRNA, 16S rRNA, COI, COIII) and nuclear (*rhodopsin, pax-6, ODH*) genes (Fig. 3.2).

Both studies indicate a shallow-water, Southern Ocean origin for clades of these taxa which invaded the deep sea. Phylogenetic analyses indicate that extant isopods inhabiting the deep sea of the Southern Ocean have arisen multiple times from shallow water lineages. Divergence-time estimates on the origin and subsequent diversification of the deep-sea octopod clade correspond with the initiation ($\sim$33 Ma) and subsequent intensification ($\sim$15 Ma) of the global thermohaline circulation respectively (Fig. 3.2). These events may have provided oceanographic

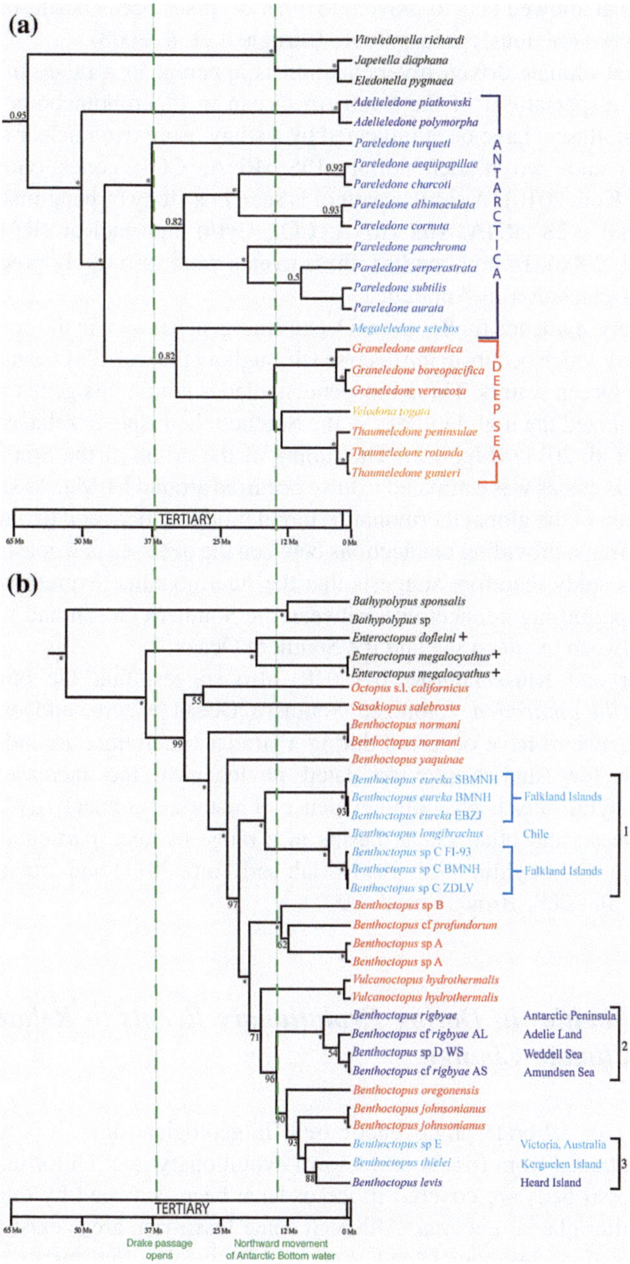
(a)
Vitreledonella richardi
Japetella diaphana
Eledonella pygmaea
Adelieledone piatkowski
Adelieledone polymorpha
Pareledone turqueti
Pareledone aequipapillae
Pareledone charcoti
Pareledone albimaculata
Pareledone cornuta
Pareledone panchroma
Pareledone serperastrata
Pareledone subtilis
Pareledone aurata
Megaleledone setebos
Graneledone antartica
Graneledone boreopacifica
Graneledone verrucosa
Velodona togata
Thaumeledone peninsulae
Thaumeledone rotunda
Thaumeledone gunteri
ANTARCTICA
DEEPSEA
0.95
0.92
0.93
0.82
0.5
0.82
TERTIARY
65 Ma
50 Ma
37 Ma
25 Ma
12 Ma
0 Ma
(b)
Bathypolypus sponsalis
Bathypolypus sp
Enteroctopus dofleini +
Enteroctopus megalocyathus +
Enteroctopus megalocyathus +
Octopus s.l. californicus
Sasakiopus salebrosus
Benthoctopus normani
Benthoctopus normani
Benthoctopus yaquinae
Benthoctopus eureka SBMNH
Benthoctopus eureka BMNH
Benthoctopus eureka EBZJ
Benthoctopus longibrachus
Benthoctopus sp C FI-93
Benthoctopus sp C BMNH
Benthoctopus sp C ZDLV
Benthoctopus sp B
Benthoctopus cf profundorum
Benthoctopus sp A
Benthoctopus sp A
Vulcanoctopus hydrothermalis
Vulcanoctopus hydrothermalis
Benthoctopus rigbyae
Benthoctopus cf rigbyae AL
Benthoctopus sp D WS
Benthoctopus cf rigbyae AS
Benthoctopus oregonensis
Benthoctopus johnsonianus
Benthoctopus johnsonianus
Benthoctopus sp E
Benthoctopus thielei
Benthoctopus levis
Falkland Islands
Chile
Falkland Islands
1
Antarctic Peninsula
Adelie Land
Weddell Sea
Amundsen Sea
2
Victoria, Australia
Kerguelen Island
Heard Island
3
55
99
93
97
62
71
54
96
90
93
88
TERTIARY
65 Ma
50 Ma
37 Ma
25 Ma
12 Ma
0 Ma
Drake passage opens
Northward movement of Antarctic Bottom water

conditions that allowed taxa to move into other deep-sea ocean basins of the world, areas that were previously inhospitable (Strugnell et al. 2008).

Additional climate-driven diversification is apparent in a range of taxa. Speciation and dispersal out of the Southern Ocean in Pleurobranchomorpha (opisthobranch molluscs) have been indicated by a study using two nuclear (18S rRNA, 28S rRNA) and two mitochondrial (16S rRNA, COI) genes (Göbbeler and Klussmann-Kolb 2010). A similar pattern is seen in a study of penguins employing mitochondrial (12S rRNA, 16S rRNA, COI, Cyt*b*) and nuclear (RAG-1) genes (Baker et al. 2006). In both studies, these events were also likely precipitated by the onset of glaciation in Antarctica.

Conversely, evidence for Southern Ocean emergence is seen in the octopod genus *Benthoctopus* which occurs in the deep sea throughout the world's oceans in addition to Southern Ocean waters. Three independent clades within this genus were shown to have colonized the high latitudes in the Southern hemisphere relatively recently (Strugnell et al. 2011) (Fig. 3.2). The timing of the origin of the Southern Ocean *Benthoctopus* clades was estimated to have occurred around 14 Ma. As stated above, intensification of the global thermohaline circulation is understood to have occurred in this time frame providing connections between the deep sea and coastal Antarctic waters. This study therefore suggests that the 'thermohaline expressway' was bi-directional permitting connectivity between the Southern Ocean and the deep sea and also between the deep sea and the Southern Ocean.

Göbbeler and Klussmann-Kolb (2010) also suggest that the opisthobranch *Bathyberthella antarctica* colonized Southern Ocean waters, and estimate this polar emergence to have occurred during a similar timeframe, around 20 Ma.

Although few studies provide dated phylogenies, the increase in global molecular phylogenies is providing evidence of historical connectivity between the Southern Ocean and other ocean basins in a range of taxa, particularly in echinoderms (e.g. O'Loughlin et al. 2011; Mah and Foltz 2011) and crustaceans (e.g. Raupach et al. 2009; Browne et al. 2007).

3.3.2 Difficulties in Dating Evolutionary Events to Relate Them to Climate Change

To constrain or calibrate phylogenetic trees in geological time, it is necessary to use dates obtained from fossils or a known evolutionary rate. Unfortunately many Antarctic fossil beds are covered in ice or have been damaged by the erosion of sediments after glacial maxima. Although some fossil-rich areas exist (e.g. the La Meseta formation, Seymour Island, Antarctic Peninsula), few fossils of Southern Ocean taxa appear suitable for divergence time estimation. For example, the single fossil eleginopid fish on which notothenioid-fish divergence times are based on is controversial (Near 2004; Janko et al. 2007, 2011) and may not even be a notothenioid.

Because molecular evolutionary rates can differ between taxa and lineages, applying an evolutionary rate (estimated from a broader taxonomic group) under the assumption of a 'strict' molecular clock, is only valid where data have been shown to be clock-like. Where suitable fossil calibrations have been available for Southern Ocean taxa, some studies have assumed a strict molecular clock and have been forced to remove non-clock-like data first, for example in Foraminifera (Darling et al. 2004 *partim*) and sea urchins (Lee et al. 2004). In the absence of suitable fossils, other studies have applied evolutionary rates from broader taxonomic groups such as echinoderms (Wilson et al. 2007) and teguline gastropods (Wilson et al. 2009) under the assumption of strict clock. This can give misleading divergence time estimates (e.g. Yoder and Yang 2000). Applying rates from broader taxonomic groups might also be misleading if substitution rates are slower in polar poikilotherms. This has been refuted for 16S rRNA in Antarctic serolid isopods (Held 2001) but has received no further attention.

Other studies have employed relaxed-clock approaches to allow variable rates among lineages without having to remove data. Some of these have fixed fossil dates to particular nodes, for example in Foraminifera (Darling et al. 2004 *partim*), and therefore give apparently precise, but not necessarily accurate, dates without confidence intervals. Others have applied fossil constraints as prior distributions to reflect the uncertainty of the fossil record, for example in octopods (Strugnell et al. 2008). This produces divergence times with broad confidence intervals, making it difficult to link to specific climatic and oceanographic events with certainty.

Future studies should target taxa with multiple, accurately-dated, well-characterized fossils with representatives just above and below the lineages to be dated. The most promising taxa for this are those with hard parts, such as echinoid, gastropod and bivalve families. Alternatively, the application of ancient DNA methodologies to historical museum specimens (collected by early Antarctic explorers) or specimens frozen in situ as was done for Adélie penguins (Lambert et al. 2002) will permit the calculation of relevant evolutionary rates.

3.3.3 Cosmopolitan Species

Connectivity between the Southern Ocean and other ocean basins is found in reports of cosmopolitan and 'bipolar' species (i.e. those found in the Southern Ocean and in Arctic waters). There are many such reports, but few reports are backed up by genetic data. For example, Park and Ferrari (2009) considered the distribution range of all deep-water calanoid copepods found south of the Polar Front. Of these 127 species, 24 were considered endemic to this region, while 84 species had a range that extended beyond the subtropical convergence. Of these 84 species, 29 had ranges thought to extend to sub-Arctic waters whereas another 13 have been reported from the high Arctic. However, increasing numbers of phylogenetic studies in recent years have been highlighting the presence of cryptic lineages (see later), and molecular data are needed to confirm such widespread distributions.

One of the few studies which provides molecular evidence of truly cosmopolitan species with a range extending from the Arctic Ocean to south of the Antarctic Polar Front was conducted on Foraminifera. Rapidly evolving ITS sequences of three species of benthic foraminiferan species (*Epistominella exigua, Cibicides wuellerstorfi, Oridorsalis umbonatus*) were shown to be almost identical between the Weddell Sea and the Arctic Ocean (Pawlowski et al. 2007). Very similar sequences of the first two species were also obtained from individuals collected from the North Atlantic Ocean, suggesting that each of these species probably exists as massive global populations.

Interestingly, asymmetric gene flow was detected in *E. exigua* with a much stronger signal in the direction from the Southern Ocean to the Arctic, than in the opposite direction (Brandt et al. 2007b). Similar reports of identical SSU rRNA sequences in pelagic foraminifera from the Arctic and the Southern Ocean are also indicative of recent genetic exchange (Darling and Wade 2008).

3.3.4 Bipolar Species

Of the deep-water calanoids thought to occur in both the Arctic and Southern Oceans (Park and Ferrari 2009), six have a reported disjunct distribution. This may be due to lack of sampling, but some authors consider 'bipolar' species to be relatively common. For example Stepanjants (2006) recognised 46 bipolar radiolarian species and 23 bipolar medusozoan species. Although some morphological studies present evidence for bipolar distributions of species, for example in bryozoans (Kuklinski and Barnes 2010), recent claims of 200 + bipolar species (see Kuklinski and Barnes 2010) are likely to be greatly overestimated. Furthermore, detailed taxonomic studies often reveal morphological differences. Bipolarity has been refuted in the sponge species *Stylocordyla borealis* following recent morphological taxonomic work (Uriz et al. 2011) and the Southern Ocean representatives redescribed as *S. chupachups*.

Generally, molecular data for species reported to have a bipolar distribution are scarce but they are available for some taxa.

Bipolarity was examined in two diatom species: *Fragilariopsis cylindrus* and *F. nana*. Unfortunately difficulties persist in identifying these species, and although bipolarity was not confirmed, neither was it refuted.

Bipolarity has also been investigated in marine bacterial isolates. Using the mitochondrial gene, 16S rRNA, it has been shown that *Shewanella frigidimarina* occurs in the Southern Ocean (Prydz Bay) and the Arctic (Greenland Sea and Canada Basin). However, limited sampling means that this species may have a cosmopolitan distribution.

Mitochondrial COI gene sequences of the pteropod *Limacina helicina* from populations sampled in the Arctic and Southern Oceans were found to differ by

33.56 %, whilst genetic distance between individuals within each of these locations was very small (i.e. Arctic 0.15 %, Southern Ocean 0.60 %). This provides strong evidence that these are different species (Hunt et al. 2010).

Gene exchange between the poles, where no intermediate populations exist, is probably a rare phenomenon. Cosmopolitanism however, particularly amongst deep-sea species, is more likely. Although datasets do not yet exist, mostly because of the difficulties in obtaining widespread deep-sea samples, future molecular data may reveal global panmixia in some deep-sea cosmopolites, highlighting the deep-ocean connections between Antarctic waters and the rest of the globe.

3.4 Connectivity within the Southern Ocean

3.4.1 The Southern Ocean as a Biodiversity Hotspot

The Ecology of the Antarctic Sea-Ice Zone (EASIZ) programme strongly contributed to refuting the idea that the Southern Ocean was species poor (Clarke and Arntz 2006). Working on a shelf-species estimate of at least 4,100 species, Clarke (2008) concluded that the richness of the shelf fauna was comparable with that of Hawaii or north-west Europe. Following intensive taxonomic work during the Census of Marine Life, the number of known shelf species has risen to 5,800 (DeBroyer and Danis 2011), but recent molecular work suggests that many cryptic species are yet to be discovered and true estimates of Southern Ocean biodiversity will be very much higher.

3.4.2 Cryptic Species

Cryptic species have been reported in a diversity of taxa including amphipods (Lörz et al. 2009; Baird et al. 2011; Havermans et al. 2011), isopods (Leese and Held 2008), ostracods (Brandão et al. 2010), sea-spiders (Mahon et al. 2010; Krabbe et al. 2010), bivalves (Linse et al. 2007), sea slugs (Wilson et al. 2009), octopods (Allcock et al. 2011), brittle stars (Hunter and Halanych 2008), sea cucumbers (O'Loughlin et al. 2011), crinoids (Wilson et al. 2007), polychaetes (Schueller 2011) and nemerteans (Thornhill et al. 2008). These studies were all molecular based, and the majority of them employed the mitochondrial gene cytochrome oxidase I, the DNA barcoding gene. Most of these studies looked for (but did not always find) evidence of a "gap" range between the highest intraspecific and the lowest interspecific distances, or looked for breaks (based on a pre-determined threshold) in haplotype networks built using statistical parsimony or similar techniques. Brandão et al. (2010) employed a maximum likelihood approach (Pons et al. 2006) to quantitatively search for species limits and it is likely that this method will be used more frequently in the future.

Almost all studies revealing cryptic species have been based on mitochondrial sequence data which are therefore haploid and only representative of the female lineage. Some studies have included diploid markers. For example, Baird et al. (2011) included the nuclear gene ITS on a study of amphipods and found congruence with their mitochondrial data; Raupach et al. (2007) employed the nuclear gene 18S rRNA on a study of isopods and also found congruence with their mitochondrial data. Conversely neither the nuclear gene 28S rRNA in a brooding bivalve (Linse et al. 2007) nor the nuclear gene ITS in ostracods (Brandão et al. 2010) were found to be congruent with mitochondrial data derived from these species. Brandão et al. (2010) suggested that introgression or incomplete lineage sorting might account for the differences. Leese and Held (2008), in the first reported study using microsatellites on invertebrate species in the Southern Ocean, did find congruence with mitochondrial data.

For the most part, the presence of cryptic lineages appears to be associated with taxa with limited dispersal capacity. This suggests that connectivity around the Southern Ocean might have been limited for these taxa at certain times, presumably during glacial maxima when ice extents were at their peak and post glacial maxima where a lack of dispersal stage prevented the populations rapidly exchanging genes. Cryptic lineages have likely had limited time to diverge morphologically and are therefore probably indicative of recent speciation. This certainly fits with a theory of limited connectivity between populations during the last glacial maximum. If speciation was widespread as a result of glacial cycles, this helps explain why the Southern Ocean appears to be a biodiversity hotspot.

Interestingly, the inclusion of rarefaction curves in some studies has indicated an under sampling of genetic diversity (Wilson et al. 2009; Baird et al. 2011). In molecular ecology, rarefaction curves plot the discovery of new haplotypes against sampling effort and, as in their ecological application, tend towards an asymptote that indicates the number of haplotypes in the population. Routine inclusion of such analyses will likely suggest that the true number of cryptic species is higher than currently detected.

3.4.3 Eurybathy and Circumpolarity

While the presence of cryptic species suggests a lack of connectivity within the Southern Ocean, there may have been persistent and continued connectivity between widespread populations in some species. Eurybathy, having a distribution encompassing a wide depth range, and circumpolarity, having a distribution that encircles the Antarctic continental landmass, are two characteristics that have been widely reported for Southern Ocean species.

In some cases, molecular studies that have revealed the presence of cryptic lineages have shown species to be have more restricted depth and geographic ranges than originally thought. Until recently, *Glycera kerguelensis*, a polychaete worm, was thought to inhabit depths ranging from around 150 m to deeper than 4,000 m. Molecular work has now shown this taxon comprises multiple cryptic species each with a much more restricted depth distribution (Schueller 2011). Nonetheless, molecular studies have confirmed extreme eurybathy in at least two species. An undescribed amphipod species in the genus *Abyssorchomene* has been confirmed as inhabiting depths from around 300 m to below 4,000 m (Havermans et al. 2011) while a single lineage of the pycnogonid *Nymphon australe* has been confirmed from 150 m to below 1,000 m (Arango et al. 2011). Whether this species extends to its reported depth of below 4,000 m awaits confirmation. Nonetheless, even the confirmed depth range would be unusual in temperate taxa.

Similarly, Hemery et al. (2012) recognised six ‘phylogroups’ within the crinoid ‘species’ *Promachocrinus kerguelensis,* which were all found to be circumpolar, sympatric and eurybathic. *P. kerguelensis* is known from 20 to 2,100 m depth and is likely to have a large dispersal potential due to its positively buoyant lecithotrophic larvae. The deeper shelf of the Antarctic continent and periodic glaciation (over Milankovitch cycles) reducing available benthic habitat on the shelf, coupled with the rise and fall of sea levels, may be natural evolutionary drivers towards eurybathy.

Geographically widespread distributions throughout the Southern Ocean are clearly less common than once thought and are most often associated with a pelagic dispersal stage (usually larval). For example analysis of two mitochondrial (16S rRNA, COI) and one nuclear (28S rRNA) gene fragments in two species of shrimps (*Chorismus antarcticus, Nematocarcinus lanceopes*) with planktotrophic larvae showed no evidence of cryptic speciation across a circum-Antarctic range (Raupach et al. 2010). Similarly, the limpet *N. concinna*, which is a broadcast spawner, has a widespread distribution but is probably restricted from circumpolarity by lack of suitable coastal habitat and the relatively short period (<4 weeks) of larval dispersal (González-Wevar et al. 2011). Population genetic studies using AFLPs and comparing *N. concinna* with the direct-developing top shell *Margarella antarctica* concluded that population structure reflected life history (Hoffman et al. 2011).

For many species, the mode of reproduction is not actually confirmed, but rather inferred from congeners or from egg size. An unlikely circum-Antarctic distribution in *Astrotoma agassizi*, a brittle star thought to lack a larval stage (Hunter and Halanych 2008), was later satisfactorily explained when a larval DNA barcoding (i.e., sequencing of the mitochondrial gene COI) programme identified a planktonic dispersive stage for this species (Heiemeyer et al. 2010). Elucidating life-history strategies through widespread DNA barcoding of zooplankton could yield much useful information on the life history strategies of Southern Ocean fauna.

It is likely that having a dispersive stage facilitates recolonisation of newly available habitat when ice shelves retreat following glacial maxima. If species do persist during glacial maxima in small refugia on the continental shelf then larvae would be rapidly distributed around the continent by ocean currents as soon as warmer conditions prevailed. This process should leave a characteristic molecular signature that reflects a population bottleneck (low diversity) and population expansion. A star-like haplotype pattern would be predicted, with one to a few very abundant ancestral haplotypes and multiple haplotypes that differ only slightly and occur at low frequencies. Several examples of such a haplotype pattern have now been reported from the Southern Ocean, for example in the sea urchin *Sterechinus neumayeri* (Diaz et al. 2011) and in the shrimp *C. antarcticus* (Raupach et al. 2010). The latter study also detected a departure from mutation-drift equilibrium in the genetic signal—evidence of population expansion. Similar population expansions have been detected in krill (Goodall-Copestake et al. 2010) and the endemic Southern Ocean limpet *N. concinna* (González-Wevar et al. 2011) and may be indicative of population expansion following glacial maxima.

3.5 Summary and Future Directions for Molecular Work

In the Southern Ocean, with the exception of studies on vertebrates and commercially valuable species, molecular ecology and phylogenetics are in their infancy. Yet they are transforming our understanding of connectivity within the Southern Ocean itself, and between the Southern and other oceans. At least half the studies result from DNA barcode data, in most cases because of the collaboration between the Canadian Centre for DNA Barcoding at the University of Guelph with the Census of Antarctic Marine Life. This has provided momentum within the Antarctic research community. Future molecular ecology studies should focus on additional genes and markers, particularly nuclear markers, including microsatellite markers. To date, microsatellites have been developed for only a handful of invertebrate species: Antarctic krill *Euphausia superba* (Bortolotto et al. 2011), the serolid isopods *Septemserolis septemcarinata* (Leese et al. 2010) and *Ceratoserolis trilobitoides* (Held and Leese 2007; Leese and Held 2008), and three octopus species: *Pareledone turqueti* (Strugnell et al. 2009a, 2012), *P. charcoti* (Strugnell et al. 2009a) and *Adelieledone polymorpha* (Strugnell et al. 2009b).

Future molecular ecology studies should also include haplotype networks and look for network patterns that can be predicted under different refugial scenarios at glacial maxima. They should also look for congruent patterns between nuclear and mitochondrial markers and they should seek evidence for recent (post glacial maximum) population expansion in the molecular signatures. They may also be able to confirm that hypothesised historical seaways (Fig. 3.1) acted as a conduit for gene flow. Similarities in molecular data between the two areas have been noted (Linse et al. 2007; Strugnell et al. 2012), but more evidence is required.

Acknowledgments This is a contribution to the SCAR EBA programme.

References

Allcock AL, Brierley AS, Thorpe JP, Rodhouse PG (1997) Restricted geneflow and evolutionary divergence between geographically separated populations of the Antarctic octopus *Pareledone turqueti*. Mar Biol 129:97–102

Allcock AL, Barratt I, Eléaume M, Linse K, Norman MD, Smith PJ, Steinke D, Stevens DW, Strugnell JM (2011) Cryptic speciation and the circumpolarity debate: a case study on endemic Southern Ocean octopuses using the COI barcode of life. Deep-Sea Res II 58:242–249

Arango CP, Soler-Membrives A, Miller K (2011) Genetic differentiation in the circum—Antarctic sea spider *Nymphon australe* (Pycnogonida; Nymphonidae). Deep-Sea Res II 58:212–219

Baird HP, Miller KJ, Stark JS (2011) Evidence of hidden biodiversity, ongoing speciation and diverse patterns of genetic structure in giant Antarctic amphipods. Mol Ecol 20:3439–3454

Baker AJ, Pereira SL, Haddrath OP, Edge KA (2006) Multiple gene evidence for expansion of extant penguins out of Antarctica due to global cooling. Proc R Soc B 273:11–17

Barnes DKA, Clarke A (2011) Antarctic marine biology. Current Biol R451–R457

Batta-Lona PG, Bucklin A, Wiebe PH, Copley NJ, Patarnello T (2011) Population genetic variation of the Southern Ocean krill, *Euphausia superba*, in the Western Antarctic Peninsula region based on mitochondrial single nucleotide polymorphisms (SNPs). Deep-Sea Res II 58:1652–1661

Bortolotto E, Bucklin A, Mezzavilla M, Zane L, Patarnello T (2011) Gone with the currents: lack of genetic differentiation at the circum-continental scale in the Antarctic krill *Euphausia superba*. BMC Genet 12:32

Brandão SN, Suger J, Schon I (2010) Circum antarctic distribution in Southern Ocean benthos? A genetic test using the genus *Macroscapha* (Crustacea, Ostracoda) as a model. Mol Phylogenet Evol 55:1055–1069

Brandt A, de Broyer C, de Mesel I, Elinsen KE, Gooday AJ, Hilbig B, Linse K, Thomson MRA, Tyler PA (2007a) The biodiversity of the deep Southern Ocean benthos. Phil Trans R Soc B 362:39–66

Brandt A, Brökeland W, Choudhury M, Brix S, Kaiser S, Malyutina M (2007b) Deep-sea isopod biodiversity, abundance, and endemism in the Atlantic sector of the Southern Ocean—results from the ANDEEP I-III expeditions. Deep-Sea Res II 54:1760–1775

Brey T, Dahm C, Gorny M, Klages M, Stiller M, Arntz WE (1996) Do Antarctic benthic invertebrates show an extended level of eurybathy? Antarctic Sci 8:3–6

Browne WE, Haddock SHD, Martindale MQ (2007) Phylogenetic analysis of lineage relationships among hyperiid amphipods as revealed by examination of the mitochondrial gene, cytochrome oxidase I (COI). Int Comp Biol 47:815–830

Clarke A (2008) Antarctic marine benthic diversity: patterns and processes. J Exp Mar Biol Ecol 366:48–55

Clarke A, Arntz WE (2006) An introduction to EASIZ (Ecology of the Antarctic Sea Ice Zone): an integrated programme of water column, benthos and bentho-pelagic coupling in the coastal environment of Antarctica. Deep-Sea Res II 53:803–814

Clarke A, Crame JA (2010) Evolutionary dynamics at high latitudes: speciation and extinction in polar marine faunas. Phil Trans R Soc B 365:3655–3666

Clarke A, Johnston N (2003) Antarctic marine benthic diversity. Oceanogr Mar Biol Ann Rev 41:47–114

Curtis C, Stewart BS, Karl SA (2009) Pleistocene population expansions of Antarctic seals. Mol Ecol 18:2112–2121

Darling KF, Wade CM (2008) The genetic diversity of planktic foraminifera and the global distribution of ribosomal RNA genotypes. Mar Micropaleontol 67:216–238

Darling KF, Kucera M, Pudsey CJ, Wade CM (2004) Molecular evidence links cryptic diversification in polar planktonic protests to Quaternary climate dynamics. Proc Nat Acad Sci 101:7657–7662

DeBroyer C, Danis B (2011) How many species in the Southern Ocean? Towards a dynamic inventory of the Antarctic marine species. Deep-Sea Res II 58:5–17

Dell RK (1972) Antarctic benthos. Adv Mar Biol 10:1–216

Diaz A, Féral JP, Saucède T, Poulin E (2011) Evolutionary pathways among shallow and deep-sea echinoids of the genus *Sterechinus* in the Southern Ocean. Deep-Sea Res II 58:205–211

Diekman B, Kuhn G, Gersonde G, Mackensen R (2004) Middle Eocene to early miocene environmental changes in the sub-Antarctic Southern ocean: evidence from biogenic and terrigenous patterns at ODP site 1090. Global Planet Change 40:295–313

EPICA (2004) Eight glacial cycles from an Antarctic ice core. Nature 429:623–628

Fraser CI, Nikula R, Ruzzante DE, Waters JM (2010) Multigene phylogeny of the southern bull-kelp genus *Durvillaea* (Phaeophyceae: Fucales). Mol Phylogenet Evol 57:1301–1311

Göbbeler K, Klussmann-Kolb A (2010) Out of Antarctica?—new insights into the phylogeny and biogeography of the Pleurobranchomorpha (Mollusca, Gastropoda). Mol Phylogenet Evol 55:996–1007

González-Wevar CA, Nakano T, Cañete JI, Poulin E (2010) Molecular phylogeny and historical biogeography of *Nacella* (Patellogastropoda: Nacellidae) in the Southern Ocean. Mol Phylogenet Evol 56:115–124

González-Wevar CA, David B, Poulin E (2011) Phylogeography and demographic inference in *Nacella* (Patinigera) concinna (Strebel, 1908) in the western Antarctic Peninsula. Deep-Sea Res II 58:220–229

Goodall-Copestake WP, Perez-Espona S, Clark MS, Murphy EJ, Seear PJ, Tarling GA (2010) Swarms of diversity at the gene cox1 in Antarctic krill. Heredity 104:513–518

Griffiths HJ, Barnes DKA, Linse K (2009) Towards a generalized biogeography of the Southern Ocean benthos. J Biogeog 36:162–177

Havermans C, Nagy ZT, Sonet G, De Broyer C, Martin P (2011) DNA barcoding reveals new insights into the diversity of Antarctic species of *Orchomene* sensu lato (Crustacea: Amphipoda: Lysianassoidea). Deep-Sea Res II 58:230–241

Heiemeyer D, Lavery S, Sewell MA (2010) Molecular species identification of *Astrotoma agassizii* from planktonic embryos: further evidence for a cryptic species complex. J Hered 101:775–779

Held C (2000) Phylogeny and biogeography of serolid isopods (Crustacea, Isopoda, Serolidae) and the use of ribosomal expansion segments in molecular systematics. Mol Phylogenet Evol 15:165–178

Held C (2001) No evidence for slow-down of molecular substitution rates at subzero temperatures in Antarctic serolid isopods Crustacea, Isopoda, Serolidae). Polar Biol 24:497–501

Held C (2003) Molecular evidence for cryptic speciation within the widespread Antarctic crustacean *Ceratoserolis trilobitoides* (Crustacea, Isopoda). In: Huiskes AHL, Gieskes WWC, Rozema J, Schorno RML, van der Vies SM, Wolff WJ (eds) Antarctic biology in a global context, Backhuys, pp. 135–139

Held C, Leese F (2007) The utility of fast evolving molecular markers for studying speciation in the Antarctic benthos. Polar Biol 30:513–521

Hemery LG, Eléaume M, Roussel V, Améziane N, Gallut C, Steinke D, Cruaud C, Couloux A, Wilson NG (2012) Comprehensive sampling reveals circumpolarity and sympatry in seven mitochondrial lineages of the Southern Ocean crinoid species *Promachocrinus kerguelensis* (Echinodermata). Mol Ecol 21:2505–2518

Hoffman JI, Clarke A, Linse K, Peck LS (2011) Effects of brooding and broadcasting reproductive modes on the population genetic structure of two Antarctic gastropod molluscs. Mar Biol 158:287–296

Hunt B, Strugnell J, Bednarsek N, Linse K, Nelson RJ, Pakhomov E, Seibel B, Steinke D, Würzberg L (2010) Poles apart: the bipola pteropod species *Limacina helicina* is genetically distinct between the Arctic and Antarctic Oceans. PLoS ONE 5:e9835

Hunter RL, Halanych KM (2008) Evaluating connectivity in the brooding brittle star *Astrotoma agassizii* across the drake passage in the Southern Ocean. J Hered 99:137–148

Janko K, Lecointre G, DeVries A, Couloux A, Cruaud C, Marshall C (2007) Did glacial advances during the Pleistocene influence differently the demographic histories of benthic and pelagic Antarctic shelf fishes?—Inferences from intraspecific mitochondrial and nuclear DNA sequence diversity. BMC Evol Biol 7:220

Janko K, Marshall C, Musilová Z, Van Houdt J, Couloux A, Cruaud C, Lecointre G (2011) Multilocus analyses of an Antarctic fish species flock (Teleostei, Notothenioidei, Trematominae): phylogenetic approach and test of the early-radiation event. Mol Phylogenet Evol 60:305–316

Janosik AM, Mahon AR, Halanych KM (2011) Evolutionary history of Southern Ocean *Odontaster* sea star species (Odontasteridae; Asteroidea). Polar Biol 34:575–586

Krabbe K, Leese F, Mayer C, Tollrain R, Held C (2010) Cryptic mitochondrial lineages in the widespread pycnogonid *Colossendeis megalonyx* Hoek, 1881 from Antarctic and Subantarctic waters. Polar Biol 33:281–292

Kuhn KL, Gaffney PM (2006) Preliminary assessment of population structure in the mackerel icefish (*Champsocephalus gunnari*). Polar Biol 29:927–935

Kuklinski P, Barnes DKA (2010) First bipolar brooder. Mar Ecol Prog Ser 401:15–20

Lambert DM, Ritchie PA, Millar CD, Holland B, Drummond AJ, Baroni C (2002) Rates of evolution in ancient DNA from Adélie penguins. Science 295:2270–2273

Lawver LA, Gahagan LM (2003) Evolution of cenozoic seaways in the circum-Antarctic region. Palaeogeog Palaeoclim Palaeoecol 198:11–37

Lee Y-H, Song M, Lee S, Leon R, Godoy SO, Canete I (2004) Molecular phylogeny and divergence time of the Antarctic sea urchin (*Sterechinus neumayeri*) in relation to the South American sea urchins. Antarct Sci 16:29–36

Leese F, Held C (2008) Identification and characterization of microsatellites from the Antarctic isopod *Ceratoserolis trilobitoides*: nuclear evidence for cryptic species. Conserv Genet 9:1369–1372

Leese F, Agrawal S, Held C (2010) Long-distance island hopping without dispersal stages: transportation across major zoogeographic barriers in a Southern Ocean isopod. Naturwissensch 97:583–594

Linse K, Cope T, Lörz AN, Sands C (2007) Is the Scotia Sea a centre of Antarctic marine diversification? Some evidence of cryptic speciation in the circum-Antarctic bivalve *Lissarca notorcadensis* (Arcoidea: Philobryidae). Polar Biol 30:1059–1068

Lörz A-N, Held C (2004) A preliminary molecular and morphological phylogeny of the Antarctic Epimeriidae and Iphimediidae (Crustacea, Amphipoda). Mol Phylogenet Evol 31:4–15

Lörz AN, Maas E, Linse K, Coleman CO (2009) Do circum-Antarctic species exist in peracarid Amphipoda? A case study in the genus *Epimeria* Costa, 1851 (Crustacea, Peracarida, Epimeriidae). ZooKeys 18:91–128

Mah C, Foltz D (2011) Molecular phylogeny of the Forcipulatacea (Asteroidea: Echinodermata): systematics and biogeography. Zool J Linn Soc 162:646–660

Mahon AR, Thornhill DJ, Norenburg J, Halanych KM (2010) DNA uncovers Antarctic nemertean biodiversity and exposes a decades-old cold case of asymmetric inventory. Polar Biol 33:193–202

Matschiner M, Hanel R, Salzburger W (2009) Gene flow by larval dispersal in the Antarctic notothenioid fish *Gobionotothen gibberifrons*. Mol Ecol 18:2574–2587

Naish T, Powell R, Levy R, Wilson G, Scherer R, Talarico F, Krissek L, Niessen F, Pompilio M, Wilson T, Carter L, DeConto R, Huybers P, McKay R, Pollard D, Ross J, Winter D, Barrett P, Browne G, Cody R, Cowan E, Crampton J, Dunbar G, Dunbar N, Florindo F, Gebhardt C, Graham I, Hannah M, Hansaraj D, Harwood D, Helling D, Henrys S, Hinnov L, Kuhn G, Kyle P, Läufer A, Maffioli P, Magens D, Mandernack K, McIntosh W, Millan C, Morin R, Ohneiser C, Paulsen T, Persico D, Raine I, Reed J, Riesselman C, Sagnotti L, Schmitt D, Sjunneskog C, Strong P, Taviani M, Vogel S, Wilch T, Williams T (2009) Obliquity-paced pliocene West Antarctic ice sheet oscillations. Nature 458:322–328

Near TJ (2004) Estimating divergence times of notothenioid fishes using a fossil-calibrated molecular clock. Antarct Sci 16:37–44

Nikula R, Fraser CI, Spencer HG, Waters JM (2010) Circumpolar dispersal by rafting in two subantarctic kelp-dwelling crustaceans. Mar Ecol Prog Ser 405:221–230

O'Loughlin PM, Paulay G, Davey N, Michonneau F (2011) The Antarctic region as a marine biodiversity hotspot for echinoderms: diversity and diversification of sea cucumbers. Deep-Sea Res II 58:264–275

Page TJ, Linse K (2000) More evidence of speciation and dispersal across the Antarctic polar front through molecular systematics of Southern Ocean *Limatula* (Bivalvia: Limidae). Polar Biol 25:818–826

Park ET, Ferrari FD (2009) Species diversity and distributions of pelagic calanoid copepods (Crustacea) from the Southern Ocean. In: Krupnik I, Lang MA, Miller E (eds) Smithsonian at the poles: contributions to international polar year science, pp 143–180, Smithsonian Institution Press, Washington

Pawlowski J, Fahrni J, Lecroq B, Longet D, Cornelius N, Excoffier L, Cedhagen T, Gooday AJ (2007) Bipolar gene flow in deep-sea benthic foraminifera. Mol Ecol 16:4089–4096

Pearse JS, Mooi R, Lockhart SJ, Brandt A (2009) Brooding and species diversity in the Southern Ocean: selection for brooders or speciation within brooding clades? In: Krupnik I, Krupnik I, Lang MA, Miller E (eds) Smithsonian at the poles: contributions to international polar year science, pp 181–196. Smithsonian Institution Press, Washington

Pfuhl HA, McCave NI (2007) Evidence for late Oligocene establishment of the Antarctic circumpolar current. Earth Planet Sci Lett 235:715–728

Pollard D, DeConto RM (2009) Modelling West Antarctic ice sheet growth and collapse through the past five million years. Nature 458:329–333

Pons J, Barraclough TG, Gomez-Zurita J, Cardoso A, Duran DP, Hazell S, Kamoun S, Sumlin WD, Vogler AP (2006) Sequence-based species delimitation for the DNA taxonomy of undescribed insects. Syst Biol 55:595–609

Raupach MJ, Malyutina M, Brandt A, Wägele JW (2007) Molecular data reveal a highly diverse species flock within the munnopsoid deep-sea isopod *Betamorpha fusiformis* (Barnard, 1920) (Crustacea: Isopoda: Asellota) in the Southern Ocean. Deep-Sea Res II 54:1820–1830

Raupach MJ, Mayer C, Malyutina M, Wägele JW (2009) Multiple origins of deep-sea Asellota (Crustacea: Isopoda) from shallow waters revealed by molecular data. Proc R Soc B 276:799–808

Raupach MJ, Thatje S, Dambach J, Rehm P, Misof B, Leese F (2010) Genetic homogeneity and circum-Antarctic distribution of two benthic shrimp species of the Southern Ocean, *Chorismus antarcticus* and *Nematocarcinus lanceopes*. Mar Biol 157:1783–1797

Rogers AD, Morley S, Fitzcharles E, Jarvis K, Belchier M (2006) Genetic structure of Patagonian toothfish (*Dissostichus eleginoides*) populations on the Patagonian Shelf and Atlantic and western Indian Ocean sectors of the Southern Ocean. Mar Biol 149:915–924

Schueller M (2011) Evidence for a role of bathymetry and emergence in speciation in the genus *Glycera* (Glyceridae, Polychaeta) from the deep Eastern Weddell Sea. Polar Biol 34:549–564

Stepanjants SD (2006) A review of bipolarity concepts: history and examples from Radiolaria and Medusozoa (Cnidaria). Mar Biol Res 2:200–241

Strugnell JM, Rogers AD, Prodöhl PA, Collins MA, Allcock AL (2008) The thermohaline expressway: the Southern Ocean as a centre of origin for deep-sea octopuses. Cladistics 24:853–860

Strugnell JM, Allcock AL, Watts PC (2009a) A panel of microsatellite loci from two species of octopus, *Pareledone turqueti* (Joubin, 1905) and *Pareledone charcoti* (Joubin, 1905). Mol Ecol Res 9:1239–1242

Strugnell JM, Allcock AL, Watts PC (2009b) Microsatellite loci from the endemic Southern Ocean octopus *Adelieledone polymorpha* (Robson, 1930). Mol Ecol Res 9:1068–1070

Strugnell J, Cherel Y, Cooke IR, Gleadall IG, Hochberg FG, Ibanez CM, Jorgensen E, Laptikhovsky VV, Linse K, Norman M, Vecchione M, Voight JR, Allcock AL (2011) The Southern Ocean: source and sink? Deep-Sea Res II 58:196–204

Strugnell JM, Watts PC, Smith PJ, Allcock AL (2012) Persistent genetic signatures of historic climatic events in an Antarctic octopus. Mol Ecol 21:2775–2785

Thatje S, Hillenbrand C-D, Larter R (2005) On the origin of Antarctic marine benthic community structure. Trends Ecol Evol 20:534–540

Thomson MRA (2004) Geological and palaeoenvironmental history of the Scotia Sea region as a basis for biological interpretation. Deep-Sea Res II 51:1467–1487

Thornhill DJ, Mahon AR, Norenburg JL, Halanych KM (2008) Open-ocean barriers to dispersal: a test case with the Antarctic polar front and the ribbon worm *Parbolasia corrugatus* (Nemertea: Lineidae). Mol Ecol 17:5104–5117

Uriz MJ, Gili JM, Orejas C, Perez-Porro AR (2011) Do bipolar distributions exist in marine sponges? *Stylocordyla chupachups* sp. nv. (Porifera: Hadromerida) from the Weddell Sea (Antarctic), previously reported as *S. borealis* (Lovén, 1868). Polar Biol 34:243–255

Vaughan DG, Barnes DKA, Fretwell PT, Bingham RG (2011) Potential seaways across West Antarctica. Geochem Geophys Geosyst 12:Q10004. doi:10.1029/2011GC003688

Wilson NG, Hunter RL, Lockhart SJ, Halanych KM (2007) Multiple lineages and absence of panmixia in the circumpolar crinoid *Promachocrinus kerguelensis* from the Atlantic sector of Antarctica. Mar Biol 152:895–904

Wilson NG, Schröld M, Halanych KM (2009) Ocean barriers and glaciation: evidence for explosive radiation of mitochondrial lineages in the Antarctic sea slug *Doris kerguelensis* (Mollusca, Nudibranchia). Mol Ecol 18:965–984

Yoder AD, Yang ZH (2000) Estimation of primate speciation dates using local molecular clocks. Mol Biol Evol 17:1081–1090

[illegible]

Thomson MRA (2004) Geological and palaeoenvironmental history of the Scotia Sea region as a basis for biological interpretation. Deep Sea Res II 51:1467–1487

[illegible]

[illegible]

[illegible]

[illegible]

[illegible]

[illegible]

Chapter 4
Pole-to-Pole Gene Flow in Protozoan Ciliates

Graziano Di Giuseppe, Fernando Dini, Claudio Alimenti, Adriana Vallesi and Pierangelo Luporini

4.1 Backgrounds

Microorganisms represent the smallest but arguably most important component of the ocean life. They are essential to all nutrient cycles because they form the bottom of the marine food chain and outnumber all other marine species by orders of magnitude. Sampling of remote and inaccessible habitats and large-scale genomic analysis have shown how little we know about the microbial life in the oceans, and how our poor knowledge of the marine chemistry and biology is preventing us from foreseeing the detrimental effects that a too rapidly changing world has on the oceans' ecosystems. In this context, polar microorganisms are attracting particular interest because of their role in global-scale biogeochemical cycles, in particular the carbon dioxide exchange with the atmosphere (Falkowski et al. 2008).

Considering this interest, planktonic and benthic microbial communities from Arctic and Antarctic areas have become the focus of more systematic sampling

G. Di Giuseppe (✉) · F. Dini
Dipartimento di Biologia, University of Pisa, via Volta 4, 56126 Pisa, Italy
e-mail: gdigiuseppe@biologia.unipi.it

F. Dini
e-mail: fdini@biologia.unipi.it

C. Alimenti · A. Vallesi · P. Luporini
Dipartimento di Scienze Ambientali e Naturali, University of Camerino, via Gentile III da Varano, 63032 Camerino, MC, Italy
e-mail: claudio.alimenti@unicam.it

A. Vallesi
e-mail: adriana.vallesi@unicam.it

P. Luporini
e-mail: piero.luporini@unicam.it

C. Verde and G. di Prisco (eds.), *Adaptation and Evolution in Marine Environments, Volume 2*, From Pole to Pole, DOI: 10.1007/978-3-642-27349-0_4,

and rigorous analyses for their taxonomic, prokaryotic (bacterial) and eukaryotic (protist), biodiversity. A relevant result of these analyses was the finding of microbial species that, like a diverse range of plant and animal species (Lindberg 1991; Crame 1993), warrant the definition "bipolar" (or "anti-tropical"), i.e. species represented by high-latitude populations physically separated in distribution across the tropics (Darling et al. 2000; Montresor et al. 2003; Brandt et al. 2007; Pawlowski et al. 2007). This concept of species bipolarity has inherently raised the intriguing question whether co-specific Antarctic and Arctic populations evolved independently since the effective separation (approximately 10–15 million years ago) between the Arctic and Antarctic cold-water provinces, or whether a trans-tropical gene flow ensures that these polar populations maintain genetic continuity (Darling et al. 2000).

Morphological studies alone are clearly insufficient to address this question, due to recurrent phenomena of parallel or convergent morphological evolution that take place under similar environmental forces. Therefore, more solid grounds supporting the concept of species bipolarity have been obtained in some species of foraminifera and dinozoans from analysis of genetic variation in sequences of the small subunit (SSU) rRNA nuclear gene (Darling et al. 2000; Montresor et al. 2003; Brandt et al. 2007; Pawlowski et al. 2007). Nevertheless, unless the calibration of a molecular clock is supported by abundance in fossil records, as is the case in foraminifera (Pawlowski et al. 1997), also this genetic approach is impaired by the fact that the same DNA regions may evolve at different rates among closely related organisms.

A solution to overcoming this challenge is provided by the well-defined, monophyletic group of ciliates which are ideal organisms for the analysis of the breeding structure of natural microbial populations and, therefore, for obtaining data which satisfy the interbreeding criterion on which the biological (Darwinian) concept of species is founded. Ciliates govern their gene exchanges through a unique sexual phenomenon, known as conjugation (or mating), in which two functionally hermaphroditic individuals unite temporarily in pairs for a mutual exchange of a "migratory" ("male") gametic nucleus destined to fuse into a synkaryon with a "resident" ("female") gametic nucleus. In addition, ciliates more than any other group of polar microorganisms can readily provide living laboratory material in virtually unlimited amounts, since every aquatic polar habitat hosts a large variety of ciliate species that are relatively easy to collect and expand into permanent clonal cultures (i.e., descendants from single wild-type specimens).

Here we review results that have been obtained from breeding analyses of Antarctic, Fuegian and Arctic populations of the ciliate *Euplotes nobilii*. These results have provided compelling evidence that these populations are genetically interconnected by gene flow and form a unique interbreeding biological species. The description of these results is preceded by an outline of our current knowledge of ciliate biodiversity in the polar regions, and by a rationalization of the genetic mechanism of the mating types that represents the major driving force of ciliate breeding strategies.

4.2 Ciliate Biodiversity at the Poles

Diversity of ciliate species in Arctic and Antarctic ecosystems has been studied inconsistently, and most studies have essentially described new morphological species from polar habitats (e.g. Corliss and Snyder 1986; Valbonesi and Luporini 1990a, b; Agatha et al. 1990, 1993; Petz et al. 1995; Kepner et al. 1999; Petz 2005; Wilbert and Song 2005, 2008). One exception is an extensive taxonomic survey of Antarctic and Arctic limnetic ciliates (collected from numerous lakes, ponds, and streams of Livingston Island in South Shetlands, Terra Nova Bay in Victoria Land, Brøgger Peninsula and Ossian Sarsfjellet in the Svalbard Archipelago), that has deliberately been carried out to assess whether ciliate polar biogeography is characterized by endemism, or global distribution of species (Petz 2004; Petz et al. 2007). Nearly 400 distinct morphospecies were overall identified. Although the majority resulted to be species already known from non-polar environments, 20 % were identified as new to science and 13 % appeared to be equally represented in Antarctic and Arctic waters. This finding represents the first (morphological) evidence that ciliates may include species represented by populations that live physically separated at the opposite ends of the globe.

4.3 Ciliate Mating Systems

As mentioned above, conjugation is the basic sexual phenomenon on which ciliates rely to perform gene exchange and regulate the genetic distance among populations. In many species, this phenomenon involves the differentiation of genetically and physiologically distinct cell classes that have been designated as "mating types" in relation to their capacity to interact mutually, either through physic cell–cell contacts or diffusible signaling pheromones, for mating-pair formation (Dini and Nyberg 1993). These mating types may only be two as shown by most species of *Paramecium*, or tens within each species as is the case in *Euplotes* and other hypotrichous ciliates (Phadke and Zufall 2009). While the binary mating systems have functionally been associated with the duality of sex, the functions of the multiple mating systems have generally been considered to be much more comparable, for example, with the devices that are intrinsic to the self/non-self recognition mechanisms evolved in flowering plants to ensure cross pollination (Luporini et al. 2005). Ciliate species with multiple mating types are in fact usually viewed as distinguishable by eco-genetic strategies strongly committed to outbreeding, because the mating options and the chances to mate with a stranger will, at least in principle, increase for every member of a species along with an increase in the mating type number.

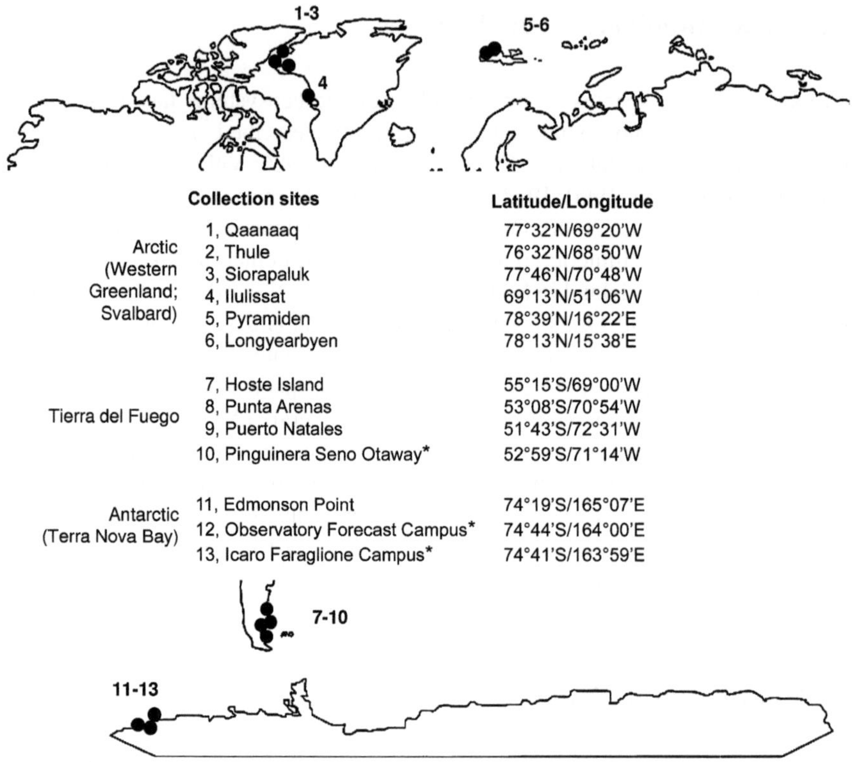

Fig. 4.1 Map of the principal sites that have been visited since 1988 to collect strains of polar species of *Euplotes*. Asterisks indicate informal names

4.4 Collection Sites and Polar *Euplotes* Species

During an interval of approximately 20 years, it has been possible to accumulate some hundreds of *Euplotes* strains as stable laboratory cultures starting from specimens isolated from various seashore sites of Antarctica, Tierra del Fuego and the Arctic (Fig. 4.1). These strains were initially analyzed morphologically to assign their species status, by measuring and counting common diagnostic parameters such as cell body dimensions, number and positional patterns of ciliary organelles, and number and geometry of cortical alveoli. Only a minority of strains have been found to represent known *Euplotes* species, such as *E. rariseta* and *E. kahli* (Valbonesi and Luporini 1990a; unpublished data), whereas the large majority appeared to represent new *Euplotes* species of which, however, only few have received official taxonomic designations so far (Valbonesi and Luporini 1990a, b).

One of the new species, *E. focardii*, is represented uniquely by Antarctic strains, and this is likely endemic to the Antarctic waters. In contrast, two other new species, i.e. *E. euryhalinus* and *E. nobilii* (Fig. 4.2), which were originally

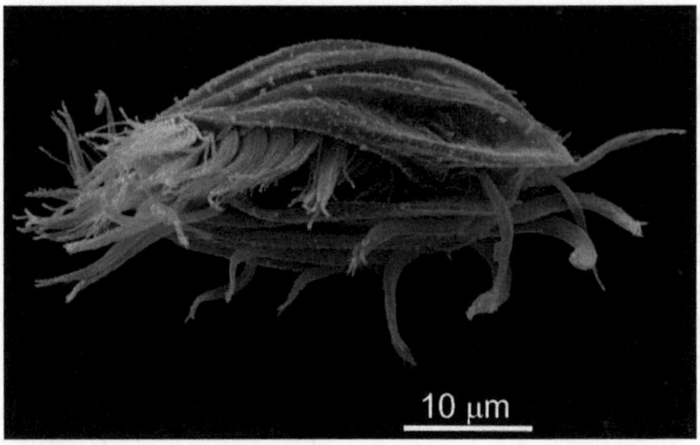

Fig. 4.2 Scanning electron micrograph of *E. nobilii* showing the cell left site with the ciliary membranelles surrounding the oral groove. On the ventral site other compound ciliary organelles (cirri) are visible whose number and positional patterns have maximal diagnostic relevance for *Euplotes* species identification

described with reference only to Antarctic strains, have subsequently been collected also from Fuegian and Arctic sites.

Because of their wider geographic representativeness and apparent incapacity to form passive dispersal stages (cysts) in response to variations in the cultivation conditions (unpublished observations), the overall complex of Antarctic, Fuegian and Arctic strains of *E. nobilii* have been chosen as experimental material to investigate, on phylogenetic and breeding grounds, whether they represented populations that are genetically discontinuous as a consequence of their spatial separation, or populations that are genetically continuous despite their spatial separation.

4.5 Phylogenetic Relationships

The phylogenetic correlations among the Antarctic, Fuegian and Arctic *E. nobilii* strains have been deduced by comparing the SSU-rRNA nuclear gene sequences determined for each of a set of 18 strains (seven Antarctic, four Fuegian and seven Arctic). As shown in Fig. 4.3, the Antarctic and Fuegian strains were found to possess completely identical sequences and, consequently, grouped altogether into the same cluster. Instead, sequence variations for one- to maximum four-nucleotide substitutions were found to be distinctive of the Arctic strains, that were thus resolved into multiple clusters distinct from the cluster including the Antarctic and Fuegian strains.

Although a single nucleotide variation between SSU-rRNA gene sequences has been reputed to be sufficient for an inter-species discrimination in other ciliates such as *Stylonychia* (Bernhard et al. 2001), more than 60 sequence variations have been found to mark the phylogenetic divergence of *E. nobilii* from its closest allied species, *E. raikovi* (Vallesi et al. 2008). In light of this degree of inter-species divergence in *Euplotes*, the one- to four-nucleotide mutations accumulated by the SSU-rRNA gene sequences of the *E. nobilii* Arctic strains have been regarded as reflecting sequence variations that have a merely intra-specific value, and not to be indicative of true genetic and evolutionary separation.

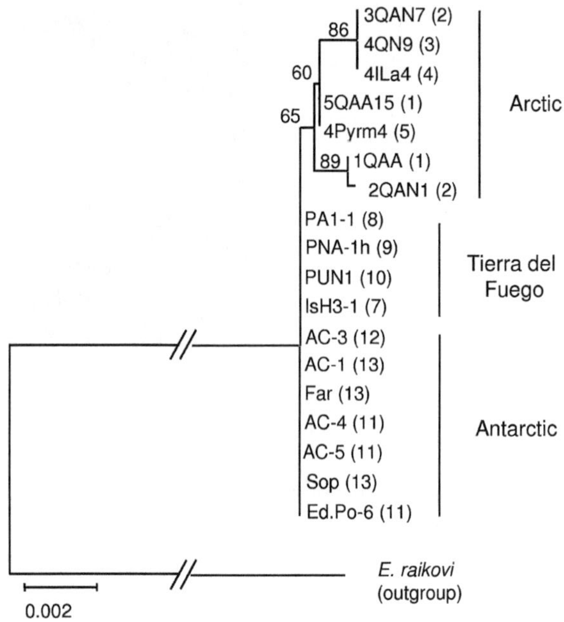

Fig. 4.3 Phylogenetic correlations among Antarctic, Fuegian, and Arctic strains derived from multiple alignment of SSU-rRNA nuclear gene sequences. The numbers at nodes of the phylogenetic tree are values estimated as percentages from 1,000 bootstrap replicates. The numbers in parentheses indicate the collection site of the strain (see Fig. 4.1). *E. raikovi*, used as outgroup, is the species most closely allied to *E. nobilii*

4.6 Mating and Breeding Interactions

Analyses of mating interactions, initially restricted to the Antarctic strains which were the first to be collected and morphologically diagnosed as representative of *E. nobilii* (Valbonesi and Luporini 1990a), provided evidence that, like other *Euplotes* species (Dini and Nyberg 1993), *E. nobilii* is represented in nature by multiple mating types that mediate their activity through diffusible, cell type-specific signaling pheromones (Felici et al. 1999). With the extension of the study of mating interactions to Fuegian and Arctic strains it was first shown that some pheromones structurally characterized from Antarctic and Arctic strains were represented by helical proteins in all members of the same homologous family (Alimenti et al. 2009; Di Giuseppe et al. 2011). Subsequently, it was observed that some Antarctic strains were fully able to form stable mating pairs not only in mixtures with Fuegian strains (with which there was sharing of identical SSU-rRNA nuclear gene sequences), but also in mixtures with Arctic strains (from which there were sequence divergences), thus implying a genetic continuity among the Antarctic, Fuegian and Arctic *E. nobilii* populations represented by these strains. This implication was verified through detailed genetic analysis of offspring clones raised from mating pairs formed in pair-wise mixtures of a selected set of three Antarctic strains (i.e., AC-1, AC-3 and AC-4 collected from Terra Nova Bay) and a selected set of three Arctic strains (i.e. 5QAA15 and 2QAN1 collected from Western Greenland, and 4Pyrm4 collected from Svalbard).

A crucial aspect of ciliate mating interactions, and *Euplotes* species in particular, is that mixing of two strains of different and compatible mating types (e.g. A

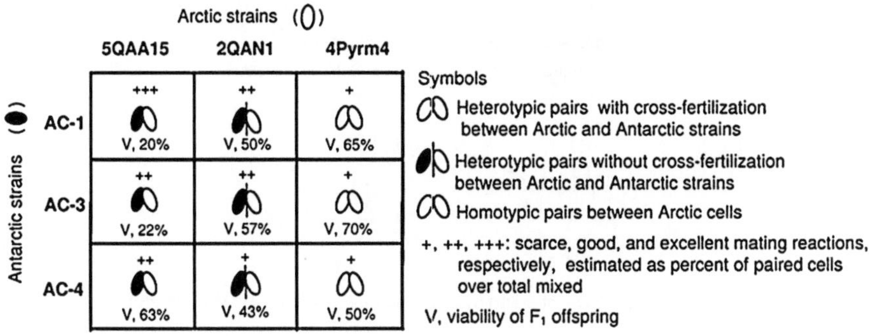

Fig. 4.4 Mating interactions between Antarctic and Arctic *E. nobilii* strains. For each pair-wise strain combination the corresponding box indicates the intensity of mating interactions given on a three-step scale, the type of cell mating pairs formed, and the viability rate computed as percentage of ex-conjugant cells that were able to develop a new nuclear apparatus and expand into fully viable progeny clones. Adapted from Di Giuseppe et al. (2011)

and B) does not necessarily result in the formation of only heterotypic pairs (AB) destined to complete cross-fertilization between mutually exchanged gametic nuclei. Homotypic pairs (AA and BB), that are obviously forced to perform self-fertilization (or autogamy in pairs), may equally be formed in addition and/or complete substitution of the heterotypic ones (Dini and Nyberg 1993). The composition in hetero- and homotypic pairs of a mating mixture (pairs that are all morphologically alike) may thus be revealed only a posteriori by analyzing offspring clones for their patterns of inheritance of the mating-type trait. However, this procedure is hardly applicable to polar ciliates, which are characterized by cell cycles that are more than four times longer than in ciliates of temperate waters, and by life cycles with immaturity (or adolescence) periods (during which cells are unable to mate) lasting months (Valbonesi and Luporini 1993). The difficulty to obtain a clear distinction between homo- and heterotypic pairs, as well as between heterotypic pairs with cross-fertilization and heterotypic pairs without cross-fertilization, has been avoided by utilizing the SSU-rRNA gene sequences (which had previously been determined in relation to the phylogenetic analyses) as strain-specific and bi-parentally inheritable nuclear markers.

All possible types of mating pairs (i.e., homotypic and heterotypic with or without cross-fertilization) were detected in the nine possible pair-wise mating combinations between the three Antarctic and three Arctic strains, and each type of pair was found capable of generating viable offspring albeit with varying survival rates (Fig. 4.4). Of decisive importance to prove gene flow between Antarctic and Arctic *E. nobilii* populations was the finding that the mixtures between the Arctic strain 5QAA15 and each of the three Antarctic strains all formed heterotypic pairs capable of completing cross-fertilization between mutually exchanged gametic nuclei. The offspring clones generated by mating pairs isolated from these three mixtures all appeared to have inherited both of the parental SSU-rRNA gene sequences, which were distinguished from one another by the presence of different nucleotides (C or T) in the sequence position 660 (Fig. 4.5). On the other hand,

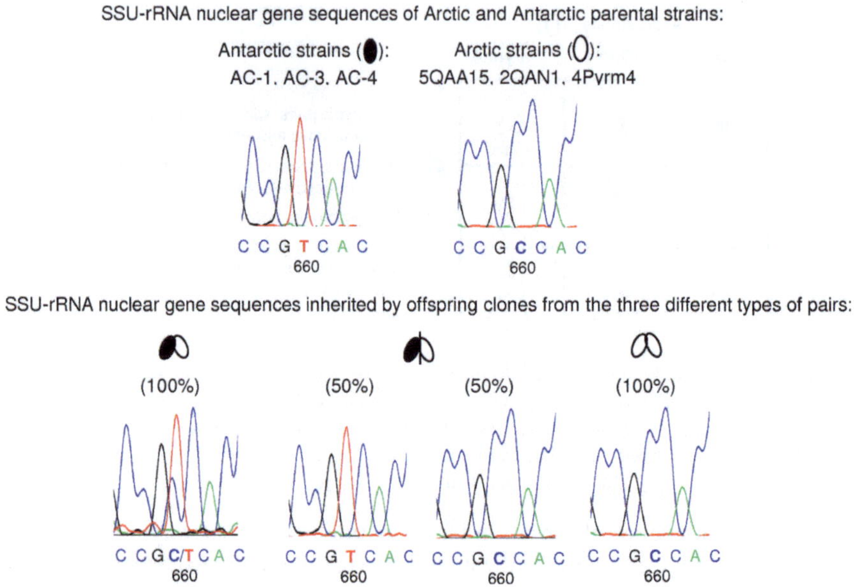

Fig. 4.5 Schematic presentation of the utilization of the strain-specific SSU-rRNA gene sequences to distinguish the different types of mating pairs (represented by symbols as in Fig. 4.4) formed by mating mixtures between Antarctic and Arctic strains. The number below the sequences indicates the position characterized by the nucleotide variation that is distinctive between the Antarctic and Arctic strains. The heterotypic pairs with cross-fertilization are unique in generating offspring clones with "hybrid" sequences characterized by a double C/T peak at the position 660, whereas the heterotypic pairs without cross-fertilization and the homotypic pairs generate offspring clones with unchanged sequences with respect to the parental strains

homotypic pairs and heterotypic pairs with self-fertilization were observed in the mixtures involving each of the three Antarctic strains with the other two Arctic strains, 4Pyrm4 and 2QAN1, thus implying that inter-strain mating compatibility is a necessary but not sufficient condition to ensure the occurrence of an effective gene exchange. The homotypic pairs were revealed through the observation that the SSU-rRNA gene sequences of the offspring clones were all identical not only to each other but also to the sequence of one of the two parental strains; the heterotypic mating pairs with self-fertilization through the observation that the two parental SSU-rRNA gene sequences were inherited in a 1:1 ratio (Fig. 4.5).

4.7 Preliminary Evidence of Pole-to-Pole Gene Flow in Nature

In light of the successful utilization of SSU-rRNA gene sequence profiles in detecting effective gene exchange in mating pairs of Antarctic and Arctic strains raised in the laboratory, preliminary attempts were carried out to obtain genetic

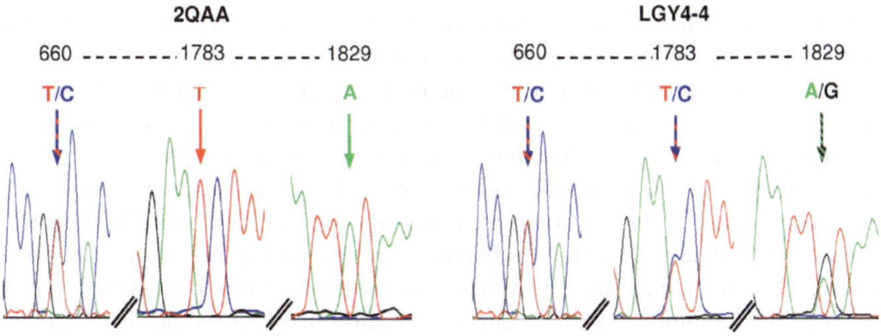

Fig. 4.6 Segments of SSU-rRNA nuclear gene sequences of the Arctic strains 2QAA and LGY4-4 showing the "hybrid" positions characterized by double T/C and A/G peaks

evidence of cross-breeding events between *E. nobilii* Antarctic and Arctic cells also in nature. These attempts were based on analyses of new Arctic *E. nobilii* strains to detect SSU-rRNA gene sequences characterized by the presence of double peaks denoting strains characterized by heterozygosity between distinct genotypes. As shown in Fig. 4.6, two strains, one (2QAA) collected from Western Greenland and one (LGY4-4) collected from Svalbard, have in fact been found (unpublished results) to contain SSU-rRNA gene sequences characterized by the presence of double peaks: only one in position 660 in strain 2QAA, and three in positions 660, 1783 and 1829 in strain LGY4-4. The double peaks in positions 1783 and 1829 appeared to be due to C/T and A/G combinations, and did not discriminate between a heterozygous condition generated by Arctic–Arctic cell cross-breeding and a heterozygous condition generated by Arctic-Antarctic/Fuegian cell cross-breeding. However, the double C/T peak in position 660 appeared to be compatible only with a heterozygous condition generated by Arctic-Antarctic/Fuegian cell cross-breeding, because every Antarctic/Fuegian SSU-rRNA gene sequence analyzed so far carries C in position 660 whereas every Arctic sequence carries T.

4.8 Concluding Remarks

Research on polar biogeography of eukaryotic microorganisms has to date largely neglected ciliates and privileged other taxa such as foraminifera and dinozoans, notwithstanding unique advantages we may derive from ciliates for assessing, directly in the laboratory, breeding performances and reproductive isolation of microbial natural populations. By crossing Antarctic, Fuegian and Arctic strains of *E. nobilii* it was possible to obtain compelling evidence that the populations represented by these strains are mating compatible and breeding fertile with one another and, therefore, share the same gene pool. How bipolar populations of *E. nobilii*, and in general microorganisms that are similarly unable to form resting cysts for passive environmental dispersal, may maintain genetic continuity despite their apparent ecological discontinuity is an intriguing question that can hardly be verified directly. It implies that trophic individuals of these populations are able to swarm, and hence to ensure a pole-to-pole gene flow from one to the opposite extremity of the globe by dwelling (and multiplying) in the permanently cold currents that cross the equatorial ocean depths. To credit this implication with experimental data, *E. nobilii* strains are being analyzed for identifying both nuclear (bi-parentally inherited) and mitochondrial (uni-parentally inherited) ribosomal gene sequences characterized by single nucleotide polymorphisms, which represent the most common type of sequence variation in genomes and have rapidly emerged as powerful genetic markers of the evolutionary history of natural populations (Brookes 1999; Sannucks 2000; Nielsen 2000; Brumfield et al. 2003). As anticipated in Sect. 4.7, two Arctic strains characterized by nucleotide polymorphisms in their SSU-rRNA nuclear gene sequences have already been identified and the interpretation of these polymorphisms includes the possibility that they reflect phenomena of natural hybridization between Arctic and Antarctic populations. After having been extensively studied in animals and plants for their driving role in speciation and evolution (Arnold 1997; Barton 2001), these phenomena have now been genetically documented also between sympatric populations of diatoms (D'Alelio et al. 2009; Casteleyn et al. 2009).

The activity of swarmers is probably not the only force underlying the genetic continuity of the bipolar *E. nobilii* populations. A synergistic force is arguably represented also by the capacity of these populations to communicate and interact, like other *Euplotes* populations, via diffusible signaling pheromones that each cell synthesizes to promote not only mating but also vegetative growth (Luporini et al. 2005). Because of their structural homology, these pheromones are mutually cross-reactive and their structural specificities are such to secure unmatched long-lasting activity and wide-range dispersal in any marine environment (Di Giuseppe et al. 2011).

Acknowledgments The authors acknowledge the financial support from the *Programma Nazionale di Ricerche in Antartide* (PNRA) and Dr. Martha Dunbar for helpful suggestions in the English revision of the text.

References

Agatha S, Wilbert N, Spindler M, Elbrächter M (1990) Euplotide ciliates in sea ice of the Weddell-Sea (Antarctica). Acta Protozool 29:221–228

Agatha S, Spindler M, Wilbert N (1993) Ciliated protozoa (ciliophora) from Arctic sea ice. Acta Protozool 32:261–268

Alimenti C, Vallesi A, Pedrini P, Wüthrich K, Luporini P (2009) Molecular cold-adaptation: comparative analysis of two homologous families of psychrophilic and mesophilic signal proteins of the protozoan ciliate *Euplotes*. IUBMB Life 61:838–845

Arnold ML (1997) Natural hybridization and evolution. Oxford University Press, New York

Barton NH (2001) The role of hybridization in evolution. Mol Ecol 10:551–568

Bernhard D, Stechmann A, Foissner W, Ammermann D, Hehn M, Schlegel M (2001) Phylogenetic relationships within the class Spirotrichea (Ciliophora) inferred from small subunit rRNA gene sequences. Mol Phylogenet Evol 21:86–92

Brandt A, Gooday AJ, Brandao SN et al (2007) First insight into the biodiversity and biogeography of the southern ocean deep sea. Nature 447:307–311

Brookes AJ (1999) The essence of SNPs. Gene 234:177–186

Brumfield RT, Beerli P, Nickerson DA, Edwards SV (2003) The utility of single nucleotide polymorphisms in inferences of population history. Trends Ecol Evol 18:249–256

Casteleyn G, Adams NG, Vanormelingen P, Debeer AE, Sabbe K, Vyverman W (2009) Natural hybrids in the marine diatom *Pseudo-nitzschia pungens* (Bacillariophyceae): genetic and morphological evidence. Protist 160:343–354

Corliss JO, Snyder RA (1986) A preliminary description of several new ciliates from the Antarctica, including *Cohnilembus grassei* n. sp. Protistologica 22:39–46

Crame JA (1993) Bipolar mollusks and their evolutionary implication. J Biogeogr 20:145–161

D'Alelio D, Amato A, Kooistra WHCF, Procaccini G, Casotti R, Montresor M (2009) Internal transcribed spacer polymorphism in *Pseudo-nitzschia multistriata* (Bacillariophyceae) in the Gulf of Naples: recent divergence or intraspecific hybridization? Protist 160:9–20

Darling KF, Wade CM, Steward IA, Kroon D, Dingle R, Leigh Brown AJ (2000) Molecular evidence for genetic mixing of Arctic and Antarctic subpolar populations of planktonic foraminifers. Nature 405:43–47

Di Giuseppe G, Erra F, Dini F, Alimenti C, Vallesi A, Pedrini B, Wüthrich K, Luporini P (2011) Antarctic and Arctic populations of the ciliate *Euplotes nobilii* show common pheromone-mediated cell–cell signaling and cross-mating. Proc Natl Acad Sci USA 108:3181–3186

Dini F, Nyberg D (1993) Sex in ciliates. In: Jones JG (ed) Advances in microbial ecology, vol 13. Plenum Press, New York, pp 85–153

Falkowski PG, Fenchel T, Delong EF (2008) The microbial engines that drive Earth's biogeochemical cycles. Science 320:1034–1039

Felici A, Alimenti C, Ortenzi C, Luporini P (1999) Purification and initial characterization of two pheromones from the marine Antarctic ciliate, *Euplotes nobilii*. Ital J Zool 66:355–360

Kepner RL Jr, Wharton RA Jr, Coats DW (1999) Ciliated protozoa of two Antarctic lakes: analysis by quantitative protargol staining and examination of artificial substrates. Polar Biol 21:285–294

Lindberg DR (1991) Marine biotic interchange between the northern and the southern hemispheres. Paleobiology 17:308–324

Luporini P, Alimenti C, Ortenzi C, Vallesi A (2005) Ciliate mating types and their specific protein pheromones. Acta Protozool 44:89–101

Montresor M, Lovejoy C, Orsini L, Procaccini G, Roy S (2003) Bipolar distribution of the cyst-forming dinoflagellate *Polarella glacialis*. Polar Biol 26:186–194

Nielsen R (2000) Estimation of population parameters and recombination rates using single nucleotide polymorphisms. Genetics 154:931–942

Pawlowski J, Bolivar I, Fahrni JF, de Vargas C, Gouy M (1997) Extreme differences in rates of molecular evolution of foraminifera revealed by comparison of ribosomal DNA sequences and the fossil record. Mol Biol Evol 14:498–505

Pawlowski J, Fahrni J, Lecroq B, Longet D, Cornelius N, Excoffier L, Cedhagen T, Gooday AJ (2007) Bipolar gene flow in deep-sea benthic foraminifera. Mol Ecol 16:4089–4096

Petz W (2004) Ciliate biodiversity in Antarctic and Arctic freshwater habitats—a bipolar comparison. Eur J Protistol 39:491–494

Petz W (2005) Ciliates. In: Scott FJ, Marchant HJ (eds) Antarctic marine protists. Australian Biological Resources Study, Canberra, pp 347–448

Petz W, Song W, Wilbert N (1995) Taxonomy and ecology of the ciliate fauna (Protozoa, Ciliophora) in the endopagial and pelagial of the Weddell Sea, Antarctica. Stapfia 40:1–223

Petz W, Valbonesi A, Schiftner U, Quesada A, Cynan Ellis-Evans J (2007) Ciliate biogeography in Antarctic and Arctic freshwater ecosystems: endemism or global distribution of species? FEMS Microbiol Ecol 59:396–408

Phadke SS, Zufall RA (2009) Rapid diversification of mating systems in ciliates. Biol J Linnean Soc 98:187–197

Sannucks P (2000) Efficient genetic markers for population biology. Trends Ecol Evol 15:199–203

Valbonesi A, Luporini P (1990a) Description of two new species of *Euplotes* and *Euplotes rariseta* from Antarctica. Polar Biol 11:47–53

Valbonesi A, Luporini P (1990b) A new marine species of *Euplotes* (Ciliophora, Hypotrichida) from Antarctica. Bull Br Mus Nat Hist Zool 56:57–61

Valbonesi A, Luporini P (1993) Biology of *Euplotes focardii*, an Antarctic ciliate. Polar Biol 13:489–493

Vallesi A, Di Giuseppe G, Dini F, Luporini P (2008) Pheromone evolution in the protozoan ciliate, *Euplotes*: the ability to synthesize diffusibile forms is ancestral and secondarily lost. Mol Phylogenet Evol 47:439–442

Wilbert N, Song W (2005) New contributions to the marine benthic ciliates from the Antarctic area, including description of seven new species (Protozoa, Ciliophora). J Nat Hist 39:935–973

Wilbert N, Song W (2008) A further study on littoral ciliates (Protozoa, Ciliophora) near King George Island, Antarctica, with description of a new genus and seven new species. J Nat Hist 42:979–1012

Chapter 5
Excess Oxygen in Polar Evolution: A Whole Organism Perspective

Hans-O. Pörtner, Kathleen Walther and Astrid Wittmann

5.1 Living Conditions in Antarctic Marine Waters

The Antarctic is characterized by more stable living conditions than the Arctic. This is due to the partial isolation of the continent and of the surrounding oceans by the Antarctic circumpolar current. Antarctic fauna has adapted to such living conditions over millions of years and the time scale of this evolutionary process has been discussed repeatedly and in the context of geological changes (see Verde et al. 2012). At the functional level the questions arise which were the drivers of Antarctic evolution and what was gained and lost over time with respect to the role and effect of these drivers. A parallel, mirror-inverted question would be which animal groups might have gained or lost in fitness during the process of environmental change. Temperature has been a prime factor considered as a driver but the question here is whether direct effects of temperature on organismic functioning are key or whether further parameters are involved such that temperature acts indirectly through such parameters. Among abiotic factors shaped by temperature, gas concentrations are key, with oxygen and carbon dioxide playing a role in life sustaining processes. In contrast, salinity is not influenced, however, water density and viscosity rise in the cold. Comparing tropical (20 °C) and Antarctic (0 °C) waters, oxygen concentrations are almost twice as high and carbon dioxide levels two times as high in Antarctic oceans. In contrast to solubility, which is enhanced

H.-O. Pörtner (✉) · K. Walther · A. Wittmann
Integrative Ecophysiology, Alfred-Wegener-Institute,
Am Handelshafen 12, 27570 Bremerhaven, Germany
e-mail: Hans.Poertner@awi.de

K. Walther
e-mail: Kathleen.Walther@awi.de

A. Wittmann
e-mail: Astrid.Wittmann@awi.de

C. Verde and G. di Prisco (eds.), *Adaptation and Evolution in Marine Environments, Volume 2*, From Pole to Pole, DOI: 10.1007/978-3-642-27349-0_5,

in the cold, the diffusibility of gases is constrained, thereby hampering gas transport, and especially oxygen uptake further. Convective transport is also constrained in the permanent cold due to enhanced water viscosity. It appears rewarding to interpret functional adjustments to these aspects in some successful key groups of the Antarctic (like notothenioid fishes), which display features not found anywhere else (e.g. Beers and Sidell 2011; Sidell and O'Brien 2006; Coppes Petricorena and Somero 2007; Garofalo et al. 2009; Pörtner et al. 2007, 2012). It is similarly rewarding to look at groups like the benthic brachyuran and anomuran crabs. These seem to have taken these adaptations to a certain limit and presently live at the doorsteps of the Antarctic or have succeeded to settle in its "warmer" water bodies, but are still excluded from life at the coldest temperatures. Similarly, the Arctic snow crab *Chiononectes opilio* penetrates into Arctic waters and tolerates water temperatures of between 4 and −1.5 °C (Burmeister and Sainte-Marie 2010), with the notion that it will experience none of its ambient temperatures permanently, but only temporarily. Egg extrusion and embryonic development occurs between 0 and 3 °C, and are reduced at lower and higher temperatures (Webb et al. 2007). Likely the species does not live permanently at temperatures below 0 °C in its natural environment, which is more variable than the Antarctic, and it may not be able to do so, but a final answer to this question still has to be provided.

5.2 A Unifying Concept: Oxygen and Capacity Limitation of Thermal Tolerance

For understanding the evolutionary process, which allows these crabs to exploit their cold-adaptation capacity, a recent concept may provide a useful mechanistic framework. The concept of oxygen and capacity limited thermal tolerance (OC-LTT) has been developed for a more comprehensive understanding of thermal tolerance and acclimation. This concept has recently also been suggested to be a suitable matrix for the integration of other stressor effects which may interfere with the levels of thermal tolerance (Pörtner 2010).

Recent years have provided evidence that the effects of ocean warming on individual species of marine ectothermic animals from various climatic zones are based on common phenomena, i.e. physiological responses that can be assumed to reflect unifying principles, common to virtually all marine animal phyla. A fundamental concept which links the various levels of biological organisation, molecule, cell, tissue, whole organism as affected by temperature and also provides a matrix for integrating various environmental stressors is the concept of OCLTT, which builds on evidence in representative species from major animal phyla. The OCLTT concept, as a general principle, explains how aerobic scope (the capacity to do work by increasing the rate of aerobic metabolism) is limited by insufficient oxygen supply at both sides of the thermal window (in temperate and tropical ectotherms). The temperature dependent balance between oxygen demand and supply and the associated functional capacity of tissues for their specific tasks in the whole

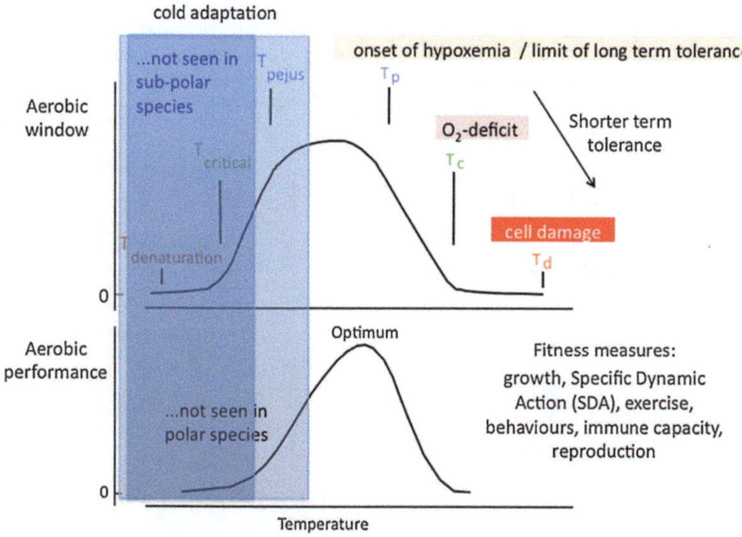

Fig. 5.1 The concept of oxygen and capacity limited thermal tolerance (OCLTT) addresses the why, how and when of thermal limitation by integrating thermal effects at various levels of biological organization. Pejus, critical and denaturation temperatures characterize the progressively more severe level of thermal limitation, with pejus levels representing the first level of thermal limitation and indicating effects at ecosystem level. Lower pejus limits are not reached in polar species and lower critical limits are not reached in sub-polar species above the freezing point of sea water (−1.9 °C)

organism shapes the performance window in animals, with an optimum close to upper pejus temperature (e.g. Pörtner and Knust 2007; Pörtner and Farrell 2008, Fig. 5.1). The OCLTT concept identifies and quantifies the physiological mechanisms that shape and limit performance (e.g. in growth, reproduction, behaviours including foraging). It thereby explains the thermal performance curve used traditionally in evolutionary biology (Huey and Kingsolver 1989; Angiletta 2009). The thermal performance window of individual specimens of a population and the minimal level of individual performance required for the sustenance of fitness delineate the borders of the thermal niche of a species (Pörtner et al. 2010). The dimensions of the thermal window and niche are specific for a species and relate to mode of life and activity levels (Pörtner 2002a, b).

Identification of the adaptive mechanisms, which shape the width and position of the performance curve on the temperature scale, started by focusing on the species-specific responses and the characteristics of the underlying physiological mechanisms. However, the OCLTT concept also reaches up to ecosystem level, where biotic interactions shape ecosystem structure and functioning. Interacting species coexist, where thermal niches overlap. Each of those interacting species will specifically be affected by changing environmental conditions. The dynamic shifts of the borders of their niches will affect the thermal and temporal range of coexistence of interacting species as well as their relative performance and thus

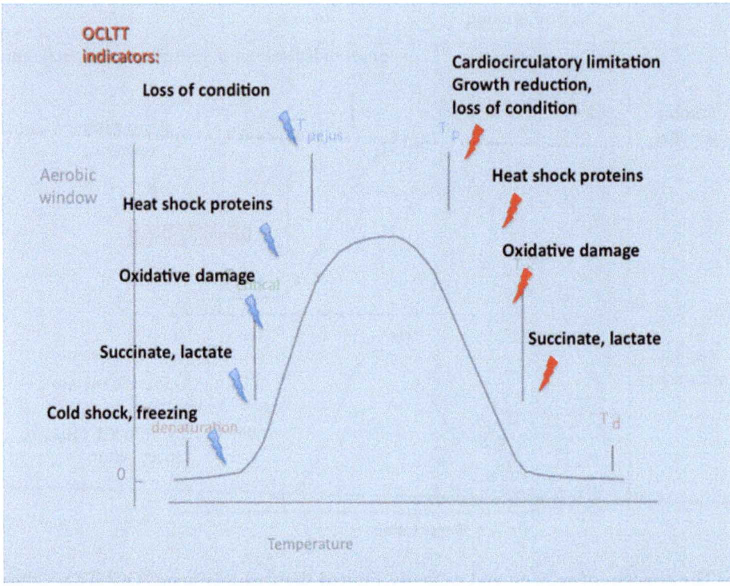

Fig. 5.2 In accordance with the thermal limits and effects at whole organism, tissue, cellular and molecular levels of biological organisation various physiological and biochemical indicators of thermal stress can be integrated into the OCLTT framework

their competitiveness. Therefore, if it comes to using the concept for an improved understanding of ecosystem level processes, studies must also reach beyond and address the consequences of thermal stress for species interactions. Comparative research should look at the changes in species-specific sensitivities and performance of interacting species with the goal to identify the mechanisms causing relevant shifts in interactions.

While further efforts need to consider multiple stressors, for example temperature, CO_2 or extreme hypoxia combined, it was suggested that integration of these stressors into a coherent picture best uses temperature as a matrix on which all of these changes occur (Pörtner 2010). In principle, all biological processes in the ocean take place at a certain temperature and are responsive to temperature change (in the sense that temperature is effective "everywhere"). Understanding the principle mechanisms of thermal adaptation and limitation will therefore provide a useful basis for integrating effects of further stressors. Temperature is the prime candidate for building such a matrix on which to comprehensively understand effects of other stressors like hypoxia, CO_2 etc.. Building on the insight that the dimensions of the thermal niche respond to the effect of additional stressors, this ultimately means that the physiological principles, which are unifying for the thermal biology of animals, are also key in integrating the effects of other stressors. A series of biochemical and physiological indicators are applicable to analyse the relevant phenomena (Fig. 5.2).

Responses to temperature changes and the shifts of response thresholds with acclimation include functional traits on various levels of biological organisation. These range from transcriptomic changes to patterns of ion and acid–base regulation and other whole animal functions with the consequence of shifting thermal tolerance limits. Such conclusions have been substantiated by comparisons of species and phyla from temperate, Arctic and Antarctic ecosystems (Pörtner 2006 for review). Interpretation of extinction patterns during earth history benefits from the respective knowledge of physiological traits identified in extant species (Knoll et al. 2007). These are framed in the OCLTT concept, which also integrates relevant environmental factors into a comprehensive picture. Studying the differential specialisation of animals on climate regimes and conversely, their sensitivity to climate therefore leads to improved understanding of ongoing and past ecosystem change. This should support higher reliability and certainty in projections of future scenarios.

For example, accumulating CO_2 causes disturbances in acid–base status. One overarching hypothesis is that CO_2 effects on physiological performance are mediated through associated shifts in extracellular pH (Pörtner et al. 2004, 2005; Pörtner and Farrell 2008). Resilience to ocean acidification may be reflected in the capacity to compensate for these disturbances or their secondary effects. Such capacity depends on temperature: ion and pH regulation comprise thermally sensitive active and passive transfer processes across membranes, with membrane structure and composition being shaped by temperature. Specific responses of ion transporter genes and their products to temperature and CO_2 were found in fish, crustaceans and bivalves. However, compensation may cause unfavourable shifts in energy budget and beyond that hamper cellular and mitochondrial metabolism, which are directly linked to the animal's aerobic performance window. In crabs, fishes and oysters, a narrowing of the thermal window is caused by increases in CO_2 levels as projected for 2100 and beyond (Pörtner and Farrell 2008; Walther et al. 2009; Munday et al. 2009; Lannig et al. 2010). In parallel, a decrease in the efficiency of energy production and a shift in energy allocation may occur and affect growth, ontogeny and fitness. Different sensitivities of life history stages indicate the existence of physiologically sensitive bottlenecks during the life cycle of marine organisms.

Overall, available evidence suggests that the concept of OCLTT comprehensively defines the physiological constraints of species and integrates the effects at underlying levels of biological organisation in response to various stressors. It provides causality and quantifies the levels and changes of performance and resistance, and supports more realistic estimates of species and ecosystem sensitivities to environmental change. The emerging picture of differential sensitivities across animal phyla is in line with existing categorisations of sensitivities from palaeo-observations during the Permian–Triassic mass extinctions (Knoll et al. 2007).

5.3 Antarctic Challenges: Physiological Pathways of Adapting to Cold

In temperate climate zones sensitivity to cold appears as a very important characteristic in shaping community composition. Accordingly, winter exposure of fish larvae is seen as a key selective factor in setting the productivity of a species. The occurrence of cold stress is probably effective in setting limits to the biogeography of cold-sensitive species. Low temperatures during winter may increase mortality, either because temperatures fall outside of the thermal window or because energy reserves become limiting, especially in smaller individuals that have relatively fewer reserves compared to larger con-specifics (Post and Evans 1989; Sogard 1997). Extreme winter events cause reductions in species abundance and ecosystem changes, for example in the German Wadden Sea (Woodhead 1964). Overcoming the sensitivity to cold, therefore, has been a prime challenge in Antarctic evolution. The functional consequences of this process go beyond those of acclimation to winter cold in temperature zone species, especially in those which reproduce in summer times.

5.3.1 Marine Crustaceans

After initial findings in sipunculids (Zielinski and Pörtner 1996), annelids (Sommer et al. 1997) and fish (van Dijk et al. 1999), crustaceans have been instrumental in developing and complementing the OCLTT concept and illustrating the relationship between temperature and oxygen supply capacity (Frederich and Pörtner 2000; Metzger et al. 2007; Walther et al. 2009; Wittmann et al. 2012).

In the warm-temperate spider crab *Maja squinado*, the highest values of arterial haemolymph oxygen partial pressures (PO_2) reached are constrained to a limited temperature range and thereby reveal the optimum of the physiological thermal tolerance range of crabs (Frederich and Pörtner 2000, Fig. 5.3). Arterial haemolymph PO_2 drops at temperatures beyond the upper thermal threshold as well as beyond the lower thermal threshold (pejus temperatures, Frederich and Pörtner 2000). Below 8.9 °C, both heart rates and haemolymph PO_2 decrease in parallel. This observation is related to a mismatch between oxygen supply and demand. The capacity of the cardiorespiratory system becomes progressively insufficient to meet the oxygen demand of tissues and especially to provide extra oxygen for functions supporting fitness, like foraging, growth and reproduction. The constraints set in at temperatures which this crab usually does not encounter in its natural environment. This optimum performance range of *M. squinado* is between 8.9 and 17.8 °C, which demonstrates a warm-temperate physiology of this crab (Frederich and Pörtner 2000). By further cooling, the lower critical temperature was reached at 0.7 °C, associated with an onset of anaerobic metabolism (Frederich and Pörtner 2000).

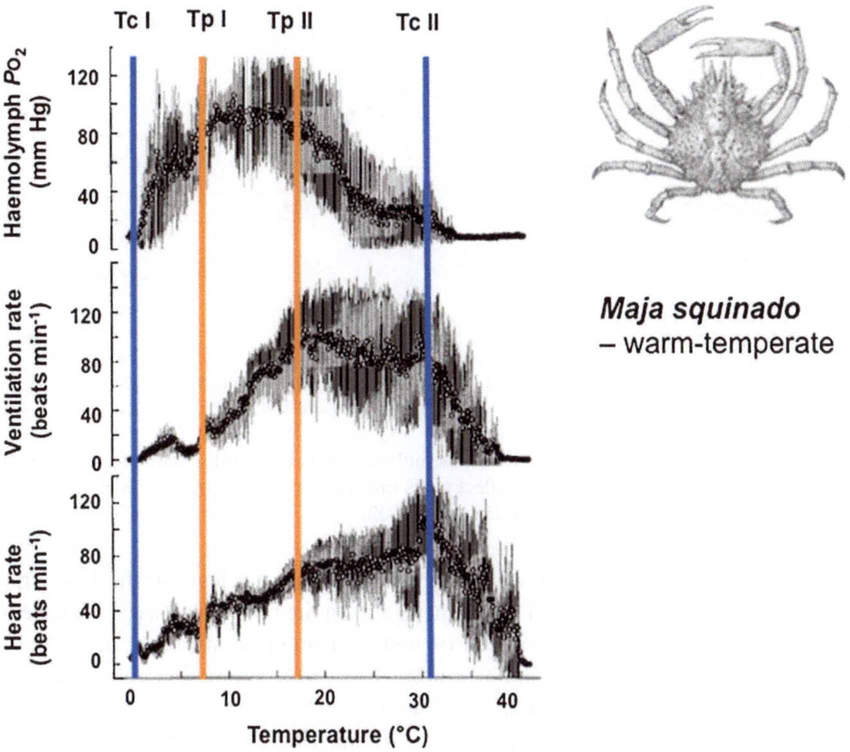

Fig. 5.3 Temperature dependent profile of haemolymph PO_2 in relation to temperature dependent patterns of ventilation and heart rates in the warm temperate spider crab, *Maja squinado* (data are means ± SD by Frederich and Pörtner 2000). Vertical lines indicate the position of relevant thermal limits (Fig. 5.1)

In contrast to *Maja squinado*, the spider crab *Hyas araneus* (collected around the island Helgoland in the Southern North Sea) shows the highest haemolymph PO_2 values in the cold (0–6 °C, Walther et al. 2009) (Fig. 5.4).

Arterial haemolymph PO_2 remained at a high level even at the lowest experimental temperatures of 0 °C. Accordingly, the optimum performance range of *H. araneus* would be found between 0 and 6 °C, indicating that this crab is cold adapted compared to the temperate *Maja squinado*. At 0 °C this species is still able to perform righting activity (to right itself after being turned on its back) at similar speeds as at 7 °C (Frederich et al. 2000). This suggests that this species is neither limited in its aerobic nor its anaerobic capacity at 0 °C, a temperature which it is usually not exposed to in the North Sea (the temperature near Helgoland ranges from 3 to 18 °C, and can peak at 20 °C in summer; Wiltshire and Manly 2004).

With rising temperatures *H. araneus* haemolymph PO_2 decreases while heart rate increases beyond pejus limits. This pattern of progressive oxygen limitation in the warmth appears uniform across warm and cold-adapted invertebrates, eurytherms and stenotherms, e.g. the Antarctic bivalve *Laternula elliptica* (Peck et al.

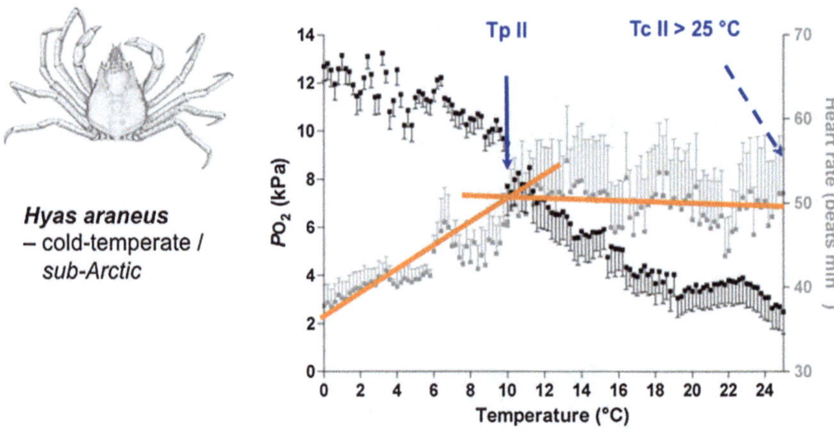

Fig. 5.4 Temperature dependent profile of haemolymph PO_2 (*black*) in relation to temperature dependent patterns of heart rate (*grey*) reflect pejus limits in the cold temperate spider crab, *Hyas araneus* (data are means ± SE by Walther et al. 2009)

2002; Pörtner et al. 2006), as well as cold-temperate species like the blue crab *Callinectes sapidus* (Mauro and Mangum 1982). In the blue crab, oxygen uptake is low at 5 °C, which indicates reduced oxygen consumption of tissues in the cold. With rising metabolic rate upon warming, heart and ventilation rates increase while PO_2 decreases from 5 to 25 °C (Mauro and Mangum 1982).

The cardiovascular system of crustaceans distributes dissolved oxygen and oxygen bound to haemocyanin to tissues and cells by convection. Heart rate correlates with ventilation rate through scaphognathite activity (Frederich and Pörtner 2000) and with the rate of oxygen consumption (Frederich et al. 2000). Haemolymph PO_2 mirrors the efficiency of oxygen transport in relation to oxygen demand. All parameters are temperature dependent and are modified once the animal undergoes acclimation to temperature (Mauro and Mangum 1982; Zainal et al. 1992; Frederich and Pörtner 2000; Frederich et al. 2000; Walther et al. 2009). An increase in temperature causes a reduction of haemolymph PO_2 and an increase in heart rate, scaphognathite rate and oxygen consumption (Ahsanullah and Newell 1971; Mauro and Mangum 1982; Zainal et al. 1992; Frederich and Pörtner 2000; Frederich et al. 2000; Walther et al. 2009).

All of these findings suggest that a cold-induced reduction of oxygen demand and functional capacity of cells and tissues is the main reason for high PO_2 values in the haemolymph of *H. araneus* and other cold adapted crustaceans. The cold-enhanced solubility of oxygen in water and body fluids (Wells 1986) facilitates oxygen supply and also favours elevated oxygen partial pressures in the haemolymph. This benefit of enhanced oxygen solubility in the cold is partially hampered by a cold-induced reduction in diffusion velocity (the diffusion constant of the Fick equation). This makes the changes in oxygen demand in relation to oxygen availability key in setting some of the functional features observed at various latitudes including body size (Pörtner 2002b) and diversity (Verberk et al. 2011).

Clearly, adaptation to variable or permanent cold temperatures and their metabolic consequences also need to be distinguished. Cold eurytherms (subpolar, usually Northern hemisphere) among invertebrates and fishes display higher metabolic rates than cold stenotherms living in a stable temperature regime (high polar, especially Antarctic, Pörtner 2006). Among crustaceans, cold compensation also exists. For example, observations by Storch et al. (2009), demonstrated that cold-adapted crab (*Taliepus dentatus*) larvae at the Chilean coast show higher cardiac performance and higher oxygen consumption rates compared to warm-adapted individuals in the cold.

For temperate *H. araneus* acclimated to 5 °C Frederich et al. (2000) reported low oxygen consumption rates in the cold (−1.5 to 3 °C) which were stimulated strongly by warming to 9 °C, following a very high Q_{10} value of 11.4. These data may reflect the cold-acclimation capacity of the spider crab, combined with energy savings in the cold, the latter being characteristic for cold-adapted species. Similarly, the lugworm *Arenicola marina* is able to switch between energy saving stenothermy during winter cold (with high Q_{10} values) and high energy turnover eurythermy during summer (Wittmann et al. 2008). High Q_{10} values over a small temperature range allow these facultative cold stenotherms to respond rapidly to small rises in temperature (see Chown et al. 2003) and stimulate seasonal acclimation for spring and summer.

In contrast, invertebrates and fishes at high latitudes and low environmental temperatures have permanently low metabolic rates (Peck et al. 2002; Pörtner 2006). In the cold, the oxygen taken up by ventilation and distributed by circulation is progressively in excess of that used by tissues, leading to an increase in steady-state haemolymph PO_2 values. Similarly, in an Antarctic bivalve, *Laternula elliptica*, haemolymph PO_2 results high and oxygen consumption low between 0–6 °C, while at higher temperatures (9 °C) haemolymph PO_2 falls due to rising metabolic rate (Peck et al. 2002). These relationships may describe the making of a polar stenotherm. With fully functional ventilation and cardiocirculation capacities, the falling energy demand in the cold causes excess oxygen availability. This in turn will lead to energy savings for oxygen supply systems (cf. Mark et al. 2002). On evolutionary timescales this allows a parallel reduction in workload and in performance, and a fall in oxygen demand by the heart and scaphognathites, and thereby, even further reductions in energy turnover, and then heart (Fig. 5.4) and ventilation rates at low temperatures (Frederich et al. 2000). However, warming then causes oxygen limitation to set in earlier (at lower temperatures) as these organs are no longer able to substantially increase their performance and therefore fail to maintain PO_2 constant over wide temperature ranges during warming.

Oxygen limitation at extreme temperatures in crustaceans might also involve the transport of oxygen by haemocyanin. The affinity of haemocyanin to oxygen is dependent on pH and temperature (Angersbach and Decker 1978; Burnett et al. 1988; Jokumsen et al. 1981; Morris and Bridges 1989; Sanders and Childress 1990). At low temperatures the affinity for oxygen is maximised. Hence, oxygen remains bound and any release to target cells and tissues is reduced. Less oxygen reaches the heart and its function may become constrained, in line with an oxygen and capacity limitation. The high pressure head of PO_2 in *H. araneus* in the cold implies a neglible contribution of haemocyanin. In contrast, the PO_2 decrease in *Maja squinado* in the cold may well invoke a role for haemocyanin to delay oxygen limitation, depending on the PO_2 reached during cooling.

Adaptation to cold temperatures occurs at the expense of tolerance to warmer temperatures, and vice versa, as the thermal window can shift in either direction. However, acclimation capacity is limited in either direction. In *H. araneus*, the changes of PO_2 and heart rate during warming reflect that capacity limitation sets in early, at around 10 °C, when PO_2 continues to decrease and the increase in heart rate levels off (Fig. 5.4). This threshold is equivalent to the upper pejus limit. Upon more severe oxygen limitation during further warming, heart rate remains stable, but then drops at the critical temperature of 25 °C (Walther et al. 2009) when anaerobic metabolism likely sets in (cf. Frederich and Pörtner 2000). The geographical distribution of this species ranges from the southern North Sea to the Arctic region around Svalbard where the species lives close to its cold-adaptation limits (Christiansen 1969, Walther et al. 2010). In the South, cold winter periods may be essential for *H. araneus* to sustain its presence in this area. The temperature of the southern North Sea around the Island Helgoland ranges from 3 to 18 °C, and up to 20 °C in the summer months (Wiltshire and Manly 2004). The pejus range is wide (between 10 and 25 °C) such that the species can also sustain warmer summer temperatures but at reduced performance levels. These periods may become longer and problematic with future warming. A detailed analysis of North Sea temperature data over the last decades showed that the lowest winter temperatures have increased over time (Wiltshire and Manly 2004). Especially with regard to future warming by up to 4 °C until 2100 in the North Sea, spider crabs may have reached their warm acclimation limit and may have to retreat to cooler waters further north. The cold adaptation characters of the (still cold-eurytermal) spider crab resemble those of polar invertebrates and may thus reflect a transition phase to the permanent cold. The stereotypes of cold adapted eurytherms and cold adapted stenotherms (see Pörtner 2006) may thus not be the only functional groups characterising the sub- and high-polar Northern and Southern hemispheres.

5.3.2 At the Doorstep to Antarctica: Sub-Antarctic Stone Crabs

Lithodid crabs are the only group of reptant decapod crustaceans that have been found in some areas of the Antarctic oceans and that may have increased in abundance or expanded recently (Smith et al. 2011). Their southernmost habitat is the continental shelf and slope of the western Antarctic Peninsula in the Bellingshausen Sea, where temperature usually remains above 0 °C (Klages et al. 1995; Thatje et al. 2008). With climate change they would be the first group of skeleton-crushing crustaceans to conquer currently uncolonised areas and dramatically change ecosystem structure through predation (Thatje et al. 2005; Smith

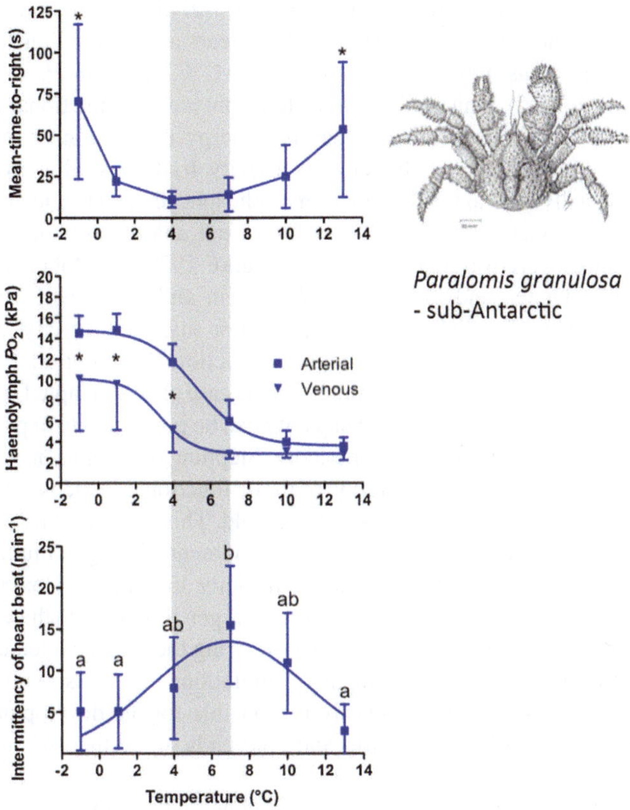

Fig. 5.5 Mean-time-to-right (*upper panel*, *significantly slower than at 4 °C, $p < 0.05$), resting haemolymph PO_2 (kPa, *middle panel*, *significantly different between arterial and venous haemolymph) and intermittency of heart beat rate at rest (min^{-1}) of the sub-Antarctic lithodid crab *Paralomis granulosa* acclimated to 4° C (*lower panel* different letters denote significant differences). Note that maximum arterial PO_2 does not reflect optimal performance, possibly due to enhanced involvement of haemocyanin oxygen transport in the warmth. Intermittent cardiovascular activity may foster the use of haemocyanin during warming. Shaded area: optimal temperature range (data are means ± SD from Wittmann et al. 2012, drawing of crab from http://carcinuss.deviantart.com/art/Paralomis-granulosa-211005359)

et al. 2011). This group would therefore be a rewarding object of study to identify the physiological limits preventing them from entering those cold-water areas.

The physiological patterns of adaptation to cold in the sub-Antarctic stone crab *Paralomis granulosa* have recently been investigated (Wittmann et al. 2012, Fig. 5.5) and resemble those seen in *Hyas araneus*. PO_2 values is both arterial and venous haemolymph are high at temperatures as low as −1 °C (Wittmann et al. 2012).

The existence of an arteriovenous difference in PO_2 implies that tissues are respiring and taking up oxygen, and that oxygen supply is not limiting in the cold. Again, a mismatch in oxygen supply and demand seems to be precluded by the reduction in metabolic rate, which leads to progressively increasing oxygen tensions in haemolymph and these in turn support low heart and ventilation frequencies. Righting activity was however impaired at −1 °C in both *H. araneus* and *P. granulosa* (Fig. 5.5), which might indicate a constraint in functional capacity of the tissues or of both aerobic and anaerobic energy production or in the speed of recovery of phosphagens. Challenging activity trials, resulting in fatigue of crustaceans, usually lead to the degradation of phosphagen and the accumulation of anaerobic end products, probably due to functional anaerobiosis and insufficient aerobic capacity (Carlsson and Gäde 1986; Burke 1979; McMahon et al. 1979; Ellington 1983; Onnen and Zebe 1983; Weinstein and Full 1998). Even during moderate challenges (like the righting response) or sustained activity, crustaceans rely on both aerobic and anaerobic energy production (Booth et al. 1982). Since functional anaerobiosis seems to be unlikely in the case of cold-adapted specimens, a thermal limitation on tissue functional capacity can be postulated to explain reduced righting speeds and fatigue at low temperature. Stepping back and considering these aspects in a wider context, cold limitation of tissue functional capacity sets in prior to oxygen limitation in in the polar and sub-polar cold. This indicates that in temperate species capacity limitation thus causes the development of oxygen limitations as in *Maja squinado* in the cold. Similarly, the capacity limitation of oxygen supply systems, e.g. cardiocirculation, sets the onset of oxygen limitation in the warmth (see Frederich and Pörtner 2000). As a corollary, studying the cold-adapted crustaceans has provided further evidence that capacity limitations of tissues, including those involved in oxygen supply are primarily responsible for the development of performance limitations at both the warm and cold ends of the thermal window of a species.

Large crustaceans possess giant muscle cells with diameters greater than 500 μm and mitochondria primarily at the periphery of the cells (Lang et al. 1970; Boyle et al. 2003). This may relate to the modest aerobic scope of crustacean muscle (Booth et al. 1982; Booth and McMahon 1992). Whereas oxygen supply of peripheral mitochondria might be granted, diffusion of ATP and the phosphagen arginine phosphate to the core of a giant muscle fibre may be limiting, even during aerobic recovery after exercise at 20 °C (where arginine phosphate takes 5 min to travel 250 μm, Kinsey and Moerland 2002; Boyle et al. 2003). Therefore, anaerobic pathways are used in the core of the fibre to aid the supply of ATP and the restoration of arginine phosphate needed to affect contractions (Boyle et al. 2003; Johnson et al. 2004; Kinsey et al. 2011). Since metabolite diffusion is reduced at low temperature (Sidell and Hazel 1987; Hubley et al. 1996), this may contribute to fatigue and reduced righting speed despite high haemolymph oxygen levels in *P. granulosa* in the cold (Wittmann et al. 2012). To our knowledge, the ultrastructure of muscle fibres in *P. granulosa* has not been studied. The existence of giant muscle fibres of up to 4-mm diameter in the confamilial *Paralithodes camtschatica* from Alaska (Lang et al. 1970) suggests that hypertrophied muscle cells also occur in *P. granulosa*. However, we are not aware of any study of the arrangement and density of mitochondria in any subpolar or polar crustacean.

These observations may contribute to but not fully answer the question why reptant decapod crustaceans do not colonise the coldest areas in the marine Antarctic. Work on temperate and sub-Antarctic crustaceans has led to the hypothesis that a low capacity for regulation of extracellular Mg^{2+} excludes reptant marine decapod crustaceans from temperatures below 0 °C and, thus, the high Antarctic (Frederich et al. 2000). This hypothesis was tested in the lithodid crab, *P. granulosa* using similar methodology as applied during the earlier studies of thermal tolerance (Wittmann et al. 2012). Mg^{2+} levels in the water and thereby, the haemolymph were modified from 30 mmol L^{-1} to a level naturally observed in Antarctic caridean shrimps (12 mmol L^{-1}). When analysing the righting speed, the study found an optimum temperature of 4 °C and falling velocities at temperatures above and below this optimum. In contrast to earlier studies in temperate and sub-polar brachyuran crabs the reduction of haemolymph [Mg^{2+}] did neither result in an increase in activity, heart or ventilation beat frequencies nor in a shift in thermal tolerance to colder temperatures. Furthermore, Antarctic isopods and the arctic-boreal brachyuran *Chionoecetes opilio* occurring at −1.5–4 °C exhibit high haemolymph [Mg^{2+}] of 35–45 mmol L^{-1} (Charmantier and Charmantier-Daures 1995; Burmeister and Sainte-Marie 2010 and references therein, Wittmann et al. 2010). These observations indicate that the inhibitory effect of Mg^{2+} may have been lost in cold-water crustaceans, but this does still not allow lithodids to inhabit the coldest polar habitats. This statement awaits confirmation from testing whether active planktotrophic versus passive lecithotrophic larvae are constrained by the Mg^{2+} effect (Wittmann et al. 2011). As larval stages are seen as bottlenecks in species fitness, it is conceivable that their sensitivity to ambient conditions sets the limits to where species actually live. Cold-adaptation constraints are especially strong at small body size, forcing larval stages more than adults into slower modes of life (Pörtner 2006). To compensate for the extended development in the cold, the evolution of abbreviated and lecithotrophic development in cold water may be preferred compared to warm water environments, where food supply is also an issue (Thorson 1950; Astorga et al. 2003; Thatje et al. 2003; Pearse and Lockhart 2004).

5.4 Perspectives

A cause and effect understanding of thermal limitation and adaptation at various levels of biological organisation is proving crucial in the elaboration of how the Antarctic climate has shaped the functional properties of extant Antarctic fauna (for review see Pörtner 2006; Pörtner et al. 2007, 2012). The skeleton-crushing crustaceans (anomurans and brachyurans) that are about to expand in the Antarctic represent a functional group where such cause and effect understanding is still in its infancy. This group is interesting, especially since it is clearly unable to settle permanently in the coldest waters of the poles, at temperature below 0 °C.

In all cold-water crustaceans looked at so far, the capacity for oxygen delivery is not acutely limiting cold tolerance. The strong reduction in metabolic rate seen in the cold may support oxygen supply systems to provide an excess of oxygen to the systemic circulation, which then again, can be exploited in energy and oxygen savings by heart and ventilation. Such maximisation of energy savings, even in cold eurythermal *Hyas araneus*, would be in line with functional characteristics attributed to Antarctic stenothermal fishes and other invertebrates (for review see Pörtner 2006; Pörtner et al. 2007, 2012). As a corollary, the crustaceans may, in similar ways as other polar and especially Antarctic fauna, display features resulting from excess oxygen availability rather than cold-induced oxygen limitation.

In line with the reasoning above, however, functional capacity and motility may nonetheless be constrained in reptant decapods in the cold to an extent that fitness and competitiveness are reduced, but these effects are not mediated by oxygen shortage. It will need to be explored in further studies whether sub-polar species show a trend to uncouple oxygen-shortage from capacity-limitation at low threshold temperatures. These conclusions contrast findings in warm temperate species where a correlation between capacity and oxygen supply exists on both sides of the thermal window.

Considering some cold adaptation features in Antarctic fishes (low heat tolerance, high membrane fluidity, cold-adapted enzyme kinetic properties (K_m and k_{cat}) and protein structural flexibility, loss of genetic information, especially for myoglobin and haemoglobin in notothenioid fish, enhanced lipid membrane densities (e.g., higher concentrations of mitochondria), large cells supporting energy savings, cold-adapted benthic brachyuran and anomuran crustaceans may display similar adaptation features as far as identified (e.g. large myocytes, low and thermally highly sensitive metabolic rates, exploitation of high oxygen solubility at low metabolic rates, enhanced oxygen supply through diffusive oxygen flux, limited oxygen supply to tissues upon warming as an early cause of functional limitation). However, starting from a slow lane position, they may have developed these features too early, i.e. at too high ambient temperatures, to still be able to push for settlement at temperatures below 0 °C and associated further mechanisms of cold adaptation. They therefore experience cold limitation at temperatures around 0 °C rather than at the freezing point of sea water.

Although less is known about adaptation to extreme cold in Antarctic crustaceans than in fishes in general, some features of cold adaptation have also been reported in polar shrimp, krill and peracarids. An increased phospholipid (main component of biomembranes) content in muscle of the Antarctic decapod shrimp *Chorismus antarcticus* compared to a temperate shrimp species (Clarke 1977) may indicate proliferation of mitochondria and parallel existing observations of high mitochondrial densities in Antarctic fishes (O'Brien et al. 2003; Sänger et al. 2005). In line with this observation the activity of citrate synthase, the key enzyme of aerobic metabolism, is five times higher in Antarctic krill (*Euphausia superba*) than in northern krill (*Meganyctiphanes norvegica*) from temperate latitudes (Buchholz and Saborowski 2000), also indicating cold induced proliferation of mitochondria. An increased content of mitochondrial membrane lipids in fish muscle is thought to compensate for the reduced diffusion velocity at low temperature in Antarctic fish (Londraville and Sidell 1990; O'Brien et al. 2003).

Peracarid crustaceans have successfully adapted to the Antarctic shelf environment and dominate the ecological niche elsewhere occupied by decapods (Brandt 1999; Aronson et al. 2007 and references therein). They also display cold-adaptation features. The isopod *Glyptonotus antarcticus* expresses a distinct myosin heavy chain isoform, which allows muscle function at low temperatures (Holmes et al. 2002). In contrast to peripheral nerves of a temperate isopod, those of *G. antarcticus* are wrapped in glial sheaths, which may increase conduction velocity in the cold (Young et al. 2006). The reduced tolerance to warming of this stenothermal species may be reflected by low circulating levels of the respiratory pigment haemocyanin, the low oxygen carrying capacity of the pigment and the low buffering capacity of the haemolymph (Whiteley et al. 1997), findings which also resemble those in cold-adapted notothenioids. However, none of the findings in the isopods would currently help to explain why anomuran and brachyuran crabs do not settle in the extreme cold.

The evolutionary forces causing low metabolic rates as a uniform character of life in Antarctic ectothermic animals as well as in the cold adapted anomurans and brachyurans studied may be linked to the requirement for high energetic efficiency as required to support higher organismic functioning in the cold. According to earlier insight, meeting this requirement allows partial compensation for cold limitation of growth (Heilmayer et al. 2004), while other functions like hatching, development, reproduction and aging are largely delayed (see Peck 2005; Pörtner et al. 2012). Energy savings may not only be reflected in the rate but also in the mode of development. Cold-adapted decapods have a lower number of larval stages. In peracarid crustaceans the development is direct and occurs in a protected brood pouch (Luxmoore 1982; Janssen and Hoese 1993; Heilmayer et al. 2008). This may contribute to the success of this group, since less energy is invested into reproduction and protected from getting lost to predators or from irregular food supply to the passive offspring.

Findings in crustacean examples from various climate regimes are in line with the hypothesis of a relaxation on oxygen supply rates and costs during cold exposure, associated with a trend towards facilitated oxygen supply in the cold. This comes with a shift from oxygen limitation to capacity limitation, a gain in energetic efficiency, and a slowed development of oxygen limitation beyond cold tolerance limits. The present integrative approach suggests that the patterns of oxygen and capacity limited thermal tolerance are linked with life history consequences and lifestyles typically seen in the permanent cold. Future research needs to address the detailed aspects of these interrelationships in the crustaceans.

Future research also needs to take into account that climate challenges do not only involve temperature changes but also the effect of additional stressors, like ocean acidification through elevated CO_2 concentrations. Approaches are available through the OCLTT concept, which allow integrating the effects of these stressors with those of temperature (see Pörtner 2010). In *H. araneus* adults from Helgoland a lower haemolymph PO_2 was seen under elevated PCO_2 at various temperatures (Walther et al. 2009). In Antarctic bivalves (*Laternula elliptica*: Cummings et al. 2011) and Arctic brittlestars (*Ophiocten sericeum*: Wood et al. 2011) elevated CO_2

concentrations cause an increase in oxygen consumption at low temperatures. This observation indicates an increased energy demand and associated workload of relevant functions under elevated CO_2 tensions, possibly of proton-equivalent ion exchange in the gills, which is key in compensating for the CO_2 induced disturbances of acid–base status. However, compensation remains incomplete (Zittier et al. 2012) and the permanent lowering of extracellular pH may cause metabolic depression in muscle and hepatopancreas (Pörtner 2008), which, however, may not compensate for the rise in energy demand of the gills. Overall, a shift in energy budget is likely, with unknown consequences on long-time scales, also for processes affected by evolutionary cold adaptation. In *H. araneus* from Helgoland the effects of high CO_2 concentrations and low temperatures develop synergistically. Crucial processes like larval development or calcification are slowed by CO_2 (Walther et al. 2010, 2011). Features of cold adaptation were seen in the Arctic *H. araneus* population and their larvae. In contrast to Helgoland larvae those from Svalbard showed no reduction in dry weight, growth was not affected in an acidified (Arctic) ocean reflecting resistance to acidification in a process that benefits from energy efficiency during cold adaptation (Walther et al. 2010). These aspects remain largely unexplored. A relevant question therefore is whether high CO_2 concentrations still exert negative effects on physiological processes that have been modified during cold adaptation. In light of these considerations our understanding of climate change impacts on polar species is far from complete.

Acknowledgments This work is a contribution to the "European Project on Ocean Acidification" (EPOCA) which received funding from the European Community's Seventh Framework Programme (FP7/2007–2013) under grant agreement n° 211384. EPOCA is endorsed by the International Programmes IMBER, LOICZ and SOLAS. This project was also supported by the German program on ocean acidification (BIOACID) funded by the BMBF and by the PACES program of the AWI and by Deutsche Forschungsgemeinschaft grants no. SA 1713/1–1 and 1–2.

References

Ahsanullah M, Newell RC (1971) Factors affecting the heart rate of the shore crab *Carcinus maenas* (L.). Comp Biochem Physiol 39:277–287

Angersbach D, Decker H (1978) Oxygen transport in crayfish blood: effect of thermal acclimation, and short-term fluctuations related to ventilation and cardiac performance. J Comp Physiol 123:105–112

Angiletta MJJ (2009) Thermal Adaptation. A theoretical and empirical synthesis. Oxford University Press, New York 320

Aronson RB, Thatje S, Clarke A, Peck LS, Blake DB, Wilga CD, Seibel BA (2007) Climate change and invasibility of the Antarctic benthos. Annu Rev Ecol Evol Syst 38:129–154

Astorga A, Fernández M, Boschi EE, Lagos N (2003) Two oceans, two taxa and one mode of development: latitudinal diversity patterns of South American crabs and test for possible causal processes. Ecol Lett 6:420–427

Beers JM, Sidell BD (2011) Thermal tolerance of Antarctic Notothenioid fishes correlates with level of circulating hemoglobin. Physiol Biochem Zool 84:353–362

Booth CE, McMahon BR (1992) Aerobic capacity of the blue crab, *Callinectes sapidus*. Physiol Zool 65:1074–1091

Booth C, McMahon B, Pinder A (1982) Oxygen uptake and the potentiating effects of increased hemolymph lactate on oxygen transport during exercise in the blue crab, *Callinectes sapidus*. J Comp Physiol B 148:111–121

Boyle K-L, Dillaman RM, Kinsey ST (2003) Mitochondrial distribution and glycogen dynamics suggest diffusion constraints in muscle fibers of the blue crab, *Callinectes sapidus*. J Exp Zool 297A:1–16

Brandt A (1999) On the origin and evolution of Antarctic Peracarida. Sci Mar 63(Suppl 1):261–274

Buchholz F, Saborowski R (2000) Metabolic and enzymatic adaptations in northern krill, *Meganyctiphanes norvegica*, and Antarctic krill, *Euphausia superba*. Can J Fish Aquat Sci 57:115–129

Burke EM (1979) Aerobic and anaerobic metabolism during activity and hypoxia in two species of intertidal crabs. Biol Bull 156:157–168

Burmeister A, Sainte-Marie B (2010) Pattern and causes of a temperature-dependent gradient of size at terminal moult in snow crab (*Chionoecetes opilio*) along West Greenland. Polar Biol 33:775–788

Burnett LE, Scholnick DA, Mangum CP (1988) Temperature sensitivity of molluscan and arthropod haemocyanins. Biol Bull 174:153–162

Carlsson K-H, Gäde G (1986) Metabolic adaptation of the horseshoe crab, *Limulus polyphemus*, during exercise and environmental hypoxia and subsequent recovery. Biol Bull 171:217–235

Charmantier G, Charmantier-Daures M (1995) Osmoregulation and salinity tolerance in zoeae and juveniles of the snow crab *Chionoecetes opilio*. Aquat Living Resour 8:171–179

Chown SL, Addo-Bediako A, Gaston KJ (2003) Forum: Metabolic cold adaptation in arthropods: a smaller-scale perspective. Funct Ecol 17:562–572

Christiansen ME (ed) (1969) Crustacea Decapoda Brachyura, vol 2. In: Marine invertebrates of Scandinavia, Universitetsforlaget, Oslo

Clarke A (1977) Lipid class and fatty acid composition of *Chorismus antarcticus* (Pfeffer) (Crustacea: Decapoda) at South Georgia. J Exp Mar Biol Ecol 28:297–314

Coppes Petricorena ZL, Somero GN (2007) Biochemical adaptations of notothenioid fishes: comparisons between cold temperate South American and New Zealand species and Antarctic species. Comp Biochem Physiol A 147:799–807

Cummings V, Hewitt J, Van Rooyen A, Currie K, Beard S, Thrush S, Norkko J, Barr N, Heath P, Halliday NJ, Sedcole R, Gomez A, McGraw C, Metcalf V (2011) Ocean acidification at high latitudes: potential effects on functioning of the Antarctic bivalve *Laternula elliptica*. PLoS ONE 6(1):e16069

Ellington WR (1983) The recovery from anaerobic metabolism in invertebrates. J Exp Zool 228:431–444

Frederich M, Pörtner HO (2000) Oxygen limitation of thermal tolerance defined by cardiac and ventilatory performance in spider crab, *Maja squinado*. Am J Physiol Regul Integr Comp Physiol 279:R1531–R1538

Frederich M, Sartoris FJ, Arntz WE, Pörtner HO (2000) Haemolymph Mg^{2+} regulation in decapod crustaceans: physiological correlates and ecological consequences in polar areas. J Exp Biol 203:1383–1393

Garofalo F, Pellegrino D, Amelio D, Tota B (2009) The Antarctic hemoglobinless icefish, fifty five years later: a unique cardiocirculatory interplay of disaptation and phenotypic plasticity. Comp Biochem Physiol A 154:10–28

Heilmayer O, Brey T, Pörtner HO (2004) Growth efficiency and temperature in scallops: a comparative analysis of species adapted to different temperatures. Funct Ecol 18:641–647

Heilmayer O, Thatje S, McClelland C, Conlan K, Brey T (2008) Changes in biomass and elemental composition during early ontogeny of the Antarctic isopod crustacean *Ceratoserolis trilobitoides*. Polar Biol 31:1325–1331

Holmes JM, Whiteley NM, Magnay JL, El Haj AJ (2002) Comparison of the variable loop regions of myosin heavy chain genes from Antarctic and temperate isopods. Comp Biochem Physiol B 131:349–359

Hubley MJ, Locke BR, Moerland TS (1996) The effects of temperature, pH, and magnesium on the diffusion coefficient of ATP in solutions of physiological ionic strength. Biochim Biophys Acta 1291:115–121

Huey RB, Kingsolver JG (1989) Evolution of thermal sensitivity of ectotherm performance. Trends Ecol Evol 4:131–135

Janssen HH, Hoese B (1993) Marsupium morphology and brooding biology of the Antarctic giant isopod *Glyptonotus antarcticus* Eights 1853 (Crustacea, Isopoda, Chaetiliidae). Polar Biol 13:145–149

Johnson LK, Dillaman RM, Gay DM, Blum JE, Kinsey ST (2004) Metabolic influences of fiber size in aerobic and anaerobic locomotor muscles of the blue crab, *Callinectes sapidus*. J Exp Biol 207:4045–4056

Jokumsen A, Wells RMG, Ellerton HD, Weber RE (1981) Haemocyanin of the giant Antarctic isopod, *Glyptonotus antarcticus*: structure and effects of temperature and pH on its oxygen affinity. Comp Biochem Physiol 70:91–95

Kinsey ST, Moerland TS (2002) Metabolite diffusion in giant muscle fibers of the spiny lobster *Panulirus argus*. J Exp Biol 205:3377–3386

Kinsey ST, Locke BR, Dillaman RM (2011) Molecules in motion: influences of diffusion on metabolic structure and function in skeletal muscle. J Exp Biol 214:263–274

Klages M, Gutt J, Starmans A, Bruns T (1995) Stone crabs close to the Antarctic Continent: *Lithodes murrayi* Henderson, 1888 (Crustacea; Decapoda; Anomura) off Peter I Island (68 51'S, 90 51'W). Polar Biol 15:73–75

Knoll AH, Bambach RK, Payne JL, Pruss S, Fischer W (2007) Paleophysiology and end-Permian mass extinction. Earth Planet Sci Lett 256:295–313

Lang F, Sutterlin A, Prosser CL (1970) Electrical and mechanical properties of the closer muscle of the Alaskan king crab *Paralithodes camtschatica*. Comp Biochem Physiol A 32:615–628

Lannig G, Eilers S, Pörtner HO, Sokolova IM, Bock C (2010) Impact of ocean acidification on energy metabolism of oyster, *Crassostrea gigas*—changes in metabolic pathways and thermal response. Mar Drugs 8(8):2318–2339

Londraville RL, Sidell BD (1990) Ultrastructure of aerobic muscle in Antarctic fishes may contribute to maintenance of diffusive fluxes. J Exp Biol 150:205–220

Luxmoore RA (1982) The reproductive biology of some serolid isopods from the Antarctic. Polar Biol 1:3–11

Mark FC, Bock C, Pörtner HO (2002) Oxygen-limited thermal tolerance in Antarctic fish investigated by MRI and 31P-MRS. Am J Physiol 283(5):R1254–1262

Mauro NA, Mangum CP (1982) The role of the blood in the temperature dependence of oxidative metabolism in decapod crustaceans. I. Intraspecific responses to seasonal differences in temperature. J Exp Zool 219:179–188

McMahon BR, McDonald DG, Wood CM (1979) Ventilation, oxygen uptake and haemolymph oxygen transport, following enforced exhausting activity in the dungeness crab *Cancer magister*. J Exp Biol 80:271–285

Metzger R, Sartoris FJ, Langenbuch M, Pörtner HO (2007) Influence of elevated CO_2 concentrations on thermal tolerance of the edible crab *Cancer pagurus*. J Therm Biol 32:144–151

Morris S, Bridges CR (1989) Interactive effects of temperature and L-lactate on the binding of oxygen by the haemocyanin of two arctic boreal crabs, *Hyas araneus* and *Hyas coarctatus*. Physiol Zool 62:62–82

Munday PL, Crawley N, Göran NE (2009) Interacting effects of elevated temperature and ocean acidification on the aerobic performance of coral reef fishes. Mar Ecol Prog Ser 388:235–242

O'Brien KM, Skilbeck C, Sidell BD, Egginton S (2003) Muscle fine structure may maintain the function of oxidative fibres in haemoglobinless Antarctic fishes. J Exp Biol 206:411–421

Onnen T, Zebe E (1983) Energy metabolism in the tail muscles of the shrimp *Crangon crangon* during work and subsequent recovery. Comp Biochem Physiol A 74:833–838
Pearse JS, Lockhart SJ (2004) Reproduction in cold water: paradigm changes in the 20th century and a role for cidaroid sea urchins. Deep Sea Res II 51:1533–1549
Peck LS (2005) Prospects for survival in the Southern Ocean: vulnerability of benthic species to climate change. Antarct Sci 17:497–507
Peck LS, Pörtner HO, Hardewig I (2002) Metabolic demand, oxygen supply, and critical temperatures in the Antarctic bivalve *Laternula elliptica*. Physiol Biochem Zool 75:123–133
Pörtner HO (2002a) Climate variations and the physiological basis of temperature dependent biogeography: systemic to molecular hierarchy of thermal tolerance in animals. Comp Biochem Physiol A 132(4):739–761
Pörtner HO (2002b) Environmental and functional limits to muscular exercise and body size in marine invertebrate athletes. Comp Biochem Physiol 133A:303–321
Pörtner HO (2006) Climate-dependent evolution of Antarctic ectotherms: an integrative analysis. Deep-Sea Res II 53:1071–1104
Pörtner HO (2008) Ecosystem effects of ocean acidification in times of ocean warming: a physiologist's view. Mar Ecol Prog Ser 373:203–217
Pörtner HO (2010) Oxygen and capacity limitation of thermal tolerance: a matrix for integrating climate related stressors in marine ecosystems. J Exp Biol 213:881–893
Pörtner HO, Farrell AP (2008) Physiology and climate change. Science 322(5901):690–692
Pörtner HO, Knust R (2007) Climate change affects marine fishes through the oxygen limitation of thermal tolerance. Science 315:95–97
Pörtner HO, Langenbuch M, Michaelidis B (2005) Synergistic effects of temperature extremes, hypoxia, and increases in CO_2 on marine animals: from Earth history to global change. J Geophys Res 110(C9):C09S10
Pörtner HO, Langenbuch M, Reipschläger A (2004) Biological impact of elevated ocean CO_2 concentrations: lessons from animal physiology and earth history. J Oceanogr 60(4):705–718
Pörtner HO, Peck LS, Hirse T (2006) Hyperoxia alleviates thermal stress in the Antarctic bivalve, *Laternula elliptica*: evidence for oxygen limited thermal tolerance. Polar Biol 29:688–693
Pörtner HO, Peck LS, Somero GN (2007) Thermal limits and adaptation: an integrative view (Antarctic ecology: from genes to ecosystems). Philos Trans R Soc B 362:2233–2258. doi:10.1098/rstb.2006.1947
Pörtner HO, Peck LS, Somero GN (2012) Mechanisms defining thermal limits and adaptation in marine ectotherms: an integrative view. In: Rogers AD, Johnston NM, Murphy EJ, Clarke A (eds) Antarctic ecosystems: an extreme environment in a changing world, First edn. John Wiley Ltd, Chichester, UK, pp 360–396
Pörtner HO, Schulte P, Wood C, Schiemer F (2010) Niche dimensions in fishes: an integrative view. Physiol Biochem Zool 83(5):808–826
Post JR, Evans DO (1989) Size-dependent overwinter mortality of young-of-the-year yellow perch (*Perca flavescens*): laboratory, in situ enclosure, and field experiments. Can J Fish Aquat Sci 46(11):1958–1968
Sanders NK, Childress JJ (1990) Adaptations to the deep-sea oxygen minimum layer: oxygen binding by the haemocyanin of the bathypelagic mysid, *Gnathophausia ingens* Dohrn. Biol Bull 178:286–293
Sänger AM, Davison W, Egginton S (2005) Muscle fine structure reflects ecotype in two nototheniids. J Fish Biol 66:1371–1386
Sidell BD, O'Brien KM (2006) When bad things happen to good fish: the loss of hemoglobin and myoglobin expression in Antarctic icefishes. J Exp Biol 209:1791–1802
Sidell B, Hazel J (1987) Temperature affects the diffusion of small molecules through cytosol of fish muscle. J Exp Biol 129:191–203
Smith CR, Grange LJ, Honig DL, Naudts L, Huber B, Guidi L, Domack E (2011) A large population of king crabs in palmer deep on the west Antarctic Peninsula shelf and potential invasive impacts. Proc R Soc B. doi:10.1098/rspb.2011.1496

Sogard SM (1997) Size-selective mortality in the juvenile stage of teleost fishes: a review. Bull Mar Sci 60:1129–1157

Sommer A, Klein B, Pörtner HO (1997) Temperature induced anaerobiosis in two populations of the polychaete worm *Arenicola marina*. J Comp Physiol B 167:25–35

Storch D, Santelices P, Barria J, Cabeza K, Pörtner HO, Fernández M (2009) Thermal tolerance of crustacean larvae (zoea I) in two different populations of the kelp crab *Taliepus dentatus* (Milne-Edwards). J Exp Biol 212:1371–1376

Thatje S, Anger K, Calcagno GA, Lovrich GA, Pörtner HO, Arntz WE (2005) Challenging the cold: crabs reconquer the Antarctic. Ecology 86:619–625

Thatje S, Hall S, Hauton C, Held C, Tyler P (2008) Encounter of lithodid crab *Paralomis birsteini* on the continental slope off Antarctica, sampled by ROV. Polar Biol 31:1143–1148

Thatje S, Schnack-Schiel S, Arntz WE (2003) Developmental trade-offs in Subantarctic meroplankton communities and the enigma of low decapod diversity in high southern latitudes. Mar Ecol Prog Ser 260:195–207

Thorson G (1950) Reproductive and larval ecology of marine bottom invertebrates. Biol Rev 25:1–45

Van Dijk PLM, Tesch C, Hardewig I, Pörtner HO (1999) Physiological disturbances at critically high temperatures: a comparison between stenothermal Antarctic and eurythermal temperate eelpouts (Zoarcidae). J Exp Biol 202:3611–3621

Verberk WCEP, Bilton DT, Calosi P, Spicer JI (2011) Oxygen supply in aquatic ectotherms: partial pressure and solubility together explain biodiversity and size patterns. Ecology 92:1565–1572

Verde C, Giordano D, Russo R, di Prisco G (2012) The adaptive evolution of polar fishes: lessons from the function of hemoproteins. In: di Prisco G, Verde C (eds) Adaptation and evolution in marine environments—the impacts of global change on biodiversity. Springer, Berlin, pp 197–213

Walther K, Anger K, Pörtner HO (2010) Effects of ocean acidification and warming on the larval development of the spider crab *Hyas araneus* from different latitudes (54° vs. 79°N). Mar Ecol Prog Ser 417:159–170

Walther K, Sartoris FJ, Bock C, Pörtner HO (2009) Impact of anthropogenic ocean acidification on thermal tolerance of the spider crab *Hyas araneus*. Biogeosciences 6:2207–2215

Walther K, Sartoris FJ, Pörtner HO (2011) Impacts of temperature and acidification on larval calcium incorporation of the spider crab *Hyas araneus* from different latitudes (54° vs. 79°N). Mar Biol 158:2043–2053

Webb JB, Eckert GL, Shirley TC, Tamone SL (2007) Changes in embryonic development and hatching in *Chionoecetes opilio* (snow crab) with variation in incubation temperature. Biol Bull 213:67–75

Weinstein RB, Full RJ (1998) Performance limits of low-temperature, continuous locomotion are exceeded when locomotion is intermittent in the ghost crab. Physiol Biochem Zool 71:274–284

Wells RMG (1986) Cutaneous oxygen uptake in the Antarctic icequab, *Rhigophila dearborni* (Pisces: Zoarcidae). Polar Biol 5:175–179

Whiteley NM, Taylor EW, Clarke A, Haj AJE (1997) Haemolymph oxygen transport and acid-base status in *Glyptonotus antarcticus* Eights. Polar Biol 18:10–15

Wiltshire KH, Manly BFJ (2004) The warming trend at Helgoland Roads, North Sea: phytoplankton response. Helgol Mar Res 58:269–273

Wittmann AC, Held C, Pörtner HO, Sartoris FJ (2010) Ion regulatory capacity and the biogeography of Crustacea at high southern latitudes. Polar Biol 33:919–928

Wittmann AC, Pörtner HO, Sartoris FJ (2012) A role for oxygen delivery and extracellular magnesium in limiting cold tolerance of the sub-Antarctic stone crab *Paralomis granulosa*? Physiol Biochem Zool 85(3):285–298

Wittmann A, Schröer M, Bock C, Steeger H, Paul R, Pörtner H (2008) Indicators of oxygen- and capacity-limited thermal tolerance in the lugworm *Arenicola marina*. Clim Res 37:227–240

Wittmann AC, Storch D, Anger K, Pörtner HO, Sartoris FJ (2011) Temperature-dependent activity in early life stages of the stone crab *Paralomis granulosa* (Decapoda, Anomura, Lithodidae): a role for ionic and magnesium regulation? J Exp Mar Biol Ecol 397:27–37

Wood HL, Spicer JI, Kendall MA, Lowe DM, Widdicombe S (2011) Ocean warming and acidification; implications for the Arctic brittlestar *Ophiocten sericeum*. Polar Biol 34: 1033–1044

Woodhead PMJ (1964) Changes in the behaviour of the sole, *Solea vulgaris*, during cold winters, and the relation between the winter catch and sea temperature. Helgol Mar Res 10:328–342

Young JS, Peck LS, Matheson T (2006) The effects of temperature on peripheral neuronal function in eurythermal and stenothermal crustaceans. J Exp Biol 209:1976–1987

Zainal KAY, Taylor AC, Atkinson RJA (1992) The effect of temperature and hypoxia on the respiratory physiology of the squat lobster, *Munida rugosa* and *Munida sarsi* (Anomura, Glatheidae). Comp Biochem Physiol 101:557–567

Zielinski S, Pörtner HO (1996) Energy metabolism and ATP free-energy change of the intertidal worm *Sipunculus nudus* below a critical temperature. J Comp Physiol B 166:495–500

Zittier ZMC, Hirse T, Pörtner HO (2012) The synergistic effects of increasing temperature and CO_2 levels on exercise capacity and acid-base balance in the spider crab, *Hyas araneus*. Mar Biol (in revision)

[illegible]

Chapter 6
Catalysis and Protein Folding in Extreme Temperature Environments

Charles Gerday

6.1 Introduction

The diversity of life partially derives from the various characteristics of natural environments encountered on planet Earth suggesting that an organism, embedded in an environment exhibiting specific properties, will benefit from physiological and biochemical adjustments tending to optimize the adequacy between this environment and the overall characteristics of the organism. Some environments are extreme meaning that as such they are not suitable for organisms such as human being or *Escherichia coli* which are commonly exposed to moderate temperatures, pH close to neutrality, atmospheric oxygen, low salinity and low pressure. In this context, the temperature parameter is crucial for survival since it considerably affects the reaction rates of the numerous chemical reactions that occur in any living organism. The relation is in fact exponential, has been defined in 1899 by the Swedish chemist Svante Arrhenius, and takes the form: $k = A.\exp^{-Ea/RT}$ in which k is the reaction rate, A is a pre-exponential term that notably depends on the activation entropy of the reaction (Collins et al. 2007), E_a is known as the activation energy, that renders the reaction possible, R is the gas constant (8.314 J mol^{-1} K^{-1}) and T is the absolute temperature in Kelvin. The dependence of reaction rate on temperature is abolished in organisms such as mammals which have chosen to keep their temperature more or less constant independently of that of the environment. This is a very costly choice since it usually consumes about 75 % of the energy absorbed by the organism under the form of food, in such a way that in winter some of these *homeotherms* cannot maintain their temperature constant due to lack of food, and enter hibernation. For

C. Gerday (✉)
Laboratory of Biochemistry, University of Liège, Institute of Chemistry,
B-4000 Liège, Belgium
e-mail: ch.gerday@ulg.ac.be

C. Verde and G. di Prisco (eds.), *Adaptation and Evolution in Marine Environments, Volume 2*, From Pole to Pole, DOI: 10.1007/978-3-642-27349-0_6,

those organisms that do not have the capacity to keep their temperature constant the problem of the environmental temperature is more acute. Indeed, low temperatures render reaction rates too slow to sustain life whereas high temperatures can accelerate them to an extent that would lead to production of unwanted or an excess of some metabolites due to the differential action of temperature on reaction rates via the activation energy.

In nature three types of organisms have been defined as a function of the temperature of their usual environment, the *psychrophiles* that thrive in environments characterized by temperatures close or below the freezing point of water, the *mesophiles,* confronted to moderate temperatures and the *thermophiles* that are exposed to temperatures which in some cases exceed that of the boiling point of water. In the context of this paper, the *mesophiles* do not pose acute problem of reaction rate and will be only consider as a reference.

6.2 The Thermophiles

On Earth, natural environments of temperatures exceeding 60 °C are always associated with volcanic activity and are either terrestrial such as geysers, hot springs and solfataric fields, or marine such as deep-sea vents. Typical temperatures in hot springs are around 80 °C (Brock 1967) whereas in hydrothermal vents the temperatures of fluids that are liberated in the cold waters of the ocean's bottom can approach 400 °C. Of course, due to the rapid dilution of these hot fluids, a local gradient of temperatures is rapidly created and cover temperatures from about 0 to 350 °C. Amazingly these environments are richly populated and display a large biodiversity of organisms from fish to hyperthermophilic bacteria; this is mainly due to the large concentration of gases and mineral nutrients present in these venting fluids. Temperatures as high as 350 °C are not compatible with the type of life we know and actually the record is held by an archaea, known as strain 121, which displays an upper growth temperature of 121 °C (Kashefi and Lovley 2003). Hyperthermophiles have been recognized in 23 genera most of them being archaea but some bacteria such as *Thermotoga maritima* or *Thermus thermophilus* also display high apparent optimum temperature close to 80 °C. At these temperatures the question of the permeability of the cytoplasmic membrane and in particular the selective proton permeability, that generates the proton gradient which usually constitutes the essential part of the metabolic energy under the form of ATP, becomes acute. It appears however that the proton permeability is rather kept constant in microorganisms characterized by different growth temperatures and this phenomenon is known as "homeo-proton permeability adaptation" (van de Vossenberg et al. 1995). This is, in general, achieved through various adaptations of the membrane composition tending to increase its rigidity in response to any increase in temperature of the environment (Gerday 2011). Other crucial components of the cell are proteins which are first very sensitive to the deleterious effect of heat on their structure and among them the enzymes which in addition have to adjust their catalytic efficiency in order to normalize reaction rates and render them more or less comparable to those that occur in mesophilic organisms.

6.2.1 The Stability Problem

Proteins are stabilized by weak bonds which have different sensitivities to temperature that depend on enthalpic and entropic terms. Van der Waals bonds are essentially electrostatic interactions that can implicate any type of atoms and that are formed between transient dipoles or transient dipoles and induced dipoles. The energy associated with these interactions is weak and strongly depends on the distance between the atoms involved. It includes an attractive and a repulsive term when atoms are located beyond or at a distance shorter than the so called van der Waals contact distance respectively. The empirical equation of the energy change is: $\Delta E = A/r^{12} - B/r^6$ in which r is the distance between groups A and B. In the case of proteins the thermodependence of these bonds varies according to the type of dipoles involved. For example the van der Waals interaction between permanent dipole and temporary dipoles is considered to be independent of the temperature (Ross and Subramanian 1981).

Hydrogen bonds occur between polar groups and can be defined as a particular type of electrostatic interactions, in which a hydrogen atom is shared between a proton donor and a proton acceptor acting as a base: $A^{\delta-}$....H....$B^{\delta+}$. The energy associated with these bonds will depend on the distance between A and B, on the geometric arrangement of the bond, a linear geometry being optimal, and on the relative acidity of A an B, the strongest bond being encountered where the $pK_a^{'}$s are similar. These bonds are formed with a negative modification of enthalpy and will therefore be sensitive to any increase of temperature.

Hydrophobic interactions are formed between lateral chains of aliphatic and aromatic amino acids. These chains are poorly soluble in water especially around room temperature. When they are transferred to water they affect the organization of the hydrogen bonds leading to significant losses in translational and rotational entropy of water molecules, a process unfavourable in terms of free energy. Clathrate-type cages are formed around the hydrophobic groups and this induces a negative modification of the entropy leading itself to a high energy level of these systems, even if the enthalpic term is negative probably due to favourable van der Waals interactions formed between the atoms of the hydrophobic groups and the solvent. They therefore tend to evolve towards a more favourable energy state in forming bonds and clusters that are mainly located in the core of the proteins. The energy change associated with the formation of these bonds can be expressed by the following equation: $\Delta G^{HPH} = \Delta H^{HPH} - T\Delta S^{HPH}$, in which HPH stands for hydrophobic. This equation can be decomposed into the following terms: $\Delta G^{HPH} = \Delta H^{vdW} + \Delta H^{Hyd} - T\Delta S^{HPH}$, in which ΔH^{vdW} corresponds to the enthalpy changes associated with the formation of van der Waals interactions, ΔH^{Hyd} to the positive enthalpy change capable to induce appropriate disorders in the organization of water molecules surrounding each individual hydrophobic groups and $T\Delta S^{HPH}$ the entropic change, at a given temperature, associated with the formation of a hydrophobic interaction, which includes a large and positive contribution ΔS^{Hyd} resulting from the release of water molecules from the individual groups. One can often read that a hydrophobic interaction is entropically driven due to the large disorder resulting from the release of water molecules, but this is forgetting that both the enthalpic and entropic terms strongly vary with the environmental temperature. At low temperature, ΔH^{Hyd} is largely positive and this is detrimental to the formation of the bond whereas at high temperature the amplitude of this term decreases and tends to zero favouring the formation of the bond. At low and moderate temperatures, $T\Delta S^{HPH}$ is largely positive due to the release of water (high value of ΔS^{Hyd}), whereas at high temperature ΔS^{Hyd} tends to zero due to large perturbation of the water molecules surrounding each individual groups in such a way that at high temperature a hydrophobic interaction is weakened and can be assimilated to a simple van der Waals interaction (Priyakumar et al. 2010). This has led to the conclusion that the hydrophobic effect is maximum at moderate temperatures and all proteins from psychrophilic, mesophilic and thermophilic organisms show a maximum thermodynamic stability close to room temperature (Kumar et al. 2002). This is clearly illustrated in Fig. 6.1 in which the thermodynamic stability curves of three α-amylases from psychrophilic, mesophilic and thermophilic organisms show that the maximum stability is close to 20 °C for the three proteins (Collins et al. 2008).

Salt bridges are formed between permanently charged side chains of opposite signs such as carboxylic and amino groups. As in the case of hydrophobic interactions, an electrostatic interaction first requires dehydration of individual charged groups, thus the energy changes associated with these interactions are rather complex and also fluctuate as a function of temperature. The general equation corresponding to the energy change associated with the formation of an electrostatic interaction between two groups of opposite sign, A^+, B^-, can be written as:

$$\Delta G^{Total} = \Delta G^{Desol} + \Delta G^{Electro} + \Delta G^{vdW} + \Delta G^{Prot}$$

in which ΔG^{Desol} correspond to the energy (unfavourable) necessary to dehydrate A^+ and B^-, $\Delta G^{Electro}$ to the favourable decrease of free energy resulting from the formation of the salt bridge, $\Delta G^{vdW}$to changes related to new van der Waals interactions induced by the formation of a salt bridge and ΔG^{Prot} to any other interactions in the protein also induced by the salt bridge. The strength of the bond also strongly depends on the temperature. Indeed, ΔG^{Desol}can be decomposed into ΔH^{Desol}-$T\Delta S^{Desol}$, ΔS^{Desol}can be considered as negligible so that ΔG^{Desol} is equal to: $\Delta H^{Dehyd\ A+} + \Delta H^{Dehyd\ B} - \Delta H^{Hyd(A+B-)}$, this term is largely unfavourable due to the energy required to dehydrate A^+ and B^-, especially at low temperature, and therefore should be compensated by a favourable energy, $\Delta G^{Electro}$, associated with the formation of the bridge, and which is equal roughly to $Z_{A+}Z_{B-}.\varepsilon^2/D.d_{A+B-}$ in which Z_{A+} and $Z_{B-}$are the charge of the ions; ε, the charge of the electron; D the dielectric constant of the medium, and d_{A+B-} the distance between A^+ and B^-.

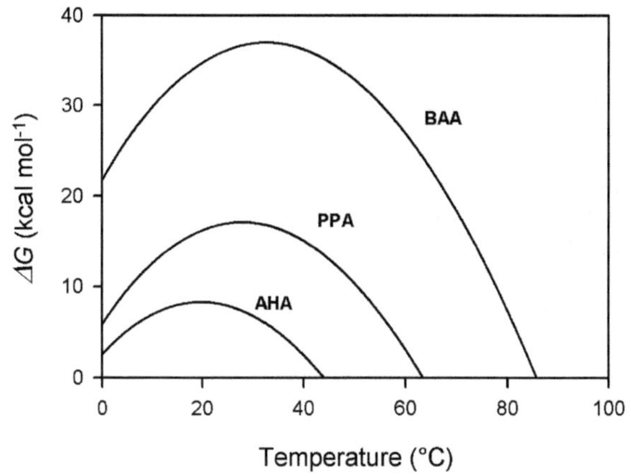

Fig. 6.1 Conformational stability of homologous α-amylases from a psychrophile (AHA, *Pseudoalteromonas haloplanktis*, Antarctic), mesophile (PPA, pig pancreatic amylase) and thermophile (BAA, *Bacillus amyloliquefaciens*) as a function of temperature. The high stability of thermophilic α-amylase is reached by lifting the stability curve (Model 1 see text) with nearly no modification of the maximum stability temperature. The cold-adapted enzyme is much less stable but its maximum stability is close to that of the thermophilic enzyme (Adapted from D'Amico et al. 2003)

So at low temperature the hydration of A^+ and B^- is favoured, the dielectric constant is high and the favourable energy change $\Delta G^{Electro}$ is minimum, inducing a weakening of the salt bridge strength. At high temperature, the energy required to desolvate A^+ and B^- decreases as well as the dielectric constant so that the formation of the salt bridge can be favoured. This is probably the reason why in thermophilic proteins numerous salt bridges, organized or not in networks, significantly participate in the stability of the molecular structure (Vieille and Zeikus 2001; Meng et al. 2008; Del Vecchio et al. 2009; Priyakumar et al. 2010).

6.2.2 Activity and Stability

To fulfil a biological activity, a protein originating from a thermophilic organism should preserve an appropriate stability at the environmental temperature but the question is: Does the maximum activity correspond to the maximum thermodynamic stability? A typical thermodynamic stability curve is shown in Fig. 6.2, in which $\Delta G_{Stab} = -RT\ln K$, K is the equilibrium constant between native N and unfolded protein U, N↔U, assuming a reversible two-state transition.

K is also equal to the ratio of two kinetic constants, $K = k_f/k_u$, where k_f is the rate constant for folding and k_u, the rate constant for unfolding, thus the stability of a protein involves equilibrium and kinetic aspects. ΔG_{Stab} is a positive value that is the difference between the free energy of the unfolded state and the free energy of

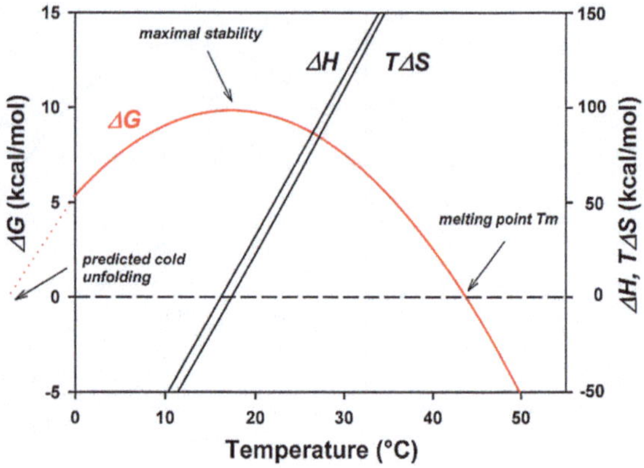

Fig. 6.2 Typical conformational stability curve ΔG, also showing the evolution of the enthalpic ΔH and entropic TΔS contribution as a function of temperature. The difference in scales show that ΔG is small, resulting from the difference between two large and opposite contributions, ΔH and ΔS (From Feller 2010)

the folded form. It also corresponds to the energy necessary to unfold the protein at a given temperature. This energy is of course composed of an enthalpic and entropic term and can be written: $\Delta G_{N-U} = \Delta H_{N-U} - T\Delta S_{N-U}$. The data can be described as a function of temperature by an altered version of the Gibbs-Helmholtz equation $\Delta G(T) = \Delta H_m(1 - T/T_m) - \Delta C_p \text{ x } [(T_m - T) + T\ln(T/T_m)]$, in which Tm is the melting temperature, ΔH_m the enthalpy change at the melting temperature and ΔC_p the heat capacity change between the native and the unfolded state. In most proteins, the maximal stability is observed around room temperature (Sawle and Ghosh 2011) whereas in the case of thermozymes the maximal activity is close to the high temperature of the environment, well on the right side of the curve shown in Fig. 6.2. This is because a certain plasticity and flexibility of the protein is required to interact with other components. In the case of a thermophilic protein this fulfilment can be easily achieved through the dissipative effect of heat, which induces the breaking of several weak bonds, contributing to increase the stabilization entropy that can ultimately lead to the unfolding of the protein. As an average, the stabilization energy of mesophilic proteins is around 40 kJ mol^{-1} [much higher values, as high as 250 kJ $mol^{-1,}$ have however been observed in the case of the co-chaperonin from hyperthermophilic microorganisms (Luke et al. 2007)], a rather low figure that corresponds to the difference between the large figures that characterize the enthalpic and entropic contributions. Therefore a small increase of the stabilization energy can transform a mesophilic protein into a thermophilic one and this increment is around 5–20 kJ mol^{-1} (Vieille and Zeikus 2001; Li and Zhou 2005; Feller 2010). This can be achieved through a reinforcement or increase in number of some weak bonds. At the temperature of maximal stability the difference between the entropies of the unfolded and native state is zero and the stability is essentially driven by the enthalpic contribution. So an increase in stability can be achieved by simply lifting the stability curve, as shown in Fig. 6.1 for the thermophilic enzyme. There are however two additional thermodynamic models that can account for the higher thermal stability of thermophilic proteins when compared to mesophilic homologues: first, the curve shown in Fig. 6.2 can be flattened to bring T_m towards higher values; in this case the lesser dependence of ΔG_{N-U} on temperature would result in a lower magnitude of ΔH_{N-U} and ΔS_{N-U} and of ΔC_p, the difference between the specific heat of the unfolded and native form respectively. This decrease, in the case of ΔC_p, can presumably originate from a more discrete exposure of hydrophobic groups on unfolding; second, the whole curve can be displaced to the right, so that the amplitude of the maximal stability of the thermophilic protein would be similar to that of the mesophilic enzyme, but this maximal value will be shifted towards higher temperatures, meaning that the thermophilic protein would be more stable at higher temperatures (Beadle et al. 1999). This is the result of a reduced loss in entropy for the folding transition. This arises from a decrease in entropy of the unfolding state, induced by increase in Pro residues and decrease in Gly residues, or by residual structures subsisting in the unfolded state. Often, a combination of increased ΔG (first model) and reduced ΔC_p (second model) is used in nature to achieve higher T_m, but more than 70 % of the thermophilic proteins rely on an increase in ΔG, independently or in combination with an increase of T_m (Razvi and Scholtz 2006). This has been somehow contradicted by a recent study on 116 proteins from mesophiles and thermophiles, that indicates that there is a significant correlation between increased T_m and reduced entropy change upon folding (third model) (Sawle and Ghosh 2011). In hyperthermophiles, a combination of the three models seems to be preferred (Luke et al. 2007). Nevertheless, knowing that only a few alterations of the amino-acid composition (inducing either additional bonds or changes in conformational or local entropy) can secure a good stability at high temperature, one can easily understand that the modifications at the level of the 3-dimensional structure will also be rather discrete. So it is rather difficult, even knowing the 3D structure, to clearly identify the parameters involved in the stabilization of a thermophilic protein. As shown in Fig. 6.2, when the dissipative effect of heat becomes too large, the ΔG stability curve crosses the zero line. This is the melting point, where the ratio between the native and unfolded form is equal to one. It is also worth mentioning that on the left side of the curve both the enthalpic and entropic contribution change sign. We will come back to that later. So, an appropriate flexibility, local or global, of the molecular edifice is necessary to secure catalytic efficiency. That is the reason why, in many cases, thermophilic enzymes show very low levels of efficiency at room temperature. By contrast, as in the case of adenylsuccinate synthetase from *Methanococcus jannaschii,* at high temperature up to 85 °C, the enzyme displays catalytic rates comparable to that of mesophilic counterparts around room temperature (Vemparala et al. 2011). Unlike its mesophilic counterpart, the catalytic function of this thermophilic enzyme only implies the movement of two loops out of five for ligand binding and catalysis. The strategy consists in a pre-arrangement of catalytic residues, an increase in polar residues and number and stability of salt bridges that induce an increased rigidity associated with fewer loop movements. Most thermophilic proteins unfold irreversibly, and so the previous equations cannot be applied. Their stability can be evaluated through the kinetic stability that expresses the rate at which the protein collapses. This is a function of the free energy of activation ΔG^* that represents the barrier which has to be overcome to eventually proceed up to denaturation. This activation energy is high in thermophilic proteins that generally unfold at slow rates. Some differences are however observed when one compared the unfolding characteristics of thermophilic proteins from archaea and bacteria. The unfolding rate of thermophilic proteins from bacteria is in general faster than that of thermophilic counterparts from archaea. Archaea and bacteria have diverged at an early state of evolution and thermophilic archaea presumably originated in hot environments, whereas thermophilic bacteria recolonized at later-stage hot environments. The difference in the unfolding rate of hyperthermostable proteins seems to reflect the characteristics of the environments from which the microorganism originates. It has indeed been proposed that proteins originating from hot environments are more compact and display a more hydrophobic core than proteins issued from bacteria that recolonized at a later stage and which are notably stabilized by salt bridges. As an example, the core of RNases from archaea such as *Sulfolobus tokodaii* and *Thermococcus kodakarensis* are richer in hydrophobic residues than that of RNases from bacteria such as *Thermus thermophilus.* The superslow unfolding of the enzymes from archaea, when compared to the faster rate of homologous enzymes from bacteria, is due to the strong effect of hydrophobic interactions on unfolding rate (Okada et al. 2010). This again underlines the fact that various strategies are used by microorganisms to adapt to an extreme environment and that the specificity of the adaptation strategy largely depends on the evolutionary history of the organisms.

6.2.3 Folding at High Temperature

Although folding of proteins can spontaneously be successful, the particular environment of the cell and the high protein concentration, around 400 mg/ml, can make folding inefficient and error-prone. In addition, as moderately elevated temperatures favour hydrophobic interactions, the folding of proteins in thermophilic organisms implies the control of aggregation processes of nascent unfolded forms and, within the cell, of a favourable equilibrium between N and U, viewed as a homeostatic control of the protein concentrations. This is achieved through the production of molecular chaperones, a multitude of proteins that act in various cellular activities such as folding, refolding of misfolded proteins, protein aggregation, multimeric protein assembly, protein hydrolysis, and protein transport. Some of these chaperones are permanent hosts of the cell, others are induced following a stress such as a heat shock, that is the reason why they are usually known as heat-shock proteins. They take any non-native conformation of proteins in charge up to their successful final destination. They can act as "holdases," in stabilizing non-native conformation, as "foldases", in assisting correct folding, and as "unfoldases", when contributing to unfold misfolded proteins (Hoffmann et al. 2010). Protein misfolding is a very important process that can occur in a cell; it is associated with more than 30 human diseases, including some of the most debilitating such as Alzheimer's, Parkinson's and Creutzfeldt–Jakob's diseases

(Tartaglia et al. 2010). For the sake of comparison between the factors involved in the folding of proteins from thermophiles and psychrophiles, we will summarize the role of some of the most studied chaperones, such as the so-called chaperonins (GroEL/GroES), practically ubiquitous in all living organisms; Dnak, known as Hsp 70 in eukaryotes, which is the major cytosolic chaperone in *E. coli*; and TF the trigger factor, the only ribosome-associated chaperone in bacteria.

6.2.3.1 GroEL/GroES

Chaperonins are subdivided into two and possibly three groups differing in structure and distribution. Group I is formed by the GroEL and GroES complexes in which GroES acts as a co-chaperone. The crystal structure of GroEL has been solved in 1994 (Braig et al. 1994). The protein is made of 60-kDa (Hsp60) subunits and consists of two heptameric rings placed back to back to form a barrel structure. They are found mainly in eubacteria, and in eukaryotic organelles such as mitochondria and chloroplasts. Roughly, the unfolded polypeptide chain is first trapped into the central cavity through hydrophobic interactions, then ATP binding induces large conformational changes that notably renders the cavity much more hydrophilic and triggers the rapid binding of the co-chaperone GroES (Hsp10) to one of the ring and that acts as a lid. The sequestration of the polypeptide chain into the hydrophilic cavity forces exposed hydrophobic residues to cluster to form a core and that is followed by protein folding. ATP hydrolysis by the ATP-bound ring is followed by the release of GroES and of the encapsulated protein. Group II is more complex than group I (Yébenes et al. 2011). Although they are also formed by two rings, GroES is replaced by a protruding rather flexible helical extension that can act as a lid essential for encapsulation of the unfolded protein. The closure of the lid is also ATP driven. The rings are made of octameric or nonameric structures formed by association of one, two or three different subunits. Group-II chaperonins, also known as thermosomes, are found in archaea and in the cytosol of eukaryotes. The crystal structure of the closed conformation of the chaperonin from *Thermoplasma acidophilum* has been solved in 1998 (Ditzel et al. 1998). Other structures and notably the crystal structure of the open state are now available (Huo et al. 2010) and a mechanism of folding chamber closure has been proposed (Zhang et al. 2010). Group-III chaperonins have been identified in carboxydotrophic bacteria; they are conserved in the genome of 11 bacteria (Techtmann and Robb 2010). They are distinct from Group-II chaperonins because their coding genes are associated with the Hsp 70 operon that includes dnaK, suggesting co-regulation and functional relationship between the two systems. They also contain a built-in lid and are composed of two octameric rings, unlike bacterial chaperonin. They form a monophyletic clade and can originate from an ancient lateral gene transfer from the archaea into an ancestral bacterium.

6.2.3.2 DnaK/DnaJ/GrpE

DnaK, known as Hsp70 in eukaryotes, is involved in various cellular processes: it prevents protein aggregation, assists in protein disaggregation and in folding and refolding of improperly folded proteins in an ATP-dependent manner. It is associated with co-factors DnaJ (Hsp40) and GrpE (nucleotide exchange factor) that respectively enhance, in *E. coli,* the ATPase rate of DnaK and significantly accelerates the nucleotide exchange. They play a role in minimizing aggregation and in enhancing the solubility of unfolded and also recombinant proteins. DnaK also plays an important role in targeting aggregated proteins to proteases. Dnak appears as monomer, dimer or higher-ordered oligomer, whereas GrpE is a homodimer and is the only component of the DnaK operon system that shows a thermal transition in the range of the temperatures experienced by the microorganism. For example, in *T. thermophilus* (growth temperature of 70–75 °C) it displays a fully reversible thermal transition at 90 °C and an irreversible one around 100 °C, compared to 48 °C and 75–80 °C in *E. coli*. As in the case of the DnaK system from *E. coli*, GrpE is considered as the thermosensor of the DnaK system that prevents thermal denaturation of substrate proteins. Its homodimeric structure has been solved by Nakamura et al. (2010). In binding to DnaK in a 2:1 stoichiometry, it accelerates ADP/ATP exchange by 80,000-fold. Amazingly, in *T. thermophilus,* $DnaJ_{Th}$ does not apparently stimulate the hydrolysis rate of ATP by DnaK nor the binding of the nucleotide, these roles being taken in charge solely by $GrpE_{Th}$ (Schlee and Reinstein 2002). In *T. thermophilus* Dnak displays high thermal stability in the absence or presence of ADP, as well as DnaJ (T_m = 100 °C). The DnaK system is present in all mesophilic archaea and only in some thermophilic archaea. The system has been recently characterized from *Methanothermobacter thermoautotrophicus* with growth temperature ranging from 40–70 °C. As in mesophilic and thermophilic bacteria, $GrpE_{Mt}$ shows a reversible thermal transition in the physiologically relevant temperature range and prevents the release of the substrate from the DnaK complex, but the significant difference is that both DnaK and Dnaj also display transition in the same temperature range. So GrpE cannot be really considered in this case as a thermosensor, since that implies that both Dnak and Dnaj would still be in a native state at the transition temperature of GrpE. The system seems to be optimized in preventing heat-induced protein aggregation (Popp and Reinstein 2009).

6.2.3.3 The Trigger Factor

The trigger factor is found in bacteria and chloroplasts and is the only chaperone associated with ribosomes. In archaea, some, not structurally related, similar factors, do also exist. It is the first chaperone, encountered by nascent polypeptide chains, that forms a stoichiometric complex with the ribosome. It is localized at the ribosomal exit of the newly built polypeptide chain. It may also act as an independent chaperone, since stable complexes with a large variety of full-length proteins were found, suggesting that the trigger factor plays an important role in the stabilization of native-like substrate prior to their incorporation into multimeric assemblies such as the ribosomal edifice (Martinez-Hackert and Hendrickson 2009). It is found in the cytosol in rather high concentration mainly in a dimeric form that can interact with folding intermediates that can be rescued by the chaperone system DnaK-DnaJ-GrpE (Liu et al. 2005). In *E. coli*, the trigger factor is a 48 kDa protein. Its 3D structure has been solved at 2.7-Ångström resolution as well as the structure of its ribosomal binding domain in complex with the large ribosomal subunit of *Haloarcula marismortui* (Ferbitz et al. 2004). The structure resembles a crouching dragon made of three domains, the N-terminal domain (1–149), or tail of the dragon, carries a specific motif (GFRxGxxP) involved in ribosome binding, the central domain (aa246–432) or carboxyl-terminal part is the chaperone module binding nascent polypeptide chains and the head of the dragon-like structure (aa150–245) has an activity of peptidyl-prolyl cis/trans isomerase. The PPIase forms a first module, whereas a second module is made of the N- and C-terminal domains, forming a crevice that provides a protective envelope over the peptide exit site on the ribosome. It has been proposed that a hydrophobic pocket in the PPIase domain assists the folding of the nascent protein bound to the chaperone site in a substrate dependent manner (Liu et al. 2010). The three dimensional structure of the N- and C-terminal domain of the trigger factor from *Thermotoga maritima* has been solved by X-ray crystallography and oligomerization and interactions have been studied (Martinez-Hackert and Hendrickson 2007). Despite low sequence identity with the *E. coli* homologues [22 % for the N-terminal domain (TF_N) and 12 % for the C-terminal domain (TF_C)], the secondary-structure elements match up surprisingly well. The structure of the Tm TF_N is also very similar to those observed in other eubacteria such as *V. cholerae* and *D. radiodurans* as well as to the structure of Hsp33s and other small heat-shock proteins. This domain exists in solution as a mixture of monomers and dimers. The structure of the T_m TF_C domain is also very similar to that of *E. coli*, but significantly diverges from the homologous domain from *V. cholerae*. Its structure is also close to that of the periplasmic chaperone SurA and a protein called mpn555. The T_m TF_C domain can form dimers as in TF_C from *E. coli*. Heteromeric assemblies between the N- and C-domains of the trigger factor from *T. maritima* were also observed (Martinez-Hackert and Hendrickson 2007). Independently of binding to ribosomes and ribosome-nascent chains, the trigger factor is thought to also interact with cytosolic proteins, favouring complex assemblies. The dimeric form of the trigger factor could represent an inactive storage form of unfolded substrates (holdase function). The peptidyl-prolyl cis/trans isomerase activity of TF accelerates proline-limited refolding, but TF also displays an in vitro chaperone activity in a concentration dependent manner. It apparently cooperates with the two ATP-dependent chaperones, the DnaK and GroeS systems that are parts of the stress-induced misfolding and aggregation of cellular proteins. According to a generally accepted sequence of events, nascent polypeptide chains first interact with ribosome-bound TF then, upon release, some chains spontaneously fold whereas some other, about one third, need further assistance by the GroeL and DnaK systems.

6.2.4 Partial Conclusion

As recently stated (Tartaglia et al. 2010), the maintenance of protein solubility that prevents misfolding and aggregation is essential for an appropriate function within the cell. Proteins can be classified into three groups as a function of their propensity to aggregate; class I has a lower aggregation propensity than class II and class III which are largely chaperone dependent. The number of highly soluble proteins decreases from class I to class III and there is an inverse relationship between cytosol abundance and GroEL requirement. Class I and class III have a low Dnak requirement in contrast to class II; this is correlated with the strong hydrophobic character of exposed regions in class II, that use the DnaK system for folding, whereas class III uses Dnak for preventing aggregation and GroEL for folding. Class-III proteins are characterized by high β-sheet content that contributes to display residual structured elements on unfolding that seem to be preferred by the GroEL system. Out of 1,158 cytosolic proteins investigated, 59 % belong to class I and are GroEL independent, 25.6 % can be grouped in class II, the others belong to class III and only contains a weak percentage (2 %) of essential proteins, in agreement with their propensity to misfold. When compared to the other cytosolic proteins, class-III proteins have lower degree of hydrophobicity and higher flexibility. These characteristics impose a high selective pressure on the amino-acid sequence of proteins and it has been possible to predict, from the sequence with a significant accuracy of 90 %, the respective requirement for the GroEl or DnaK systems (Tartaglia et al. 2010).

6.3 The Psychrophiles

Psychrophiles are organisms living in permanently cold habitats and that are exposed to temperatures often well below the freezing point of water, such as in permafrost zones or in the brine veins of sea water (Gilichinski et al. 2007; Deming 2007). Cold environments are also by far the most abundant on Earth since they include polar regions, deep sea (below a depth of 1,000 m), glaciers, sea and freshwater ice, permanent snows and permafrost zones. All these habitats have been colonized by microorganisms, as well as by invertebrates and lower vertebrates in the case of the oceans for example. So, in contrast to what was initially expected, the diversity of life in these regions is high and one can raise the question of how these organisms maintain appropriate metabolic rates despite the negative effect of low temperatures on reaction rates. Low temperatures have in addition the tendency to freeze some macromolecular structures and we have seen that a certain flexibility of these edifices is necessary to secure a defined biological function. This can also be important in the case of protein folding implicating the various chaperones which have to be produced in appropriate rates. Within the other physico-chemical constraints that can impair reaction rates, one has also to mention the spectacular increase of water viscosity at low temperature and the decrease of molecular diffusion rates. Another important parameter is related to protein stability, and one can see from Figs. 6.1 and 6.2 that, as at high temperatures, proteins can be significantly destabilized at low temperatures up to unfolding. As discussed above, this is due first to the weakening of the hydrophobic forces through a favourable effect of low temperatures on hydration of the hydrophobic groups and also to weakening of the electrostatic interactions implicating charged groups of opposite sign, due to the importance of the solvation penalty at low temperature (dehydration needed of individual groups and unfavourable enthalpy) that cannot be overcome by the low heat in the environment. The temperature has therefore a strong effect on the growth of psychrophilic microorganisms. Recently, typical growing curves have been obtained on seawater Antarctic *Pseudoalteromonas haloplanktis* (Piette et al. 2011b) (Fig. 6.3).

Maximal cell densities are observed between 4 and 18 °C and, although the generation time is the lowest around 22 °C, this high temperature leads to drastic decrease in cell densities at the stationary phase indicating that the bacterium is under severe stress in this range of temperatures. This type of data has also been recently observed in other psychrophilic and mesophilic strains (Margesin 2009) with the conclusions that "the temperature at which microbial growth was fastest was in no case the temperature at which the investigated strains produced the highest amount of biomass. All tested psychrophilic bacteria and yeast strains produced highest amounts of cells at 1 °C. Thus cultivation temperatures close to the maximum growth temperature are not appropriate for studying psychrophiles". These observations are crucial for the correct analysis of psychrophiles, since too often authors are still convinced that the temperature showing the shortest generation time is the best suitable temperature for the microorganism. It is just only an apparent optimum temperature that has nothing to do with the well-be of the bacterium. Unlike thermophiles, the first challenge of which is to secure appropriate stability at the environmental temperature, the main problem for psychrophiles is to secure an appropriate rate of the numerous reactions occurring in the organism despite the low temperature of the environment. We indeed know that the rate of a chemical reaction is exponentially depending on temperature, $k = A\ e^{-Ea/R}$. One can easily deduced from this equation that the thermo-dependence of a chemical or enzymic reaction is strongly dependent on the value of Ea, the activation energy with is also equal to: $Ea = \Delta H^{*} + RT$ so Ea is in fact a measure of the activation enthalpy that can be calculated from an Arrhenius plot: lnk as a function of 1/T. This gives in a certain range of temperature a straight line of slope—Ea/R (Collins et al. 2007). So, a low value of Ea will make the reaction less dependent on temperature since, with $Ea = 0$, the exponential term is equal to 1 and the reaction is almost independent of temperature (not quite, since the pre-exponential term is directly dependent on temperature). To be complete, one has

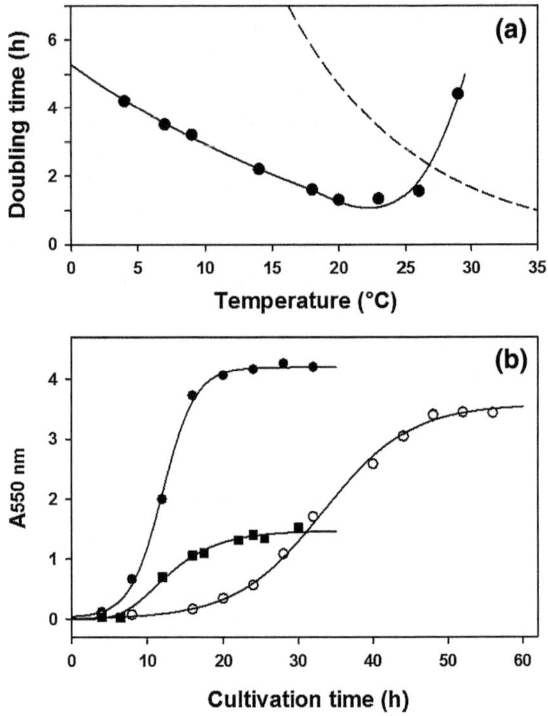

Fig. 6.3 Growth of the Antarctic psychrophilic bacterium *Pseudoalteromonas haloplanktis* as a function of temperature. **a** shows the thermodependence of the generation time compared to that of mesophilic *E. coli* RR1 (*dashed curve*). **b** shows the growth curve at 4 °C (*open circles*), 18 °C (*black dots*) and 26 °C (*black squares*). The shortest doubling time also corresponds to the temperature that gives the lowest cell density (From Piette et al. 2011b)

also to mention that the rate of an enzymic reaction is also dependent on an important term, the transmission coefficient κ, expressed in the temperature dependence of the rate of catalysis given by an equation proposed by Eyring and similar to the Arrhenius law: $k_{cat} = \kappa\ k_B T/h\ e^{-\Delta G^*/RT}$ (Marx et al. 2006; Feller 2010) or its more generalized expression $\gamma_{(T)}$ (Garcia-Viloca et al. 2004), that expresses the possibility that the activated state has some probability to return to its ground state (re-crossing the activation barrier). Often neglected, this term is not equal to one in the case of reactions occurring at low temperature, and notably depends on the viscosity of the medium.

6.3.1 Enzyme Activity at Low Temperatures

In living organisms, most reactions occurring in an organism are catalyzed by enzymes and from the discussion above one can see that one way to cope with the negative effect of low temperatures would be to lower the activation energy of the enzyme, either by decreasing the activation enthalpy or by increase the activation entropy since $\Delta G^* = \Delta H^* - T\Delta S^*$. The activation parameters have been calculated for a few psychrophilic enzymes and compared to those of mesophilic counterparts (D'Amico et al. 2003; Marx et al. 2006; Coquelle et al. 2007; Bjelic et al. 2008). In all cases, and in agreement with the higher catalytic efficiency, a lower activation energy ΔG^* was recorded for psychrophilic enzymes approximately in the range of a decrease of 2–10 %. This lower activation energy systematically results from a large decrease of the activation enthalpy often higher than 20 % however compensated by a much more negative activation entropy. The lower activation enthalpy has been attributed to the lower number of weak bonds, enthalpy-driven, that have to be broken to reach the activated state. The lower activation enthalpy also depicts the lower temperature dependence of the catalytic activity of these enzymes. With regards to the much more negative value of the activation entropy the figures are also consistent with the idea that the ground state of the cold-adapted enzyme has a rather loose structure, that needs a larger re-ordering to be transformed into a well-organized transition state (Lonhienne et al. 2001; Feller 2010). The consequence is that the lower activation enthalpy is counterbalanced by a less favourable entropy change on activation, so that the activation energy is higher than what would be predicted by the difference in the activation enthalpy. That is not really a problem since for example in the case of α-amylase from an Antarctic bacterium a decrease of activation energy by only 2 % allows to increase the rate of catalysis three fold at 10 °C when compared to the mesophilic counterpart (Collins et al. 2007). The lower stability and the more negative activation entropy suggests, as stated above, that cold-adapted enzymes display higher flexibility when compared to their mesophilic counterparts and intuitively this can indeed counteract the freezing effect of low temperatures on the 3D structure of macromolecules and preserve some plasticity for accommodation of substrates and release of products. Numerous techniques have been applied for the evaluation of the relative flexibility of protein structure (for a more detailed discussion see Collins et al. 2008) such as fluorescence quenching, neutron scattering and molecular dynamics. Fluorescence quenching has been successfully applied to several psychrophilic and mesophilic enzymes: Ca^{2+}–Zn^{2+} proteases (Chessa et al. 2000), xylanases (Collins et al. 2003), DNA ligases (Georlette et al. 2003), α-amylases (D'Amico et al. 2003), cellulases (Sonan et al. 2007) and zinc metalloproteases (Xie et al. 2009) and from these experiments one can conclude with a certain degree of confidence that psychrophilic enzymes display a more flexible structure. As recently stated (Bjelic et al. 2008), this flexibility is not probably localized systematically at the level of the active sites, in general highly conserved in thermal homologues, but rather in other important sites more or less remote from the active site, but which, in any case, allow to decrease the activation energy of the enzyme by facilitating access to the active site. As shown by the large difference in thermal stability existing between the two main domains of a cold-adapted phosphoglycerate kinase (Bentahir et al. 2000), flexibility does not indeed concerns the whole structure but only crucial parts of it. Later it was demonstrated (Zecchinon et al. 2005) that the two stability domains of this cold-adapted phosphoglycerate kinase do not match the well known N- and C-domains, since the heat-stable domain is made of only 80 residues of the C-domain, including the nucleotide binding site. The remaining part of the protein forms the heat-labile domain and it was proposed that some rigidity of a crucial part of the protein was necessary to preserve appropriate affinity of the cold-adapted enzyme for its substrate, since these intracellular enzymes are probably working far from substrate saturation. An important characteristic of cold-adapted enzymes that concerns the activity is that thermal inactivation always precedes any structural changes that can be detected either by fluorescence or by microcalorimetry. Figure 6.4 indeed concerns, on the left side, α-amylases from psychrophilic, mesophilic and thermophilic counterparts, and on the right side psychrophilic, mesophilic and thermophilic glycoside hydrolases. In mesophilic and thermophilic enzymes, the unfolding thermal transitions strictly match the apparent optimum of activity whereas, in cold-adapted enzymes, the loss of activity largely precedes any apparent changes in the 3D structure. These data can have two explanations. First, the active site of cold-adapted enzymes can be altered to some extent without significant modification of the structure; it has indeed been shown in cold-adapted α-amylase that the active site is the first structural element that unfolds. Second, it is also possible that the enzyme–substrate complex is the first structure to be altered by heat simply by weakening of the interactions between the two partners. In any case these data favour the idea that indeed the low stability of cold-adapted enzymes is undoubtedly related to increase of the flexibility of crucial domains, counterbalancing in this way the freezing effect of low temperatures on structure and reaction rates. This can have additional consequences at least in some cases, e.g. better accessibility of the active site that can also induce broadening of the specificity at least in multi-substrate enzymes (Tsigos et al. 1998; Smalås et al. 2000).

6.3.2 Folding at Low Temperatures

As discussed above, in thermophiles protein folding is often assisted by several cellular tools that prevent aggregation, prepare the protein for proper folding and eliminate misfolded assemblies. The folding of proteins at low temperature can also be impaired by an improper adaptation of the participating enzymes, by possible decrease of hydrophobic and electrostatic interactions (salt bridges) at low temperatures, and by physical constrains such as the important increase in viscosity of the aqueous intracellular space. It has been shown that the intrinsic folding rate constants of psychrophilic, mesophilic and thermophilic proteins are similar (Piette et al. 2011a) at a given temperature, meaning that at low temperature psychrophilic proteins should fold at much slower rates than mesophilic or thermophilic counterparts at their usual environmental temperatures. We have seen that in thermophiles many proteins require assistance for proper folding under the form of chaperones and, from the analysis of the genome sequence of a high number of psychrophiles (Casanueva et al. 2010), it is clear that psychrophiles also have the potential to express all types of chaperones, but the question is to know whether they really express all of them or only specific chaperones at low temperatures. To answer these questions, a limited number of studies have recently

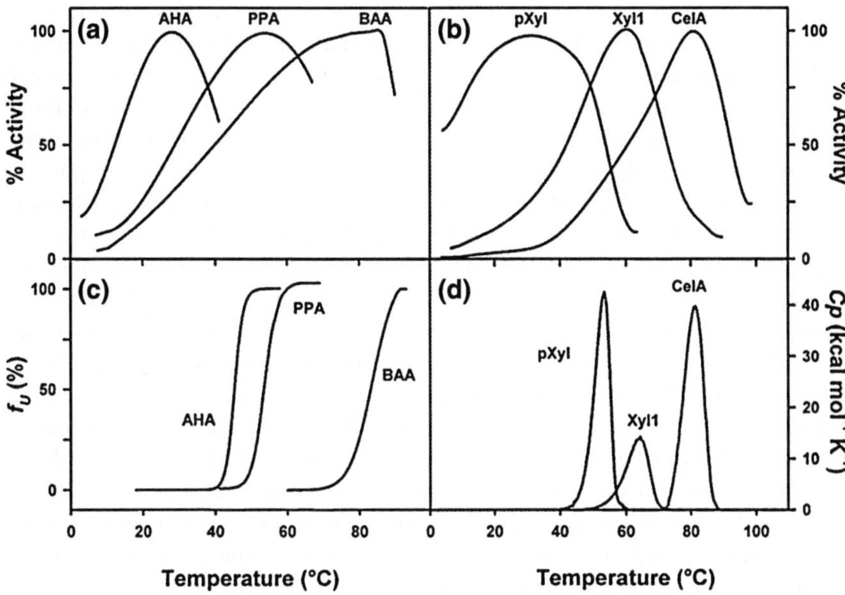

Fig. 6.4 Inactivation (**a**) and unfolding (**c**) of psychrophilic (AHA), mesophilic (PPA), and thermophilic (BAA) α-amylases. Unfolding, followed by fluorescence as a function of temperature. Inactivation (**b**) and unfolding (**d**) of psychrophilic (pXyl) xylanase from Antarctic *P. haloplanktis*, mesophilic (Xyl1) xylanase from *Steptomyces* sp. S38, and thermophilic (CelA) endoglucanase from *Clostridium thermocellum*. Unfolding, followed by microcalorimetry. In psychrophiles, unlike mesophiles and thermophiles, the temperature of inactivation of cold-adapted enzymes largely precedes any significant structural modification. Fluorescence (**c**); microcalorimetry (**d**) (From Marx et al. 2006)

focused on the differential expression of these chaperones as a function of growth temperature of cold-adapted microorganisms. These analyses have been carried out on *Shewanella* sp. SIB1 (Suzuki et al. 2004), *Methanococcoides burtonii* (Goodchild et al. 2004), *Exiguobacterium sibiricum* (Qiu et al. 2006; Rodrigues et al. 2008), *Psychrobacter arcticus* (Zheng et al. 2007), *Shewanella livingstonensis* (Kawamoto et al. 2007), *Spingopyxis alaskensis* (Ting et al. 2010) and *P. haloplanktis* (Piette et al. 2010). Depending on the strain, certain chaperones are overexpressed or down-regulated at low temperature, making general conclusions rather hazardous. What appears rather constant at low temperatures is the overexpression of the enzyme peptidyl-prolyl cis/trans isomerase. Prolyl isomerisation is a rate limiting step in the folding of protein because it is a slow process. The trans-cis transformation is also energetically unfavourable and is certainly further impaired by the low environment temperature in psychrophiles. Some attenuation of this unfavourable process occurs in psychrophiles since they generally contain a lower amount of prolyl residues. By contrast, mesophiles and thermophiles usually contain a higher amount of Pro residues that decreases the entropic changes on folding rendering the process easier and the protein more stable. The important chaperone GroEL is down regulated in some cold-adapted species as well as the associated factors GroES and GrpE, a similar trend is also observed in the case of the DnaK system. As in peptidyl-prolyl cis/trans isomerase, the trigger factor is also generally overexpressed except in for *P. arcticus.* Not forgetting the possibility that the apparent differential expression of chaperones in organisms living at low temperatures could simply reflect the evolutionary history, it is also possible that the differences observed in the temperature regulation of the expression of chaperones as a function of temperature could be due to differences in methodology. In some cases indeed, some authors use, as upper growth temperature, the temperature of optimum growth rate. But these temperatures, as stated above, induce a severe stress, since maximal growth rates are just kinetic effects which have nothing to do with the physiological state of the bacterium (Margesin 2009).

It seems clear that, in psychrophiles, the trigger factor that interacts with nascent polypeptide chains is essential for securing proper folding of cold-adapted proteins, probably in association with peptidyl-prolyl cis/trans isomerase (Piette et al. 2011a). The down-regulation of the other chaperones can possibly be related to the fact that low temperatures reduce to some extent the risk of aggregation and of misfolding. In support of this, it has been also observed that overexpression of a cold trigger factor represses expression of other chaperones when *E. coli* is grown at low temperatures (Kandror and Goldberg 1997).

6.4 Conclusions

The rate of growth of a microorganism is related to the rate of the metabolic reactions, most of them being catalyzed by enzymes that require proper folding and stability. That does not mean that the temperature of maximum growth rate is the best possible temperature for the microorganism, since these temperatures, due to the deleterious effect of heat, induce partial unfolding or overproduction of unwanted metabolites due to the differential effect of heat on enzyme activity through largely different values of activation energies. A delicate balance has therefore to be found between conflicting parameters. In thermophiles, although high temperatures favour reaction rates, they also induce unfolding of molecular structures or misfolding notably due to uncontrolled hydrophobic forces. Proteins in thermophiles have therefore slightly modified their structure to resist to the high temperature of their environments, while the folding of a high number of their essential proteins is assisted by a large set of folding helpers named chaperones. The maximum stability of a protein is however not suitable in living organisms since a certain plasticity and flexibility of the molecular structures are required to secure the necessary interactions between partners in a very crowded cell. In psychrophiles, one can consider that the main problem is to secure appropriate metabolic fluxes by acting on the activity or relative abundance of enzymes involved in catalysis. In the history of evolution, the strategy to produce higher amounts of catalysts is not the cheapest way in terms of energy costs, so in general the reaction-rate problem has been solved through progressive evolution towards enzymes displaying lower activation energy and in consequence lower thermal dependence of the activity that allows the organisms to be transiently exposed to unusually low or high temperatures. Therefore a continuum in the adaptation of cold-adapted microorganisms, depending on their evolutionary history, is expected, and subdivisions such as those introduced in the form of terms such as psychrotolerants or psychrotrophs appear devoid of any interest. Tentatively, also the apparent simplification or limitation of the tools rendering appropriate the folding of proteins at low temperature is correlated to reduction of the physico-chemical constrains that can counteract the folding process and possibly also to the energetic care to limit protein synthesis to the minimum required.

References

Beadle BM, Baase WA, Wilson DB, Gilkes NR, Shoichet BK (1999) Comparing the thermodynamic stabilities of a related thermophilic and mesophilic enzyme. Biochemistry 38:2570–2576

Bentahir M, Feller G, Aittaleb M, Lamotte-Brasseur J, Himri T, Chessa J-P, Gerday C (2000) Structural, kinetic and calorimetric characterization of the cold-active phosphoglycerate kinase from the Antarctic *Pseudomonas* sp. TACII18. J Biol Chem 275:11147–11153

Bjelic S, Brandsdal BO, Aqvist J (2008) Cold adaptation of enzyme reaction rates. Biochemistry 47:10049–10057

Braig K, Otwinowski Z, Hedge R, Boisvert DC, Joachimiak A, Horwich AL, Sigler PB (1994) The crystal-structure of the bacterial chaperonin groel at 2.8-Ångstrom. Nature 371:578–586

Brock TD (1967) Microorganisms adapted to high temperatures. Nature 214:882–885

Casanueva A, Tuffin M, Cary C, Cowan DA (2010) Molecular adaptations to psychrophily: the impact of 'omic' technologies. Trends Microbiol 18:374–381

Chessa J-P, Petrescu I, Bentahir M, Van Beeumen J, Gerday C (2000) Purification, physico-chemical characterization and sequence of a heat labile alkaline metalloprotease isolated from a psychrophilic Pseudomonas species. Biochim Biophys Acta 1479:265–274

Collins T, Meeuwis MA, Gerday C, Feller G (2003) Activity, stability and flexibility in glycosidases adapted to extreme thermal environments. J Mol Biol 328:419–428

Collins T, D'Amico S, Marx J-C, Feller G, Gerday C (2007) Cold-adapted enzymes. In: Gerday C, Glansdorff N (eds) Physiology and Biochemistry of extremophiles. ASM Press, Washington, pp 165–170

Collins T, Roulling F, Piette F, Marx J-C, Feller G, Gerday C, D'Amico S (2008) Fundamentals of cold-adapted enzymes. In: Margesin R, Schinner F, Marx J-C, Gerday C (eds) Psychrophiles, from biodiversity to biotechnology. Springer-Verlag, Berlin, pp 211–227

Coquelle N, Fioravanti E, Weik M, Vellieux F, Madern D (2007) Activity, stability and structural studies of lactate dehydrogenases adapted to extreme thermal environments. J Mol Biol 374:547–562

D'Amico S, Marx J-C, Gerday C, Feller G (2003) Activity-stability relationships in extremophilic enzymes. J Biol Chem 278:7891–7896

Del Vecchio P, Elias M, Merone L, Graziano G, Dupuy J, Mandrich L, Carullo P, Fournier B, Rochu D, Rossi M, Masson P, Chabriere E, Manco G (2009) Structural determinants of the high thermal stability of SsoPox from the hyperthermophilic archaeon *Sulfolobus solfataricus*. Extremophiles 13:461–470

Deming JW (2007) Life in ice formation at very low temperatures. In: Gerday C, Glansdorff N (eds) Physiology and biochemistry of extremophiles. ASM Press, Washington, pp 133–144

Ditzel L, Löwe J, Stock D, Stetter KO, Huber H, Huber R, Steinbacher S (1998) Crystal structure of the thermosome, the archaeal chaperonin and homolog of CCT. Cell 93:125–138

Feller G (2010) Protein stability and enzyme activity at extreme biological temperatures. J Phys Condens Matter 22:32101–321018

Ferbitz L, Maier T, Patzelt H, Bukau B, Deurling E, Ban N (2004) Trigger factor in complex with the ribosome forms a molecular cradle for nascent proteins. Nature 431:590–596

Garcia-Viloca M, Gao J, Karplus L, Truhlar DG (2004) How enzymes work: analysis by modern rate theory and computer simulations. Science 303:186–195

Georlette D, Damien B, Blaise V, Depierreux E, Uversky VN, Gerday C, Feller G (2003) Structural and functional adaptations to extreme temperatures in psychrophilic, mesophilic and thermophilic DNA ligases. J Biol Chem 278:37015–37023

Gerday C (2011) Life at the extreme of temperature. In: Storz G, Hengge R (eds) Bacterial stress response. ASM Press, Washington, pp 425–444

Gilichinski D, Vishnivetskaya M, Petrova M, Spirina E, Mamykin V, Rivkina E (2007) Bacteria in permafrost. In: Margesin R, Schinner F, Marx J-C, Gerday C (eds) Psychrophiles: from biodiversity to biotechnology. Springer-Verlag, Berlin, pp 83–102

Goodchild A, Saunders NF, Erlan H, Raftery M, Guilhaus M, Curmi PM, Cavicchioli R (2004) A proteomic determination of cold adaptation in the Antarctic archaeon, *Methanococcoides burtonii*. Mol Microbiol 53:309–321

Hoffmann A, Bukau B, Kramer G (2010) Structure and function of the molecular chaperone, trigger factor. Biochim Biophys Acta 1803:650–661

Huo Y, Hu Z, Zhang K, Wang L, Zhai Y, Zhou Q, Lander G, Zhu J, He Y, Pang X, Xu W, Bartlam M, Don Z, Sun F (2010) Crystal structure of group II chaperonin in the open state. Structure 18:1270–1279

Kandror O, Goldberg AL (1997) Trigger factor is induced upon cold shock and enhances viability of Escherichia coli at low temperatures. Proc Natl Acad Sci U S A 94:4978–4981

Kashefi K, Lovley DR (2003) Extending the upper temperature limit for life. Science 301:934

Kawamoto J, Kurihara T, Kitagawa M, Kato I, Esaki N (2007) proteomic studies of an Antarctic cold-adapted bacterium, *Shewanella livingstonensis* Ac 10, for global identification of cold-inducible proteins. Extremophiles 10:819–826

Kumar S, Tsai C-J, Nussinov R (2002) Maximal stabilities of reversible two-state proteins. Biochemistry 41:5359–5374

Li WF, Zhou PL (2005) Structural features of thermozymes. Biotechnol Adv 23:271–281

Liu CP, Perrett S, Zhou JM (2005) Dimeric trigger factor stably binds folding-competent intermediates and cooperates with the DnaK-DnaJ-GrpE chaperone system to allow refolding. J Biol Chem 280:13315–13320

Liu CP, Zhou QM, Fan DJ, Zhou JM (2010) PPIase domain of trigger factor acts as auxiliary chaperone site to assist the folding of protein substrates bound to the crevice of trigger factor. Int J Biochem Cell Biol 42:890–901

Lonhienne T, Gerday C, Feller G (2001) Psychrophilic enzymes: revisiting the thermodynamic parameters of activation may explain local flexibility. Biochim Biophys Acta 1543:1–10

Luke KA, Higgins CL, Wittung-Stafshede P (2007) Thermodynamic stability and folding of proteins from hyperthermophilic organisms. FEBS J 274:4023–4033

Margesin R (2009) Effect of temperature on growth parameters of psychrophilic bacteria and yeasts. Extremophiles 13:257–262

Martinez-Hackert E, Hendrickson WA (2007) Structure of and interactions between domains of trigger factor from *Thermotoga maritima*. Acta Crystallogr Sect D63:536–547

Martinez-Hackert E, Hendrickson WA (2009) Promiscuous substrate recognition in folding and assembly activities of the trigger factor chaperone. Cell 138:923–934

Marx J-C, Collins T, D'Amico S, Feller G, Gerday C (2006) Cold-adapted enzymes from marine Antarctic microorganisms. Mar Biotechnol 9:293–304

Meng G, Xia-Yu X, Xian-Ming P (2008) Salt bridges in the hyperthermophilic protei Sso10b are resilient to temperature increases. J Biol Chem 283:31690–31696

Nakamura A, Takumi K, Miki K (2010) Crystal structure of a thermophilic GrpE protein: insight into thermosensing function for the DnaK chaperone system. J Mol Biol 396:1000–1011

Okada J, Okamoto T, Mukaiyama A, Tadokoro T, You D-J, Chon H, Koga Y, Takano K, Kanaya S (2010) Evolution and thermodynamics of the slow unfolding of hyperstable monomeric proteins. BMC Evol Biol 10:207–218

Piette F, D'Amico S, Struvay C, Mazzuchelli G, Renaut J, Tutino ML, Danchin A, Leprince P, Feller G (2010) Proteomics of life at low temperatures: trigger factor is the primary chaperone in the Antarctic bacterium *Pseudoalteromonas haloplanktis* TAC 125. Mol Microbiol 76:120–132

Piette F, D'Amico S, Mazzucchelli G, Danchin A, Leprince P, Feller G (2011b) Life in the cold: a proteomic study of cold-repressed proteins in the Antarctic bacterium *Pseudoalteromonas haloplanktis* TAC125. Appl Environ Microbiol 77:3881–3883

Piette F, Struvay C, Feller G (2011a) The protein folding challenge in psychrophiles: facts and current issues. Environ Microbiol 13:1924–1933

Popp SL, Reinstein J (2009) Functional characterization of the DnaK chaperone system from the archaeon *Methanothermobacter thermoautrophicus ΔH*. FEBS Lett 583:573–578

Priyakumar UD, Ramakrishna S, Nagarjuna KR, Reddy SK (2010) Structural and energetic determinants of thermal stability and hierarchical unfolding pathways of hyperthermophilic proteins, Sac7d and Sso7d. J Phys Chem B 114:1707–1718

Qiu Y, Kathariou S, Lubman DM (2006) Proteomic analysis of cold adaptation in a Siberian permafrost bacterium *Exiguobacterium sibiricum* 255–15 by two-dimensional liquid separation coupled with mass spectrometry. Proteomics 6:5221–5233

Razvi A, Scholtz JM (2006) Lessons in stability from thermophilic proteins. Protein Sci 15:1569–1578

Rodrigues DF, Ivanova N, He Z, Huebner M, Zhou J, Tiedje JM (2008) Architecture of thermal adaptation in an *Exiguobacterium sibiricum* strain isolated from 3 million year old permafrost: a genome and transcriptome approach. BMC Genomics 9:547

Ross PD, Subramanian S (1981) Thermodynamics of protein association reactions: forces contributing to stability. Biochemistry 20:3096–3102

Sawle L, Ghosh K (2011) How do thermophilic proteins and proteosomes withstand high temperature? Biophys J 101:217–227

Schlee S, Reinstein J (2002) The DnaK/ClpB chaperone system from *Thermus thermophilus*. Cell Mol Life Sci 59:1598–1606

Smalås AO, Leiros HK, Os V, Willassen NP (2000) Cold adapted enzymes. Biotechnol Annu Rev 6:1–57

Sonan GK, Receveur-Bréchot V, Duez C, Aghajari N, Czjzek M, Haser R, Gerday C (2007) The linker region plays a key role in the adaptation to cold of the cellulose from an Antarctic bacterium. Biochem J 407:293–302

Suzuki Y, Haruki M, Takano K, Morikawa M, Kanaya S (2004) Possible involvement of an FKBP family member protein from a psychrotrophic bacterium, *Shewanella* sp. SIB1 in cold adaptation. Eur J Biochem 271:1372–1381

Tartaglia GG, Dobson CM, Hartl FU, Vendruscolo M (2010) Physicochemical determinants of chaperone requirements. J Mol Biol 400:579–588

Techtmann SM, Robb FT (2010) Archaeal-like chaperonins in bacteria. Proc Natl Acad Sci U S A 107:20269–20274

Ting L, Williams TJ, Cowley MJ, Lauro FM, Guilhaus M, Raftery MJ, Cavicchioli R (2010) Cold adaptation in the marine bacterium, *Shingopyxis alaskensis* assessed using quantitative proteomics. Environ Microbiol 12:2658–2676

Tsigos I, Velonia K, Smonou I, Bouriotis V (1998) Purification and characterization of an alcohol dehydrogenase from the Antarctic psychrophile *Moraxella* sp. TAE 123. Eur J Biochem 254:356–362

Van de Vossenberg JL, Ubbink-Kok T, Elferink MG, Driessen AJ, Konings WN (1995) Ion permeability of the cytoplasmic membrane limits the maximum growth temperatures of bacteria and archaea. Mol Microbiol 18:925–932

Vemparala S, Mehrotra S, Balaram H (2011) Role of loop dynamics in thermal stability of mesophilic and thermophilic adenylsuccinate synthetase: a molecular dynamics and normal mode analysis study. Biochim Biophys Acta 1814:630–637

Vieille C, Zeikus G (2001) Hyperthermophilic enzymes: sources, uses, and molecular mechanisms for thermostability. Microbiol Mol Biol Rev 65:1–43

Xie B-B, Bian F, Chen X-L, He H-L, Guo J, Gao X, Zeng Y-X, Chen B, Zhou B-C, Zhang Y-Z (2009) Cold adaptation of zinc metalloprotease in the thermolysin family from deep sea and Arctic sea ice bacteria revealed by catalytic and structural properties and molecular dynamics. J Biol Chem 284:9257–9269

Yébenes H, Mesa P, Munoz IG, Montoya G, Valpuesta JM (2011) Chaperonins: two rings for folding. Trends Biochem Sci 36:424–432

Zecchinon L, Oriol A, Netzel U, Svennberg J, Gerardin-Otthiers N, Feller G (2005) Stability domains, substrate-induced conformational changes and hinge-bending motions in a psychrophilic phosphoglycerate kinase. A microcalorimetric study. J Biol Chem 280:1307–41314

Zhang J, Baker ML, Schröder GF, Douglas NR, Reissman S, Jakane J, Dougherty M, Fuc J, Levitt M, Ludtke SJ, Frydman J, Chiu W (2010) Mechanisms of folding chamber closure in a group II chaperonin. Nature 463:379–383

Zheng S, Ponder MA, Shih JY, Yiedje JM, Thomashow MF, Lubman DM (2007) A proteomic analysis of *Psychrobacter arcticus* 273–4 adaptation to low temperature and salinity using a 2-D liquid mapping approach. Electrophoresis 28:467–488

Part III
Monitoring and Management

Chapter 7
Changing the Look on Seals from Pole to Pole with Satellite Technology

Arnoldus Schytte Blix, Lars P. Folkow and Erling S. Nordøy

7.1 Satellites in Seal Research

All seals are dependent on land or ice to breed upon, and many species of seals spend their entire life in close proximity to the coast or the edge of ice, where they can be observed and collected throughout the year. However, several species of phocid seals are pelagic outside the short breeding and moulting periods, and it was not until the invention of satellite linked dive recorders (SLDRs) some 20 years ago that we could start to learn about the whereabouts and behaviour of sev-eral ecologically and economically important species of seals outside the breeding season. Also, knowledge of the seasonal distribution and diet composition of many species of seals had until then been based on incidental sightings and occasional shipboard surveys and analysis of a limited number of stomach contents from animals shot in the pack ice. Thus, while these studies have provided knowledge of the diet of seals where they are most accessible for collection, erroneous conclusions may be drawn by extrapolating from such studies if large proportions of the stocks spend a considerable amount of time in open water, where they may pursue different prey species.

We have used satellite-linked dive recorders (SLDRs) of various designs and models produced by Wildlife Computers, Redmond, USA, from the very beginning, to determine location and diving behaviour throughout the year of a great number of seals both in the Arctic and in the Antarctic. In so doing we were

A. S. Blix (✉) · L. P. Folkow · E. S. Nordøy
Department of Arctic Biology, University of Tromsø,
NO-9037 Tromsø, Tromsø, Norway
e-mail: asb000@uit.no

L. P. Folkow
e-mail: lars.folkow@uit.no

E. S. Nordøy
e-mail: erling.nordoy@uit.no

C. Verde and G. di Prisco (eds.), *Adaptation and Evolution in Marine Environments, Volume 2*, From Pole to Pole, DOI: 10.1007/978-3-642-27349-0_7,

Fig. 7.1 Crabeater seal (*Lobodon carcinophagus*) in the pack ice of the Ross Sea, Antarctica (Photo: AS Blix)

also able to determine their haul-out patterns, which have been instrumental in converting aerial sightings into population numbers (e.g. Southwell et al. 2012).

7.2 Antarctic Seals

7.2.1 Crabeater Seal (Lobodon carcinophagus)

The first species we had the opportunity to study was the crabeater seal (Fig. 7.1), which we tagged with SLDRs in the pack ice off Queen Maud Land in Antarctica in February. It is the most abundant of all seals with population estimates varying wildly from 8 to 50 million over time (Laws 1973), while consensus today is close to the lowest of these estimates (Southwell et al. 2012).

It had long been known that during the austral summer, when the pack ice distribution is limited to the coastal waters these animals are distributed within the small patches of pack ice, sometimes in relatively high densities, while in winter little was known of their whereabouts, except for occasional observations of stragglers on the sub-Antarctic islands (King 1983). Their presence in the pack ice is certainly true for breeding and moulting during summer, but we could show that the animals hardly ever left the pack ice but moved north with it as it expanded in Austral winter, distributing themselves over a vast area (Fig. 7.2) (Nordøy et al. 1995; Nordoy and Blix 1998). They made about 150 dives per day, of which 90 % were made to less than 52 m, with a daily diving pattern that confirmed that crabeaters feed predominantly on krill (*Euphausia superba*) throughout the year. Individual maximum diving depths in this species varied between 288 and 528 m.

7.2.2 Ross Seal (Ommatophoca rossii)

The Ross seal (Fig. 7.3) has perhaps been the most elusive of all ice-breeding seals. Since its discovery in 1840 <200 sightings had been made over the next

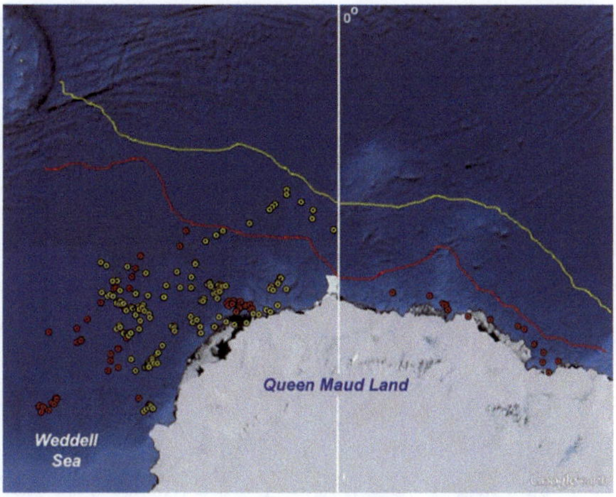

Fig. 7.2 Daily locations of 5 and 6 crabeater seals between mid-April and mid-May in 1993 (*yellow*) and 1997 (*red*), respectively, in relation to the outer edge of the pack-ice (*yellow* and *red* lines). The zero-degree longitude is indicated (Modified from Nordøy et al. 1995 and Nordoy and Blix 1998)

Fig. 7.3 Ross seal (*Ommatophoca rossii*) in the pack-ice of King Haakon VII's Sea, off Queen Maud Land, Antarctica (Photo: AS Blix)

130 years (Hoffman et al. 1973), and aside from detailed anatomical descriptions very little was known about this species 150 years after its discovery. Because of this overwhelming paucity of observations it was moreover generally assumed that the species was low in numbers and confined to heavy pack ice year around (e.g. Condy 1977; King 1983; Laws 1984). However, from one seal that we tagged off Queen Maud Land after the moult in 1997 we got the first indication of another explanation (Blix and Nordoy 1998). This inspired a study in which 10 animals were tagged off Queen Maud Land, and from which we had the unusual success of getting data from 6 animals for a full year.

Now we found that these animals did indeed head North shortly after moulting in late January-early February and entered into open water in the area south of the Antarctic Polar Front, where they remained pelagic until October, when they headed back almost due South in preparation for breeding (Fig. 7.4). From the drift pattern of the females we were also for the first time able to pin-point the nursing period to 13 days in mid-November, which would indicate that mating takes place in December (Blix and Nordoy 2007).

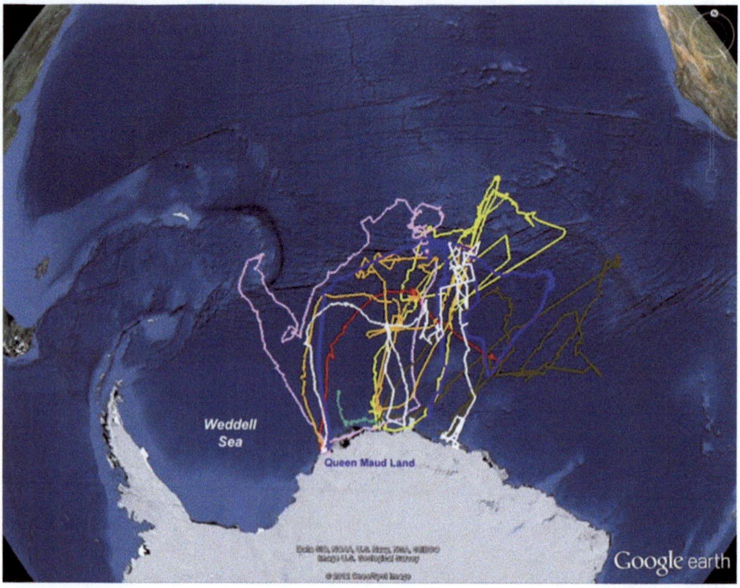

Fig. 7.4 Overall movements of nine adult Ross seals tagged with SLDR's after moulting in mid-February at two locations indicated with circles. Six of the animals provided data for about a year. *BI*, Bouvet Island. (From Blix and Nordoy 2007)

We also found that these seals made about 100 dives a day throughout the year, most of the time to depths of 100–300 m with durations of 5–15 min, the deepest dive on record being 792 m, while less than 10 of more than 200,000 dives lasted for more than 30 min. This distribution and diving pattern is at least suggestive of a diet consisting of mid-water fish such as *Pleuragramma antarcticum*, squid, and to some extent krill (*Euphausia superba*), when in the pack ice, and myctophid fish and several species of squid, when in the open ocean. The Ross seals also provided data to show that the haul-out pattern changes during the austral summer (Fig. 7.5), which is important information with regard to correction of survey data when estimating animal abundance (Southwell et al. 2008).

7.2.3 Leopard Seal (Hydrurga leptonyx)

Leopard seals are few and solitary in the Antarctic, and they are not particularly inviting for handling.

We were nevertheless able to tag two adult animals (Fig. 7.6), again off Queen Maud Land after the moult in late February. Two animals are not many, but since information on leopard seals, aside from anatomy, estimates of abundance, and a few stomach contents is almost nil, any information is important.

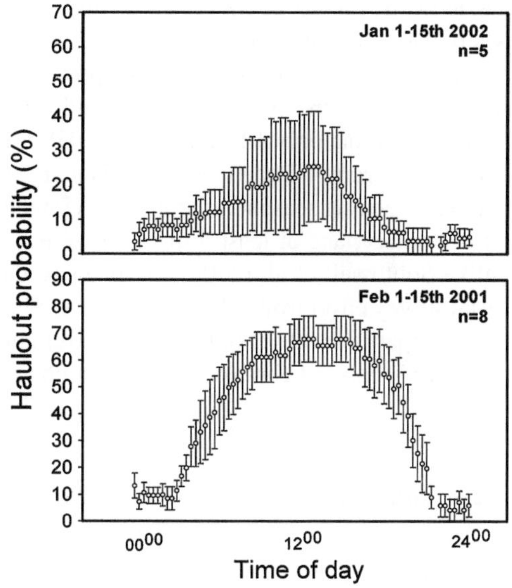

Fig. 7.5 Haul-out probability ± SEM (%) of Ross seals in relation to time of day in first half of January (*top*) and during the first half of February (*bottom*) (Modified from Blix and Nordoy 2007)

Fig. 7.6 Leopard seal (*Hydrurga leptonyx*) with a head-mounted SLDR in the pack-ice of King Haakon VII's Sea off Queen Maud Land, Antarctica (Photo: AS Blix)

It is known that these animals are mostly found within 300 m of the edge of the pack ice close to the Antarctic continent during Austral summer (Condy 1977; Bester et al. 2002), and both our seals remained within the pack-ice relatively close to the Antarctic continent, from tagging in mid-February until early May, when contact was lost with one of the seals. The one remaining, however, then migrated to the South Sandwich Islands in only 3 weeks, and thereafter headed East, always close to the outer edge of the pack ice, but with only an occasional haul out, until contact was lost at 55°S in September. Both seals made mostly short, <5 min, dives to depths of 10–50, the deepest dive recorded being 304, dives to more than 150 m being extremely unusual (Nordoy and Blix 2009). These observations may indicate that leopard seals spend the winter mainly in open water, off the edge of the pack ice, where they hunt near the surface. It is therefore likely that krill as well as penguins and a variety of fish are important prey items. Leopard seals have never been common anywhere in the Antarctic, and Southwell et al. (2008) have speculated over the fact that only 12 sightings in 5,978 km of aerial surveys were made in east Antarctica, from 4 December to 10 January, which has been supposed to overlap with the breeding season of the species. However, we examined 16 adult leopards in mid-February, and at least 4 of those were in a very early stage of moulting, and unlike Southwell et al. (2008) we made 16 definite sightings in 3 days of modest aerial survey effort from 14 to 16 February. This suggest that a significant proportion of the leopards are at sea in December and January and that the best time for surveys of leopard seals is February to early March rather than at mid-summer.

7.2.4 Spatial Segregation of Antarctic Phocid Seals

The last two of the Antarctic phocid seals, the Weddell (*Leptonychotes weddelli*) and the southern elephant seals (*Mirounga leonina*), have not been studied by us, but by others using the same general approach, and the following rather neat spatial distribution of the Antarctic species of phocid seals appears:

The Weddell seals behaviour has been known for a long time, in part because they oddly enough show no fear of man, and because these animals inhabit the rather stable pack ice over the continental shelf, close to the continent, all year round. In this fast ice they keep blow holes open, even in very thick ice, through which they often feed on bottom fish even at depths of 400 m, but mid-water fish and a variety of squid and crustaceans are also eaten (e.g. Kooyman et al. 1980).

The crabeater seals are distributed throughout the pack ice where they feed on krill close to the surface.

The leopard seals hunt mainly in rather shallow water, just outside the pack ice, where penguins, a variety of fish and krill are likely prey.

The Ross seals are mainly truly pelagic and operate at high sea close to the Antarctic Polar Front for a major part of the year, where they dive to 100–300 m and most likely prey on myctophid fish and a variety of squid.

The southern elephant seals are the world's largest seals that breed primarily on the sub-Antarctic islands, but spend more than 80 % of their lives at high sea, mostly close to the Antarctic continent and the Antarctic Polar Front, where they forage at depths of 300–400 m, while some animals, notably males, occasionally dive to depths in excess of 2,000 m (e.g. McIntyre et al. 2010). With this great variation in foraging behaviour and distribution it is quite obvious that these animals eat a variety of deepwater fish, pelagic fish and squid, while in some areas they also forage on benthic organisms.

Fig. 7.7 Harp seal (*Pagophilus groenlandicus*) in the pack ice off the east coast of Greenland (Photo: AS Blix)

7.3 Arctic Seals

In the northern hemisphere, notably in the North Atlantic Ocean, there are two species of phocid seals that by numbers and their habit of congregating in great concentrations in specific locations on the pack ice for breeding and moult have had very significant impact on the economy of coastal communities for the last 100 years. Those are the harp seals and the hooded seals. Moreover, since the North Atlantic and the adjacent Barents Sea-White Sea hold some of the world's most important fish stocks the ecological and economical implications of the millions of these seals are matters of concern.

7.3.1 Harp Seal (Pagophilus groenlandicus)

The harp seal (Fig. 7.7) is the most abundant pinniped species in the North Atlantic Ocean. The population is traditionally divided into three stocks: the North West Atlantic (off the coast of Newfoundland and in the Gulf of St Lawrence), the Greenland Sea, and the White Sea stocks. We have studied the Greenland Sea and the White Sea stocks.

7.3.1.1 Greenland Sea

The Greenland Sea stock gathers into large and dense breeding aggregations on the pack ice north/north-west of Jan Mayen Island, where they give birth between mid-March and early April.

After a lactation period of about 2 weeks the pups are abandoned and mating takes place. After that the animals disperse, and except for known aggregations of moulting animals in May and occasional spot observations of animals that haul out at the edge of the pack-ice, little was known about their whereabouts and

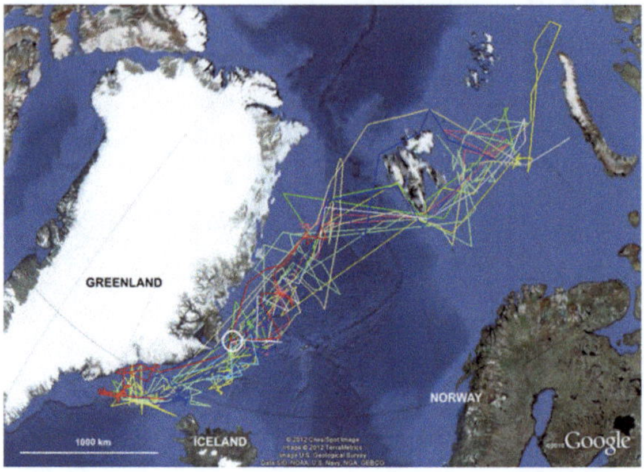

Fig. 7.8 Overall movements of 10 adult harp seals of the Greenland Sea stock tagged with SLDRs (tagging site *white circle*) after moulting in late May (Modified from Folkow et al. 2004)

behaviour for the rest of the year. We (Folkow et al. 2004) found that the seals remained near the pack-ice edge in the Greenland Sea between breeding and moulting (April/May) and during the first 7 weeks after moulting, when they were performing shallow dives of <100 m. This diving behaviour seems to fit well with diet data that show that the crustacean *Parathemisto sp.*, the squid *Gonatus fabricii*, and the polar cod *Boreogadus saida*, as well as capelin *Mallotus villosus*, are important preys during spring and early summer in the Greenland Sea. Most surprisingly, however, we found that the Greenland Sea population migrates into the Barents Sea in mid-July and stays there, together with the Barents Sea stock, until late autumn when they return home (Fig. 7.8). During their sojourn in the Barents Sea, when they utilize the entire water column (0–300 m), their distribution seemed to be closely linked to the distribution of concentrations of capelin. The maximum recorded diving depth was 568 m, the average maximum depth for 5 animals being 451 ± 76 m.

7.3.1.2 White Sea

The White Sea stock breeds on the ice in the "Gorlo" (Throat) of the White Sea in late February/early March and Moult in its "Vronka" (Funnel) in late April/early May. We (Nordoy et al. 2008) found that between breeding and moulting the adult animals were distributed along the coasts of Kola (Russia) and Finnmark (Norway), coinciding in time and space with the spawning capelin. Between moulting and breeding they encircled the entire Barents Sea, in so doing by and large reflecting the annual migrations of the capelin (Fig. 7.9). Most of this time was spent in open water, using the water column from 0 to 300, the average all time maximum diving depth for 9 animals being 410 ± 67, the deepest dive ever recorded being 510 m. Capelin is therefore assumed to be the main prey item for the White Sea stock, the presence of which being the reason for the apparent mass migration of the Greenland Sea stock into the Barents Sea during the autumn when they overlap in time and space with the native stock (Folkow et al. 2004).

Fig. 7.9 Gross annual migrations of the Barents Sea harp seal population between breeding in February–March and moulting (*black arrow*), and between moulting and breeding (*red*). The breeding area is indicated in *yellow* and the moulting area in *blue* (From Nordoy et al. 2008)

7.3.2 Hooded Seal (Cystophora cristata)

Hooded seals (Fig. 7.10) breed in late March on the pack-ice between the east coast of Greenland and Jan Mayen (the "West Ice"), off the west coast of Greenland (in the Davis Strait), in the Gulf of St Lawrence, and off the east coast of Newfoundland (the "Front"). After a very short breeding period (Bowen et al. 1985) they disappear and then reappear to moult in July.

The mature females that we tagged in the breeding rookery (Folkow et al. 1996) dispersed at high sea between breeding and moulting, distributing themselves more or less all over the Greenland and Norwegian Seas. It has long been known, and exploited commercially, that large numbers of hooded seals congregate in the Denmark Strait, and it has been assumed that animals both from the "West Ice" and Newfoundland stock are moulting there. However, another moulting area north of Jan Mayen was reported by Nansen (1890), but later ignored. At the time of moulting our seals returned to the same general area northwest of Jan Mayen, and although the sample size is small ($n = 3$) this suggests that the hooded seals that moult in the Denmark Strait belong to the Newfoundland stock, while the moulting grounds of the "West Ice" stock are located much further north.

Between the moult in July and breeding in March the ice edge along the east coast of Greenland was an important area for these animals. However, most of the seals performed long and often repeated journeys to very distant waters: Some travelled to the Faroe Islands, via both the north and the south side of Iceland,

Fig. 7.10 Hooded seal (*Cystophora cristata*) male with inflated "hood" in the pack ice off Newfoundland (Photo: G Lightfoot)

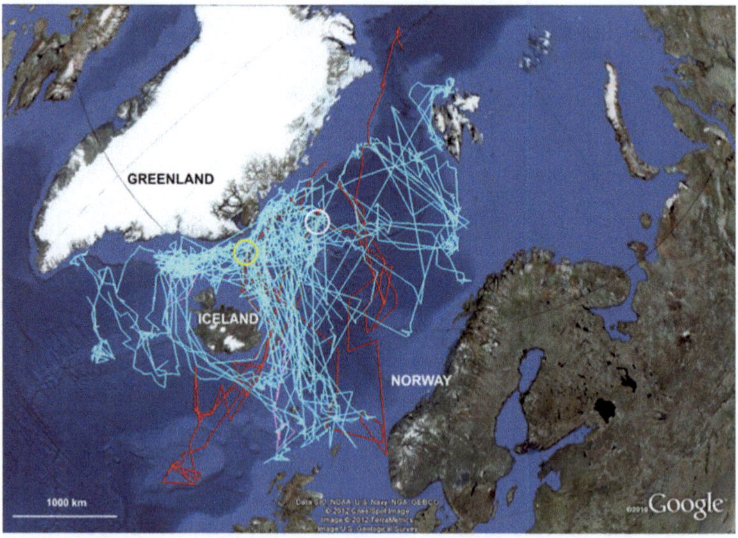

Fig. 7.11 Overall movements of 15 hooded seals of the Greenland Sea stock between tagging with SLDRs after moult in July until breeding in March the following year (*blue lines*). *Violet lines:* continued tracking of 6 of these seals after breeding in March; *red lines:* movements of 4 seals between tagging during breeding in March until moulting Moult tagging site is indicated with *white circle*; breeding tagging site is indicated with *yellow circle* (Modified from Folkow et al. 1996)

some resided for extended periods in the Irminger Sea, some went to the shelf break of the Barents Sea near Bear Island, while one seal travelled to the north side of Svalbard (Fig. 7.11) (Folkow et al. 1996). During this time at high sea, dives of 5- to 25-min duration to 100–600 m dominated (75 %), but most of the animals sometimes performed some very deep (≥1,000 m) and long (≤52 min) dives

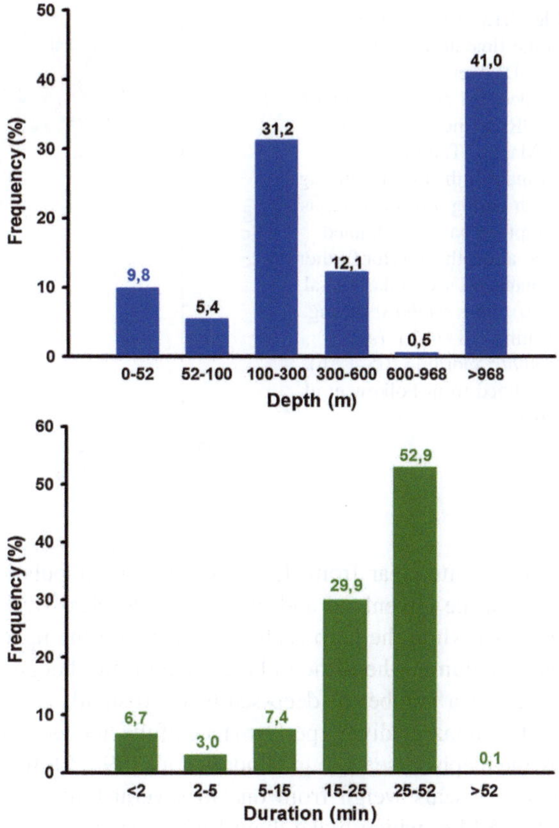

Fig. 7.12 Dive depths (*top*) and durations (*bottom*) of a sub-adult (120 kg) male hooded seal which was diving off Jan Mayen in December (From Folkow and Blix 1995)

(Fig. 7.12), and diving at high sea was continuous through day and night, with an estimated 90 % of the time spent submerged (Folkow and Blix 1999). The at-sea excursions occurred at all times of the year, but were not synchronized in time, and hooded seals therefore do not display any general seasonal migration pattern. Moreover, the dispersal pattern after the moult appears to be random, but makes more sense when the locations are related to the bathymetry of the oceans.

A large proportion of the at-sea locations is confined to waters along the continental shelf breaks, submarine ridges and sea mounts, where upwelling presumably causes high biological productivity.

In comparing the known distribution of prey in time and space, Greenland halibut (*Reinhardtius hippoglossoides*), redfish (*Sebastes spp.*) polar cod (*Boreogadus saida*), herring (*Clupea harengus*), squid (*Gonatus fabricii*) and blue whiting (*Micromesistius poutassou*) are likely prey. The average duration of the excursions was 47 ± 22, the longest being 99 days. The record speed of travel is held by one seal that travelled almost 800 km in 7 days. One male and one female with still active transmitters returned to the breeding site in March the following year.

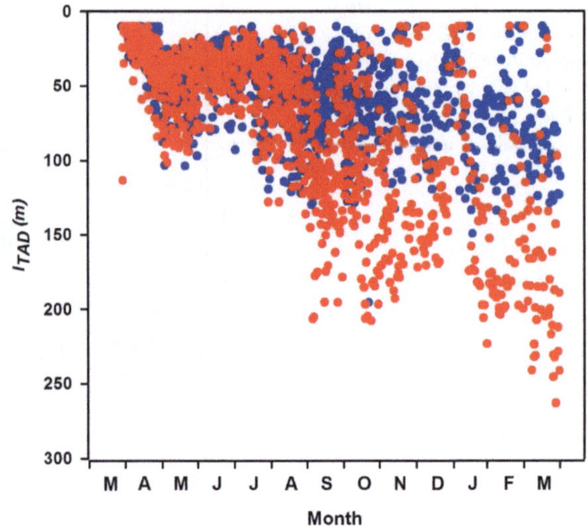

Fig. 7.13 Temporal changes in the time-at-depth index (ITAD) over a year for 7 hooded seal pups tagged with SLDRs at the age of 1 week in March. ITAD is an estimate of the average diving depth during a 6-h period, as computed based on binned time-at-depth data (for further explanation see Folkow et al. 2010). *Blue symbols:* during evening (18–24 h), *red symbols*: morning (06–12 h) (Modified from Folkow et al. 2010)

It is quite clear from these data that the population of harp seals and hooded seals in the Greenland and the Norwegian Seas overlap in geographical distribution, but while the harp seals are occupying the upper few hundred meters of the water column where they chase capelin, the hooded seals are exploiting deeper water for a number of deep-sea fish and squid.

The amazing diving performance of the hooded seal in so doing is only matched by the elephant seals (e.g. Hindell et al. 1991; McIntyre et al. 2010), but while the elephant seals weigh from one to several tons, our hooded seals weighed only 100–150 kg, which make them truly exceptional.

Recently, we (Folkow et al. 2010) have shown that the pups of the hooded seal start diving at the age of about 7 days and that they thereafter disperse over vast areas of the Greenland and the Norwegian Seas in a manner similar to adults. Already within 3 weeks of age they reached depths of more than 100 m and dived for more than 15 min (Fig. 7.13) and throughout the autumn and winter dives of depths of more than 700 m and 30 min duration were recorded.

7.4 Adaptations to Diving

How is this amazing diving performance of the hooded seal possible? Burns et al. (2007) have shown that the hooded seals have the greatest oxygen stores of all mammals, and it has for long been known that the hooded seal like other seal species have the ability to direct almost the entire cardiac output to the brain, while the rest of the body must rely on oxy-myoglobin and anaerobic metabolism when the animal decides to go for long dives (Ramirez et al. 2007). It has moreover been shown that seals have the ability to selectively cool their brains during diving (Blix et al. 2010), and that the shivering that otherwise would have occurred is then inhibited (Kvadsheim et al. 2005). Finally, it has been suggested that cellular metabolism is down-regulated in many tissues during particularly long dives (Ramirez et al. 2007; Folkow and Blix 2010).

References

Bester MN, Ferguson JWH, Jonker FC (2002) Population densities of pack ice seals in the Lazarev Sea, Antarctica. Antarct Sci 14:123–127

Blix AS, Nordoy ES (1998) Ross seal diving behaviour and distribution: a reassessment? NZ Nat Sci 23(Suppl):14

Blix AS, Nordoy ES (2007) Ross seal (*Ommatophoca rossii*) annual distribution, diving behaviour, breeding and moulting, off Queen Maud Land, Antarctica. Polar Biol 30:1449–1458

Blix AS, Walloe L, Messelt EB, Folkow LP (2010) Selective brain cooling and its vascular basis in diving seals. J Exp Biol 213:2610–2616

Bowen WD, Oftedal OT, Boness DJ (1985) Birth to weaning in 4 days: remarkable growth in the hooded seal, *Cystophora cristata*. Can J Zool 63:2841–2846

Burns JM, Lestyk K, Folkow LP, Hammill MO, Blix AS (2007) Size and distribution of oxygen stores in harp and hooded seals from birth to maturity. J Comp Physiol B 177:687–700

Condy PR (1977) Results of the fourth seal survey in the King Haakon VII Sea, Antarctica. S Afr J Antarct Res 7:10–13

Folkow LP, Blix AS (1995) Distribution and diving behaviour of hooded seals. In: Blix AS, Walloe L, Ulltang Ø (eds) Whales, seals, fish and man. Elsevier Science BV, Amsterdam, pp 193–202

Folkow LP, Blix AS (1999) Diving behaviour of hooded seals (*Cystophora cristata*) in the Greenland and Norwegian Seas. Polar Biol 22:61–74

Folkow LP, Blix AS (2010) Air breathers under water: diving mammals and birds. In: Nilsson GE (ed) Respiratory physiology of vertebrates. Cambridge University Press, Cambridge, pp 221–264

Folkow LP, Mårtensson P-E, Blix AS (1996) Annual distribution of hooded seals (*Cystophora cristata*) in the Greenland and Norwegian Seas. Polar Biol 16:179–189

Folkow LP, Nordoy ES, Blix AS (2004) Distribution and diving behaviour of harp seals (*Pagophilus groenlandicus*) from the Greenland Sea stock. Polar Biol 27:281–298

Folkow LP, Nordoy ES, Blix AS (2010) Remarkable development of diving performance and migrations of hooded seals (*Cystophora cristata*) during their first year of life. Polar Biol 33:433–441

Hindell MA, Slip DJ, Burton HR (1991) The diving behaviour of adult male and female southern elephant seals, *Mirounga leonina, (Pinnipedia, Phocidae)* Austr. J Zool 39:595–619

Hoffman R, Erickson A, Siniff D (1973) The ross seal (*Ommatophoca rossii*) In: Seals. IUCN publ new series, suppl paper No 39. IUCN, Morges. pp 129–139

King JE (1983) Seals of the World. Oxford University Press, Oxford

Kooyman GL, Wahrenbrock EA, Castellini MA, Davis RA, Sinett EE (1980) Aerobic and anaerobic metabolism during voluntary diving in Weddell seals: evidence of preferred pathways from blood chemistry and behaviour. J Comp Physiol 138:335–346

Kvadsheim PH, Folkow LP, Blix AS (2005) Inhibition of shivering in hypothermic seals during diving. Am J Physiol 289:R326–R331

Laws RM (1973) The current status of seals in the southern hemisphere. In: Seals. IUCN Suppl. Paper No. 39. pp 144–161

Laws RM (1984) Seals. In: Laws RM (ed) Antarctic ecology. Academic Press, London, pp 621–715

McIntyre T, de Bruyn PJN, Ansorge IJ, Bester MN, Bornemann H, Plötz J, Tosh CA (2010) A lifetime at depth: vertical distribution of southern elephant seals in the water column. Polar Biol 33:1037–1048

Nansen F (1890) Paa ski over Grønland (The First Crossing of Greenland). Aschehoug, Kristiania

Nordoy ES, Blix AS (1998) Seasonal distribution, dive behaviour and temperatures at depth of crabeater seals. NZ Nat Sci 23(Suppl):137

Nordoy ES, Blix AS (2009) Movements and dive behaviour of two leopard seals (*Hydrurga leptonyx*) off Queen Maud Land, Antarctica. Polar Biol 32:263–270

Nordøy ES, Folkow L, Blix AS (1995) Distribution and diving behavior of crabeater seals (*Lobodon carcinophagus*) off Queen Maud Land. Polar Biol 15:261–268

Nordoy ES, Folkow LP, Potelov V, Prischemikhin V, Blix AS (2008) Seasonal distribution and dive behaviour of harp seals (*Pagophilus groenlandicus*) of the White Sea-Barents Sea stock. Polar Biol 31:1119–1135

Ramirez J-M, Folkow LP, Blix AS (2007) Hypoxia tolerance in mammals and birds: from the wilderness to the clinic. Annu Rev Physiol 69:113–143

Southwell C, Bengtson J, Bester, M, Blix AS, Bornemann H, Boveng P, Cameron M, Forcada J, Laake J, Nordoy E, Plöz J, Rogers T, Southwell D, Steinhage D, Stewart BS, Trathan P (2012) A review of data on abundance, trends in abundance, habitat use and diet of ice-breeding seals in the southern ocean. CCAMLR Science (in press)

Southwell CJ, Paxton CGM, Borchers DL, Boveng PL, Nordoy ES, Blix AS, De la Mare WK (2008) Estimating population status under conditions of uncertainty: the Ross seal in East Antarctica. Antarctic Sci 20:123–133

Chapter 8
Environmental Processes, Biodiversity and Changes in Admiralty Bay, King George Island, Antarctica

Lúcia S. Campos, Carlos A. M. Barboza, Manuela Bassoi, Marcelo Bernardes, Sandra Bromberg, Thaïs N. Corbisier, Roberto F. C. Fontes, Paula F. Gheller, Eduardo Hajdu, Helena G. Kawall, Priscila K. Lange, Andre M. Lanna, Helena P. Lavrado, Gabriel C. S. Monteiro, Rosalinda C. Montone, Tatiana Morales, Rafael B. Moura, Cristina R. Nakayama, Thayane Oackes, Rodolfo Paranhos, Flávio D. Passos, Monica A. V. Petti, Vivian H. Pellizari, Carlos E. Rezende, Mariane Rodrigues, Luiz Henrique Rosa, Eduardo Secchi, Denise R. Tenenbaum and Yocie Yoneshigue-Valentin

8.1 Introduction

The isolation of Antarctica from South America during the Oligocene ($\sim$35 mya) formed the Southern Ocean, generated the northward flow of the Atlantic Antarctic Bottom Water, and numerous unique geological and oceanic processes (e.g. an active

L. S. Campos (✉) · C. A. M. Barboza · M. Bassoi · P. K. Lange · A. M. Lanna · H. P. Lavrado · T. Morales · R. B. Moura · T. Oackes · R. Paranhos · M. Rodrigues · D. R. Tenenbaum · Y. Yoneshigue-Valentin
Instituto de Biologia, Universidade Federal do Rio de Janeiro, Av Carlos Chagas Filho 373, Rio de Janeiro-RJ 21941-902, Brazil
e-mail: campos-lucia@biologia.ufrj.br

C. A. M. Barboza
e-mail: carlosambarboza@gmail.com

M. Bassoi
e-mail: manu.bassoi@gmail.com

P. K. Lange
e-mail: prilange@gmail.com

A. M. Lanna
e-mail: andrebioufrj@gmail.com

H. P. Lavrado
e-mail: hpasseri@biologia.ufrj.br

C. Verde and G. di Prisco (eds.), *Adaptation and Evolution in Marine Environments, Volume 2*, From Pole to Pole, DOI: 10.1007/978-3-642-27349-0_8,

spreading centre in the Bransfield Strait, ridge trench collision, gas hydrates on modern sediments, and complex circulation) along the northern end of the Antarctic Peninsula in particular (Barker and Burrell 1982; Pearse et al. 2001; Barker and Thomas 2004; Thomson 2004; Turner et al. 2009).

T. Morales
e-mail: thmorales@hotmail.com

R. B. Moura
e-mail: lytechinusvariegatus@gmail.com

T. Oackes
e-mail: t.oackes@gmail.com

R. Paranhos
e-mail: rodolfo@biologia.ufrj.br

M. Rodrigues
e-mail: marianeers@gmail.com

D. R. Tenenbaum
e-mail: deniser@biologia.ufrj.br

Y. Yoneshigue-Valentin
e-mail: yocie@biologia.ufrj.br

C. A. M. Barboza · M. Bernardes
Universidade Federal Fluminense, Outeiro de São João Batista s/no,
Niterói-RJ 24020-141, Brazil
e-mail: uff.bernardes@gmail.com

S. Bromberg · T. N. Corbisier · P. F. Gheller · H. G. Kawall · G. C. S. Monteiro ·
R. C. Montone · M. A. V. Petti · V. H. Pellizari
Instituto Oceanográfico, Universidade de São Paulo, Praça do Oceanográfico 191,
São Paulo-SP 05508-120, Brazil
e-mail: bromberg@usp.br

T. N. Corbisier
e-mail: tncorbis@usp.br

P. F. Gheller
e-mail: paulafgheller@gmail.com

H. G. Kawall
e-mail: helena.kawall@gmail.com

G. C. S. Monteiro
e-mail: gabrielmonteiro@usp.br

R. C. Montone
e-mail: rmontone@usp.br

M. A. V. Petti
e-mail: mapetti@usp.br

V. H. Pellizari
e-mail: vivianp@usp.br

R. F. C. Fontes
Campus Experimental do Litoral Paulista, Universidade Estadual Paulista,
Praça Infante Dom Henrique s/no, São Vicente-SP 11330-900, Brazil
e-mail: rcfontes@clp.unesp.br

Limited by the Antarctic continent to the South and Polar Front to the North, the Southern Ocean has been considered one of the best defined marine ecosystems on Earth (Griffiths 2010 and authors therein). The Polar Front has remained a natural barrier to many organisms. As deep as 1000 m, it represents a distinct biogeographical discontinuity with only a few epipelagic and benthic taxa found within and beyond the Southern Ocean (Clarke and Johnston 2003; Linse et al. 2006; Griffiths et al. 2008, 2009; Griffiths 2010). Exceptions are migratory birds and mammals, some mid-water and deep-sea organisms to whom the front does not act as an effective barrier (Brandt et al. 2007; Acevedo et al. 2007; Aguayo-Lobo et al. 2011; Allcock et al. 2011). In general, life has flourished in abundance and diversity therein, despite the fact that its environment has been characterized by glaciations, strong currents, and the sub-zero temperatures close to the seawater freezing point (Clarke and Johnston 2003; Peck et al. 2006; Pörtner et al. 2007; Pugh and Convey 2008). These features have led to speciation and a unique biota with a limited capacity to adapt to the recent environmental changes (Rogers et al. 2012). As the Southern Ocean warms more rapidly to greater depths than the

E. Hajdu
Museu Nacional, Universidade Federal do Rio de Janeiro, Quinta da Boa Vista, s/no, Rio de Janeiro-RJ 20940-040, Brazil
e-mail: eduardo.hajdu@gmail.com

R. B. Moura
Programa de Pós-Graduação em Biologia Animal, Universidade Federal de Pernambuco, Av Prof Moraes Rego 1235, Recife-PE 50670-420, Brazil

C. R. Nakayama
Instituto de Ciências Ambientais, Químicas e Farmacêuticas, Universidade Federal de São Paulo, Rua Artur Riedel 275, Diadema-SP 09972-270, Brazil
e-mail: crnakayama@unifesp.br

F. D. Passos
Instituto de Biologia, Universidade Estadual de Campinas, Rua Monteiro Lobato 255, Campinas-SP 13083-970, Brazil
e-mail: flaviodp@unicamp.br

C. E. Rezende
Laboratório de Ciências Ambientais, Universidade Estadual do Norte Fluminense Darcy Ribeiro, Av. Alberto Lamego 2000, Campos dos Goytacazes-RJ 28013-602, Brazil
e-mail: crezende@uenf.br

L. H. Rosa
Instituto de Ciências Biológicas, Universidade Federal de Minas Gerais, Av Antônio Carlos, 6627, Belo Horizonte-MG 31270-901, Brazil
e-mail: lhrosa@ufop.br

E. Secchi
Instituto de Oceanografia, Universidade Federal do Rio Grande, Av Italia, Km 8, Rio Grande-RS 96201-900, Brazil
e-mail: edu.secchi@furg.br

global ocean average, ongoing dramatic changes in physical and biogeochemical processes have become life-threatening (Rintoul et al. 2012). Therefore, understanding how these processes affect Antarctic organisms and their diversity in different spatial and temporal scales is imperative, and essential to better comprehend the functioning of the Earth system (Turner et al. 2009; Gutt et al. 2011).

The West Antarctic Peninsula region is subject to one of the fastest rates of climate change on the planet (Turner et al. 2009). Northwest from the peninsula, the South Shetland Islands include King George Island, the biggest and busiest of the region. This island is located in the boundary of the marine ice extension at the Antarctic atmospheric front's edge (Hofmann et al. 2004). It is basically subject to a maritime climate characterized by a frequent succession of low pressure centres, where the cyclonic activity is high in the summer as a result from the migration of low pressure centres from the Antarctic Polar Circle (Setzer et al. 2004). The high pressure over the Antarctic Peninsula generates humid and mild air temperature from North to Northwest towards the Shetland Islands (Braun et al. 2001), which could explain the mild air temperature over King George (Setzer et al. 2004).

Long term meteorological data have indicated a fast increase (four times higher than the global average) in the atmospheric temperature of the Antarctic Peninsula region over the last 50 years (Turner et al. 2009). Associated to this regional warming, King George Island has lost 7 % of its ice cap in this period. In fact, the largest ice volume losses observed in the planet have occurred only 350 km away from King George Island with the retreat of more than 7000 km^2 since the beginning of the twenty-first century, i.e. the Larsen shelves (Gutt et al. 2011).

Unlike general trend, atmospheric measurements at King George have shown relative temperature stability, with a decrease of −0.6 °C over the past 15 years (Setzer et al. 2004, 2010). Conversely, sea-water temperature in the region has increased significantly (Clarke et al. 2005). This supports the warming hypothesis, a threat to the characteristics of Antarctic shelf communities (Gutt et al. 2011). Temperature rise may remove physiological barriers allowing fast moving, durophagous (skeleton-crushing) predators such as sharks and crabs, excluded from Antarctic waters for millions of years due to low water temperatures (Aronson et al. 2007), to reinvade the shelf and alter the structure of endemic communities.

Admiralty Bay is located at the centre of King George Island, and has had some of the longest term environmental and marine biological data, collected by several nations including Poland and Brazil (Trivelpiece et al. 1987; Setzer et al. 2004; Sicinski et al. 2011). Monitoring this area has become essential in order to detect and understand its changes over time, and distinguish natural from anthropogenic impacts (Montone et al. 2012). The International Polar Year 2007–2008 gave the opportunity to sample the less studied zones of this bay and compare new with past biodiversity data. Also, new tools and modelling were used to evaluate oceanic processes in the area and their effect on marine life. A general overview on our main results from IPY is provided here, together with some of the main processes that may affect biodiversity. A better understanding of ecosystem functioning in the context of climate change is critical, and building up knowledge from the best known areas may provide essential clues to the effect of environmental changes on

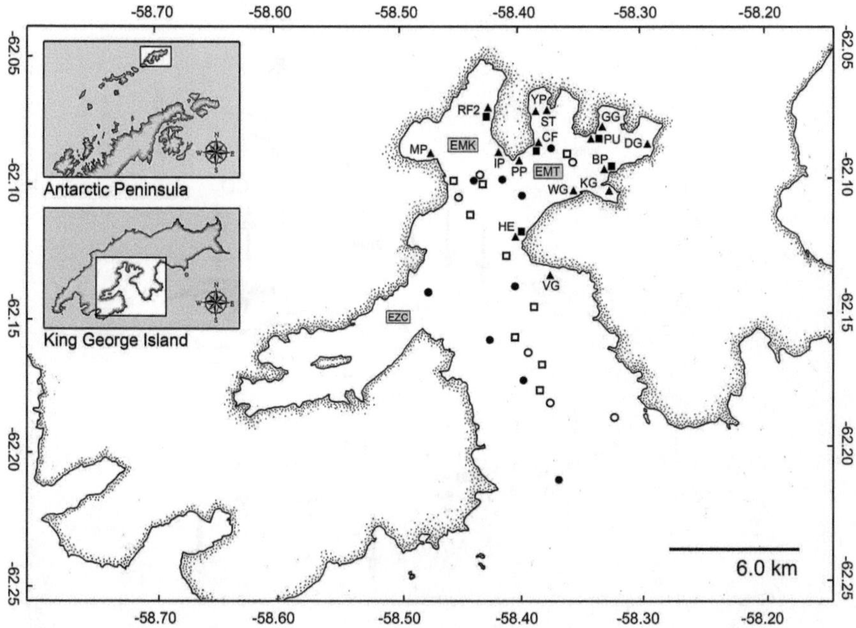

Fig. 8.1 Main sampling stations in Admiralty Bay at King George Island: (▲) Imaging stations; (●) stations sampled with a box-corer (2008–2009); (○) stations sampled with a box-corer (2009–2010); (■) stations sampled with dredge (2008–2009); (□) stations sampled with dredge (2009–2010); *EZC* Ezcurra Inlet; *EMK* Mackelar Inlet; *EMT* Martel Inlet (modified from Moura 2009)

terrestrial and marine biodiversity, and contribute to reducing uncertainties from forecast models related to this still fairly pristine region.

8.2 Environmental Processes

Environmental assessment for monitoring and management has been applied in several sub-Antarctic islands, but there are fewer marine than terrestrial and atmospheric studies (Campos et al. 2012). Since 2002 and throughout the International Polar Year (IPY), a multidisciplinary effort to better understand the processes affecting the structure of biological communities within Admiralty Bay has been intensified in the context of the Census of Antarctic Marine Life (Schiaparelli et al. 2012) and the SCAR programme 'Evolution and Biodiversity in the Antarctic' (e.g. Nakayama et al. 2011a; Sicinski et al. 2011). Sampling stations have been established to study physical, chemical and biological features (Fig. 8.1).

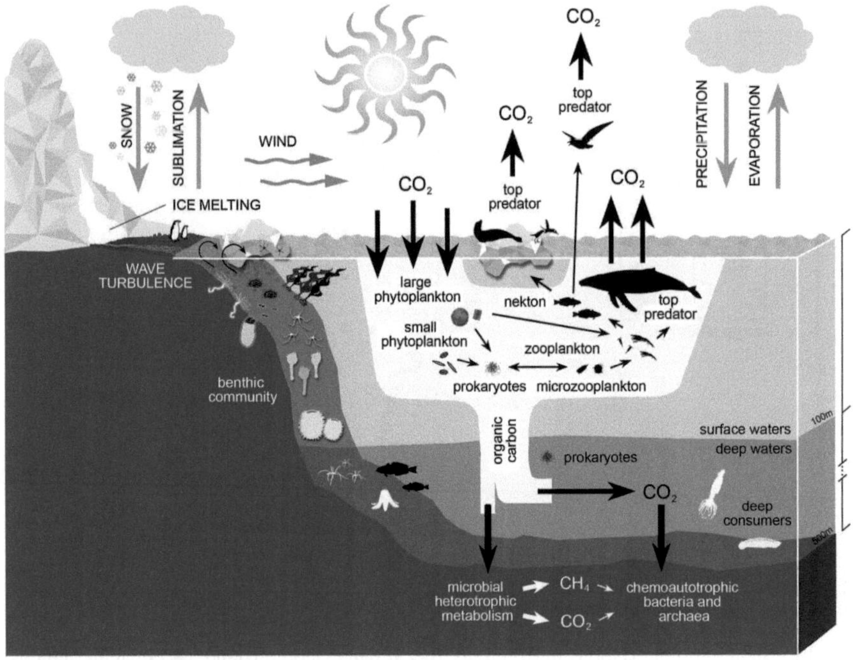

Fig. 8.2 Admiralty Bay at King George Island: scheme of the main environmental and biological processes (adapted from Chisholm 2000)

These data are being gathered in order to model the natural processes within Admiralty Bay against anthropogenic ones (Yoneshigue-Valentin et al. 2011, Campos et al. 2012). Several aspects of the atmospheric, terrestrial and marine environment have been considered in the light of main known geophysical and oceanographic features within Admiralty Bay (Fig. 8.2), and also taking into account the potential for human impacts in the area (see Montone et al. 2012).

8.2.1 Atmosphere

Brazil has maintained a long-term climate series at Admiralty Bay recording solar radiation, wind speed/direction, air temperature and atmospheric phenomena at different levels (Correia et al. 2011; Marani and Alvalá 2011; Pinheiro et al. 2011; Setzer et al. 2010). The Sun is the main source of energy and responsible for life on Earth, but the atmosphere filters the solar radiation (mainly X-rays and ultraviolet), which in turn may damage terrestrial and marine life (Correia et al. 2011).

In recent years, the ozone layer and ground intensity of UV-A and UV-B variability has been accompanied by changes in the ionosphere, which in turn is controlled by solar radiation, varying in close association with the decreasing activity of the 23rd solar cycle (Correia 2011). During winter, the ionosphere (mainly its lowest portion) is strongly influenced by meteorological processes underneath, especially gravity and planetary waves, with strong coupling between all atmospheric layers (Correia et al. 2011). These were monitored in order to understand the energy exchange between layers, better characterize the influence of the Sun-Earth interaction in climate change, and how this affects terrestrial and marine environments. Bageston et al. (2009) have shown that gravity waves observed at the Brazilian Ferraz Station are consistent with wave parameters at Rothera Station.

The dynamics and thermal structure of the mesosphere and lower thermosphere over King George Island have been studied at Admiralty Bay for the past 30 years. The bay has a mean temperature of −1.8 °C, an annual precipitation of 366 mm (more abundant in March and April) and relative humidity always above 80 % influenced by cyclones from the Drake Passage (Rakusa-Suszczewski 1993; Setzer et al. 2004). The Brazilian National Space Institute (www.cptec.inpe.br/antartica) shows that annual mean wind speed is 5.9 m/s ($\sim$21 km/h), but there are frequent gusts between 15 and 20 m/s, which can easily reach 49 m/s (176 km/h). Admiralty Bay has well marked features of periglacial climate (Setzer et al. 2004).

Meteor radars have been installed in order to observe wind structure at 80–100 km altitudes (Bageston et al. 2009). Mariano et al. (2011) have shown that atmospheric SO_2 at Ferraz had no correlation with solar radiation, wind speed and O_3 total column; they appear related to anthropic activities at the station. Fossil fuel and organic waste incineration have impacted the atmosphere only locally within 200 and 400 m around Ferraz, possibly due to persistent strong local winds and atmospheric stability pattern, which are important for gas and particulate matter dispersion originated in Ferraz (Sodré 2011; Montone et al. 2012).

8.2.2 Terrestrial Environment

Antarctic terrestrial habitats are generally best represented in coastal zones, especially in the Maritime Antarctica (Convey 2010), which includes King George Island and Admiralty Bay. Usually these habitats are considered low in species richness, and higher level taxa are not well represented, but as our knowledge on microbial life improves, this perspective may change and reveal thousands of species in ice, soils, lakes, and associated with plants (Rosa et al. 2009, 2010; Teixeira et al. 2010; Vaz et al. 2011; Santiago et al. 2012; Gonçalves et al. 2012).

The ice free zone of Admiralty Bay hosts microorganisms, plants, invertebrates (including Acari, Collembola, Nematoda, and two insects of the order Diptera).

The permafrost occurs away from the littoral zone as the soil temperature during summer is sometimes higher than 5 °C causing melting near shore (Setzer et al. 2004). Even though precipitation is low, it exceeds evaporation, and the soils become saturated for the whole year inducing hydromorphism. These features favour the existence of moss, e.g. *Polytrichum* spp., *Syntrichia* spp. and *Bryum* spp., and lichens found mainly in outcrop rocks and large rock fragments after defrost (Pereira et al. 2010). Also, warming and changes in precipitation patterns may increase biological production due to lower ice-cover duration and water-column- mixture increase in lakes elsewhere (Hodgson et al. 2010). Even though there have been few studies on the lakes at Admiralty Bay, this is also likely to occur there with the additional threat of potentially invasive insects and other microorganisms. The fungal communities from lakes of Admiralty Bay include cosmopolitan and endemic species (Vaz et al. 2011; Gonçalves et al. 2012).

Despite Antarctic isolation, it is possible that certain microbial groups do not follow this condition. Some microbial taxonomic units have a wide distribution between Admiralty Bay and South America (Lemos et al. 2011). This may be due to the fact that these organisms may have high potential for atmospheric transfer (Convey 2010), or can be transported by animals, and human activities.

Differences related to runoff glacier waters, presence of bird colonies, and plants induce conspicuous changes in the environment and its biota down to the microbiological level (Simões et al. 1999; Albuquerque et al. 2004). The structure of microbial communities from soils exposed by glacier retreat and ice samples from Baranowski Glacier revealed an ecological succession, and heterotrophic bacteria occurred all through the process (Duarte 2010). Over time, the increase of ice-free zones and soil association with plant cover, especially bryophytes and *Deschampsia*, together provide a darker horizon surface, which in turn modify the albedo effect and contribute to increasing temperatures (Setzer et al. 2004).

Directly or indirectly, human activities interfere with the terrestrial environment communities (Costa and Alves 2007, 2008; Krüger and Petry 2011). The bird populations have been monitored for over 30 years (Trivelpiece et al. 1987). Persistent organic pollutants (POPs) and polycyclic aromatics hydrocarbons (PAHs) were found in preserved blubber samples from skuas (*Catharacta spp.*), three species of penguins (*Pygoscelis adeliae*, *P. papua* and *P. antarctica*), terns (*Sterna vittata*), snowy sheathbill (*Chionis alba*) and blue-eyed shag (*Phalacrocorax atriceps*) from the vicinity of Ferraz and Arctowski stations (Taniguchi et al. 2009). Although PAHs are well metabolized and readily excreted, they can be detected in fat depots a short period after uptake; the detection of POPs, particularly in penguins, having high fidelity to the Antarctic environment, is evidence of global dispersion of these compounds. Conversely, absence of nitrogen, high content of petroleum hydrocarbon and low microbial diversity was found in soil samples near Ferraz as a result of oil contamination due to tank rupture in the mid 80's in addition to little spills and use of motor vehicles (Rosado et al. 2011).

The terrestrial ecosystem vulnerability to change is high in the context of species distribution, colonization and potential invasion of non-native species (Convey 2010). This is particularly true in the bay considering increased ice-free zone, proximity with South America, and potential for human impact. An interaction of physical (e.g. temperature/humidity increase, cloud cover, thinning of ozone layer) and biological factors (e.g. competition, predation) may affect the biota, and modify the food web (Peck et al. 2006). The native biota is not well equipped to respond to rapid changes, lacking well-developed competition abilities; it is likely that functional and structural aspects of the ecosystem will inevitably be altered as new ecological niche and trophic functions are established with the entrance of species that were not part of this environment before (Convey 2010).

Additionally, research facilities in Admiralty Bay, including associated logistics, may induce chronic and acute local impacts, such as dispersion of chemical pollutants (Santos et al. 2005; Martins et al. 2010), dust and direct damage to plant communities, soils and fresh water (Sodré et al. 2011), which need to be carefully monitored after the recent fire. Unfortunately, many decades are necessary for recovery of Antarctic plant communities and soil that dramatically suffer from these types of disturbances (Turner et al. 2009 and authors threrein; Convey 2010).

8.2.3 Marine Environment

The marine environment at Admiralty Bay (pelagic and benthic realms) has been surveyed over the years by several nations. It is very heterogeneous and characterized as a fjord system with a 550 m deep central basin. Its habitats and their biodiversity, pelagic (surface to deep) and benthic (mud, rocks, algae beds), were studied during IPY within the scope of the CAML. The physical, chemical and geological aspects of the environment have been investigated and used to better understand distribution and adaptations of the biota. Past data, especially those related to the benthic system, have been gathered and synthesized in Sicinski et al. (2011). They recoded approximately 1,300 benthic species in this area. Fig. 8.3 illustrates a variety of marine organisms.

Since the summer 2002–2003, a strategy beyond BACI (Before After Control Impact) was established taking into account the sewage outfall and the oil tanks near Ferraz as one of the main contamination sources, and the minimum of 5 further reference sites within the bay spread through its inlets (Montone et al. 2012). A bathymetric gradient study (20–30 m, 50–60 m, 100 m, 300 m, 500 m, 700 m and 1100 m), always using at least 3 launching replicates at each depth, was carried out during IPY in order to sample benthic organisms from the shallowest to the deepest zones of Admiralty Bay, from Martel Inlet towards the Bransfield Strait, using a box corer and a small Agassiz trawl (cm 60 w × 40 h × 60 l). Besides these, a camera system was used to record community structure in the coastal zone down to 30 m in 14 different sampling sites especially within Martel and Mackelar Inlets. These surveys indicated that shallow-water organisms within inlets, especially at Martel, were typical from habitats with higher hydrodynamics than those predicted by models generated and based on previous assessment studies (e.g. Montone et al. 2012). The dominant megabenthic community structure (10–30 m depth) has usually been described in relation to substrate types (e.g. synthesis in Sicinski et al. 2011), but our image analysis results were then correlated to wave dynamics. Thus, here we provide an overview on processes that may affect biological communities especially in the marine environment.

Fig. 8.3 Examples of biota found in Admiralty Bay, King George Island. **a** Humpback whale (*Megaptera novaengliae*); **b** Weddell seal (*Leptonychotes weddellii*; **c** Chinstrap penguin (*Pygoscelis antarctica*); **d** macroalgae (*Adenocystis utricularis*) in the intertidal; **e** sea urchin (*Sterechinus neumayeri*); **f** solitary ascidian (*Molgula* sp.), and ophiuroid (*Ophionotus victoriae*). Photos: (a) M Bassoi; (b–d) RB de Moura; (e–f) E Hajdu

8.2.3.1 Physical Setting

Models are needed to interpolate between sparse observations, to integrate diverse observations into consistent estimates of the ocean state, detect significance of variations in time scales beyond observation duration, infer aspects of circulation not observable directly (e.g. vertical velocity), and combine circulation and biological observations in a consistent manner (Rintoul et al. 2012). More efforts to gather physical data are necessary within the bay, as its topography is

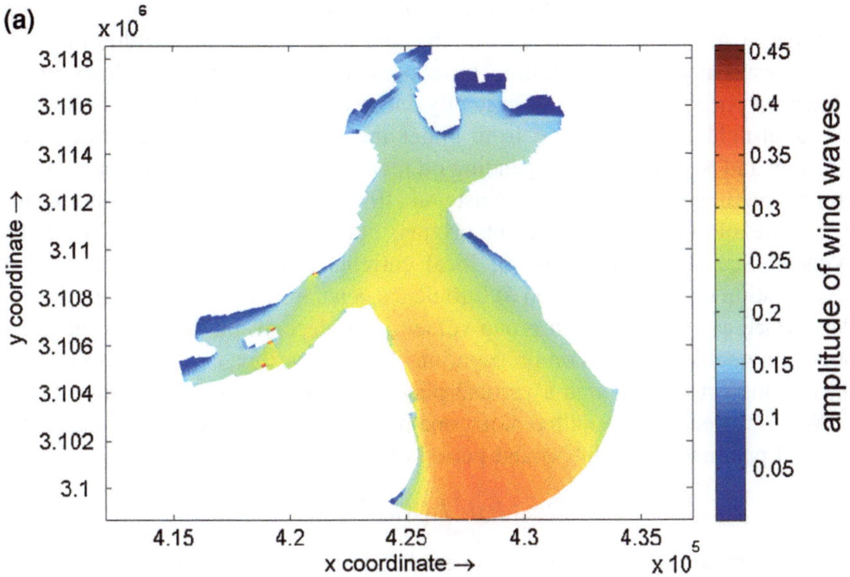

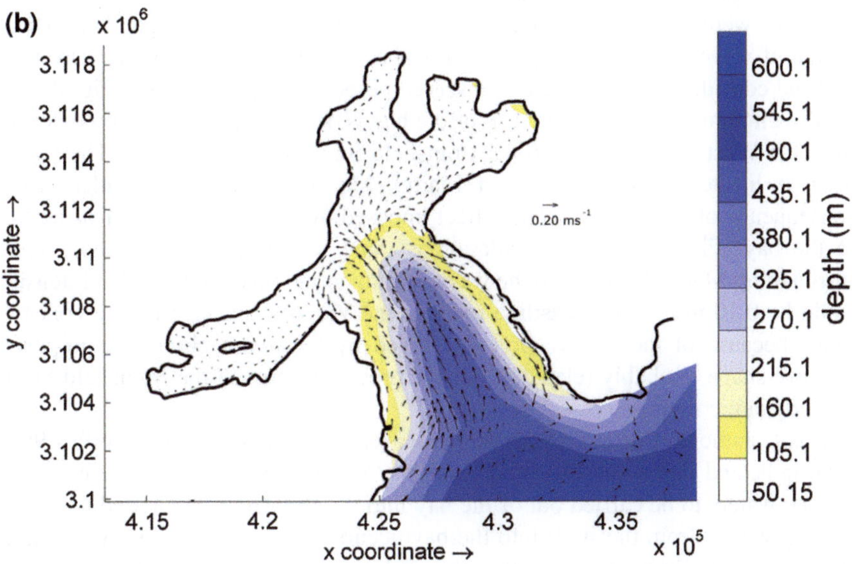

Fig. 8.4 **a** *top*: Amplitude of wind waves forced at the Admiralty Bay entrance based on wind wave modelling using wind velocity (medium intensity) = 6 m s^{-1} wind direction = N; **b** *bottom*: Main flow direction of currents within the bay

heterogeneous and also winds affects terrestrial and marine environments. So far, only sparse physical data exist, and have shown that the amplitude of wind waves is forced at the entrance of Admiralty Bay and propagates along the inner inlets (Figs. 8.4a, 8.5).

Wind forcing is a typical wind field from summer climatology described in the literature. Currents in the centre of the bay are more intense than within inlets. In situ measurements and modelling have pointed out low-energetic flows in the innermost portions of the inlets, basically due to tides. But the strongest flows from the Bransfield Strait originate from either the Weddell or Bellingshausen Seas (Gordon and Nowlin 1978), depending on regional circulation, winds and seasonal climate regime. Basterretxca and Arístegui (1999) suggest that only the surface has a mixture of water from different origins, but deep water originates from the Weddell Sea. The low energy of local currents has led us to focus on wave investigation and its mechanism as turbulence generator. Waves propagating along the surface carry energy with group velocity that depends on the shallowness and how bathymetry interferes on propagation. The energy flux, or wave power, is the mean transport rate through a vertical plane of width parallel to the wave crest.

Wave power (P) transmitted along one meter of coastline is proportional to the spectral parameters height (h) and period (T):

$$P \alpha h^2 T \, (Wm^{-1})$$

We estimated the spectral parameters height and period, from the analysis of movies simultaneously taken at 8 different locations in Admiralty Bay. These estimates were taken from a stick mark as reference for wave height. The distribution of wave quantities, proportional to the wave transmitted power, were normalized considering the minimum observed proportional power value (Fig. 8.5).

The circulation in the deep basin of the bay is defined by a return loop entering from the West and returning to the Bransfield Strait through the East. The return point is located at the entrance of Martel, Mackelar and Ezcurra Inlets, as a consequence of the hydrodynamic effect imposed by the geomorphological setting of the bay (Fig. 8.4b). The circulation is more intense in the near-shore zone surrounding Martel Inlet due to the wave dynamics and tidal effect, which intensity could be 0.10 ms^{-1}. The coastline towards the Vieville Glacier shows a concave shape because of the erosive process caused by the return circulation imposed concave shape, probably related to erosive processes through the Bransfield Strait hydrodynamic regime.

Some authors (Pruszak 1980; Robakiewicz and Rakuza-Swazcsewski 1999) have indicated that when the WSW and NWN winds predominate, these induce surface waters to be carried out of the bay into the Bransfield Strait, and an influx of deep waters from the strait into the bay occurs mainly through its SW margin. However, modelling suggested that this influx occurs independently from wind fields, being mostly related to the circulation of the Bransfield Strait.

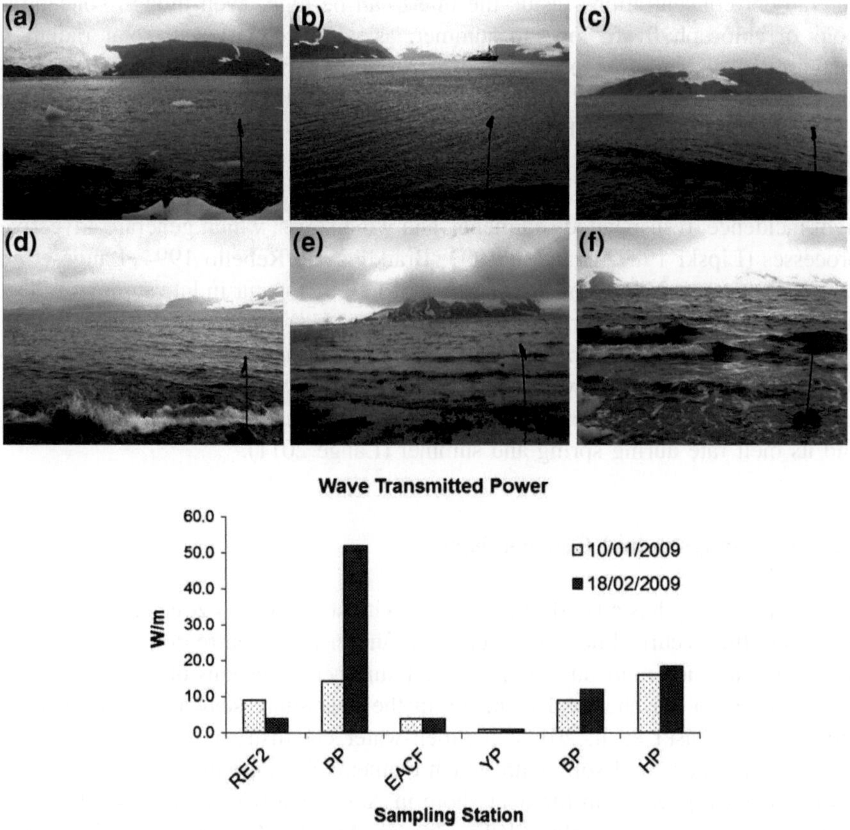

Fig. 8.5 Simultaneous wave record at eight different sampling sites spread through Admiralty Bay, King George Island. **a** Yellow Point; **b** Brazilian Station Comandante Ferraz; **c** Plaza Point; **d** Ipanema; **e** Botany Point; **f** Hennequin Point; **g** Sampling sites distribution of normalized values of wave power in Admiralty Bay (*Ref2* Refuge 2, *PP* Plaza Point, *EACF* Ferraz Station, *YP* Yellow Point, *BP* Botany Point, *HP* Hennequin Point). Measurements were estimated from the analysis of movies taken during two periods: 01/11/2009 and 02/18/2009

8.2.3.2 Hydrochemistry

Admiralty Bay can be compared to an estuarine region as the freshwater output from terrestrial ice melt induces lower salinities on a seasonal basis, especially in its inner portions, than sites near its opening to the Bransfield Strait (Lipski 1987; Lange et al. 2007). A temperature and salinity gradient could be formed between coastal and oceanic waters, shifting depending on the terrestrial freshwater input to the marine environment related to atmospheric temperature changes and ice deposition during the Austral winter (Medlin and Priddle 1990). West and Southwest high wind speeds usually occur in concomitance with intense solar radiation.

Nutrient concentrations inside the inlets can be high, even though concentrations of chlorophyll are low. In summer, as wind velocity and solar radiation increase, so do seawater temperature, nitrite and ammonium, but silicate, phosphate and nitrate decrease (Lipski 1987). However, nitrate, phosphate and silicate concentrations are even higher in the central basin than within inlets (Lange 2011).

The temporal variation on distribution of inorganic nutrients also correlates with other biogeochemical processes and physical dynamics of the bay, such as light incidence, resuspended sediments, and wind stress, which generate advective processes (Lipski 1987, Brandini 1993; Brandini and Rebello 1994; Lange et al. 2007). The lowest N:P ratios (8.5 ± 4.5 mol l^{-1}) that occur in late summer reflect highest phosphate concentrations (4.5 ± 2.5 mol l^{-1}) during this season (Lange et al. 2007). The concentrations of these nutrients are also affected by microphytoplankton in shallow waters near shore. They show considerable spatial and temporal variation depending on sea and terrestrial ice formation over the winter, and its melt rate during spring and summer (Lange 2011).

8.2.3.3 Geophysics and Geochemistry

As Admiralty Bay has a fjord-like structure with steep coastal zones in some areas such as in the Ezcurra Inlet, and deepening further at its centre down to 550 m in its main canal, ice melt during spring and summer is responsible for significant contributions to the chemical elements in the near-shore seawaters and also the sediments (Sicinski et al. 2011). Ice-melt-water chemistry is directly related to path through rocks and soils with which elements are exchanged, and this has a strong effect, especially in the near shore in Antarctica (Costa et al. 2004).

Several elements (Ca, Mg, Al, Fe, Zn, Cu, Mn, Ni, Pb, Cr, Cd, and V) have been measured at in the soil, ice melt and sediments of Admiralty Bay (Costa et al. 2004; Santos et al. 2005). Elements such as Fe and Al show concentrations 65 and 87 % higher than standards in melt waters, which usually feed the lakes in Admiralty Bay, but also run to the sea. Al, for instance shows values ten times higher than the maxima allowed by the Health World Organization, whilst Ni may be 70-fold higher than recommended for consumption (Costa et al. 2004). The ice-melt waters, as they run through the soil, fragments of rocks and sediments, absorb chemical elements and become an important exchange interface between the terrestrial and marine ecosystems (Schaefer et al. 2004).

Metals occur naturally in the Antarctic pristine environment, but can also be introduced by human activities. Those that have no biological function such as Hg, Pb, and Cd, are the most dangerous (Santos et al. 2005, 2007). Some are introduced in Antarctica through the atmosphere and deposited in the ice and further enter the marine system through ice melt (Schaefer et al. 2004; Santos et al. 2005).

Sediments also have high Al concentrations near Thomas Point compared to Botany Point and a site near Ferraz Station in Admiralty Bay (Schaefer et al. 2004). These authors have also shown that pH in interstitial water was typically moderately basic, usual for marine sediments (Table 8.1).

Table 8.1 Main marine physical factors and their features at Admiralty Bay, King George Island, including data from the literature and this work

Environmental Factor	Features	References
Mean tidal range	1.4 m, but maximum tides can reach 2.1 m	Catewicz and Kowalik (1983)
Current speeds	Main canal = 0.20–0.40 m s^{-1} Inlets = 0.02 m s^{-1}	This work
Seawater temperature	Usually -1.6-3 °C Mean values: 1–2.3 °C if origin in Bellingshausen Sea; 0.8–1 °C if origin in the Weddell Sea. During IPY: 0–1.1 °C, lowest in deepest water and highest near shore	Tokarczyk (1987) This work
Seawater freezing	At irregular intervals, e.g. during 11 winters in 20 years (from 1977 to 1996). Ice cover in recent years (1995, 2007, 2009) achieved the maximum level during winter	Kruszewski (1999); Setzer et al. (2010)
Seawater salinity	16–34	Tokarczyk (1987) Lange et al. (2007) This work
Nutrients in the water (annual variation)	Nitrite 0.06–0.19 μmol kg^{-1} Nitrate 9.7–36.7 μmol kg^{-1} Silicate 58.3–73.5 μmol kg^{-1} Phosphate 1.69–2.34 μmol kg^{-1}	Lipski (1987) Brandini and Rebello (1994); Lange et al. (2007)
Bioavailable macro- and micro-nutrients in sediments	P, K, Ca^{2+}, Mg^{2+}, Al^{2+}, H^{+}	Schaefer et al. (2004)
Bioavailable metals	Ba, B, Cd, Cr, Cu, Zn, Ni, Pb, V	Schaefer et al. (2004); Santos et al. (2005, 2007)
pH in sediments	7.7–9.7	Schaefer et al. (2004)
Interstitial seawater salinity	34–36, increasing with depth	Maciel et al. (2006)
Total organic matter in the sediments	Mean ± SD = 8.69 ± 2.67 % in front of Ferraz	Schaefer et al. (2004); Santos et al. (2005)
Total Organic Carbon	Corg = 0.39 ± 0.16 % at 20 m, 0.49 ± 0.31 % at 30 m and 0.41 ± 0.15 % at 60 m	Maciel et al. (2006)

The values of micronutrients show that coastal sediments have a strong relationship with terrestrial anomalies, for instance Cu was high in sediments near Keller Peninsula (8–14.8 mg dm^{-3}), and much lower near Thomas Point (4.4–9.0 mg dm^{-3}). Cu had a significant correlation with silt, clay, P, Ca, and organic matter, as copper easily forms organic-clay complexes. The values of iron were fairly high due to its reductive nature in sediments (rich in Fe^{2+}), but there is a considerable variability between sites within the bay (224–1,100 mg dm^{-3}). Besides high concentration of Cu near Peninsula Keller, strongly mineralised in chalcopyrite and other sulphates, these authors did not find expressive anomalies

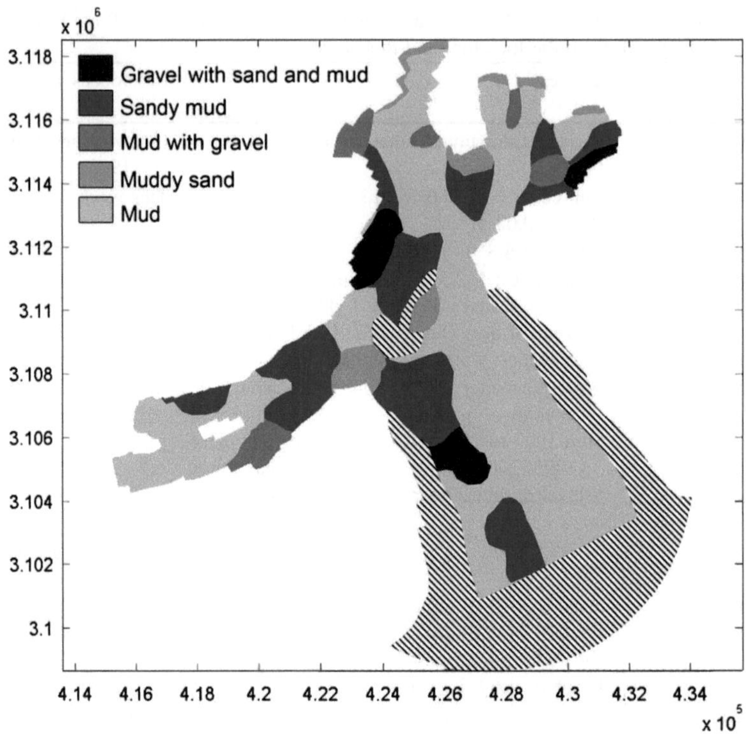

Fig. 8.6 GIS generated map distribution of sediment types within Admiralty Bay. Cross-hatched areas were not sampled (modified from Gruber 1990)

or pollution by heavy metals. There was a tendency for highest values of certain bio-available metals (e.g. Cu 4.36–14.7 mg kg^{-1}; Zn 1.85–6.48 mg kg^{-1}; V 1.50–5.78 mg kg^{-1}). However, Santos et al. (2005) found evidence of metal contamination in front of Ferraz surface sediments enhanced by paints, sewage and petroleum [B, Mo and Pb (>90 %); V and Zn (70–80 %); Ni, Cu, Mg and Mn (30–40 %)], although their low bioavailability, probably triggered by iron sulphide, represents small risk to the marine life.

The sediments in Admiralty Bay are heterogeneous with a large grain-size distribution (Sicinski et al. 2011), but patterns corroborate with the interaction between bottom topography and oceanic circulation, for instance, the concave shape of the coastline towards the Vieville Glacier. Also, in areas where strongest effects of turbulence occur or the substrate is steeper, such as within Ezcurra and surrounding Thomas Point, the clay values are lower with more contribution of sands (Schaefer et al. 2004; Sicinski et al. 2011). Samples taken from the central basin of Admiralty Bay showed that sediments are predominately composed of silt and clay and generally classified as fine to very fine silt, but occasional pebbles and rubbles may be mixed in (Fig. 8.6). Also, ice scour and anchor ice effects are more common and greatly contribute to the geomorphology near shore, especially at Mackelar and Martel inlets (Rodrigues et al. 2003).

Measurements of interstitial seawater salinity showed that it increases with depth (Maciel et al. 2006). The total organic matter was higher in front of Ferraz (mean ± SD = 8.69 ± 2.67 %) than in Machu Picchu, Botany or Hennequin points (Schaefer et al. 2004; Santos et al. 2005). The actual concentrations of total organic carbon (Corg) in the proximity of Ferraz were comparatively low, but showed a slight increase with increasing depth (Table 8.1).

Using organic geochemical markers, Martins et al. (2005) showed that the sewage from Ferraz reaches as far as 400 m away from discharge. A latest environmental assessment indicated that faecal sterols and petroleum hydrocarbon concentrations normally increase slightly throughout the summer near Ferraz, but values had been very low indicating that their contribution had not been significant. Montone et al. (2010) suggested that faecal sterols have been associated to marine-mammal activities near shore rather than anthropogenic.

8.2.3.4 Marine Life

The marine life in Admiralty Bay is rich and diverse, and community structures are greatly influenced by geophysical and chemical processes, besides natural biological interactions (e.g. Brandini and Rebello 1994; Costa and Alves 2007; Sicinski et al. 2011). All Antarctic trophic levels are well represented in the area, from microorganisms to top predators such as seabirds, seals and whales (Figs. 8.2 8.4). As humpback whales travel from western Antarctic Peninsula to the Weddell Sea (Dalla Rosa et al. 2008), they occasionally enter the bay during the summer, and other marine mammals and birds use the bay as resting, feeding or even breeding ground (e.g. Trivelpiece et al. 1987; Costa and Alves 2008).

Admiralty Bay is within the Antarctic continental shelf area, therefore greatly influenced by seasonal sea-ice cover and primary production, factors that may ultimately favour strong benthopelagic coupling, as reported elsewhere on the shelf in the Peninsula region (e.g. Smith et al. 2006). Corbisier et al. (2004) described bentho-pelagic coupling between plankton and suspension feeders such as *Laternula elliptica* (bivalve) and the fish *Chaenocephalus aceratus*, whilst some mollusks (*Nacella* and *Yoldia*) and nematodes showed a close relationship with the microphytobenthos; many organisms display mixed diets.

Resuspension of benthic microalgae is possibly the main reason for high phytoplankton abundance and biomass in coastal areas of the bay (Brandini and Rebello 1994). But Lange et al. (2007) showed that pennate diatoms were dominant close to shore in the 2002/3 summer, suggesting this was also consequent of ice melting. Strong correlation between phytoplankton dominant species in near-shore waters and atmospheric indexes that reflect climate changes, such as the Antarctic Oscillation—AAO (e.g. dominance of diatoms during negative AAO polarities and dinoflagellates during AAO positive polarities) suggests that phytoplankton may be the most immediate biological indicator of climate change in this area (Lange 2011). However, the chemical parameters that assist phytoplankton-ice coupling need further investigation, as well as the implications of phytoplankton shifts (and consequently its cell size and nutritional content) in the food web.

Even though the first phytoplankton sampling in Admiralty Bay was carried out during the 'Pourquoi-Pas?' Expedition in 1909, only after the 1970s studies on these organisms advanced considerably, mainly do to Polish and Brazilian scientists (e.g. Rakusa-Suszczewski 1980; Rakusa-Suszczewski et al. 1993; Ligowski and Kopczÿnska 1993; Brandini and Rebello 1994; Lange et al. 2007; Tenenbaum et al. 2011; Tenório et al. 2011). But observations of phytoplankton biomass, species distributions and primary production are still needed to relate environmental variability to biological activity (Rintoul et al. 2012).

It has been observed that shelf seafloor is nourished by fresh organic matter not only from phytoplankton, but also from algae that grow in sea ice, as these become freely available upon ice melting during summer (Hofmann et al. 2004). So far, 140 species of phytoplankton (mainly diatoms and a few dinoflagellates and silicoflagellates) have been identified (Tenenbaum et al. 2011). Amongst the diatoms 21 genera were planktonic and 22 benthic (epiphytic, epilithic, episamic and eponthic), all found in the water column, supporting hypotheses related to resuspension and/or resulting from ice melt (Tenenbaum et al. 2011). It is important to highlight that, usually, blooms that occur in the near-shore zone do not reach the bay interior (Kopczynska 2008). This supports the theory that there is a front between coastal waters subjected to terrestrial and sea-ice melt, where benthic phytoplankton species are more frequent, and central waters derived from the Bransfield Strait, where centric pelagic diatoms are dominant (Lange 2011). And this is also possibly related to the circulation within the bay (Fig. 8.4).

Molecular ecology assessments of Bacteria and Archaea domains showed high diversity with the presence of cosmopolitan taxonomic groups in shallow sediments (up to 20 m), but also detected groups of archaea that have so far only been described in Antarctica (Nakayama et al. 2011a). Differences in the structure of archaeal communities along a 100–1100 m depth gradient were detected by denaturing gradient gel electrophoresis (DGGE), suggesting that communities at 700 and 1100 m were clearly distinct from those in shallower sediments (unpublished data). From a different perspective, faecal pollution bacteria *Escherichia coli* and *Clostridium perfringens* were found mainly near the sewage outfall off Ferraz. *C. perfringens*, from anthropic or animal origin, was detected also in Botany and Ullman points (Nakayama et al. 2011b). A hypothesis has been raised regarding the potential wind induced circulation influence of Ferraz towards those areas based on atmospheric data by Sodré (2011), showing the dominant gas emission plume from the station over the bay. This led us to believe that the circulation processes were possibly underestimated in benthic-community-structure analyses.

In general, natural processes, such as anchor ice and ice scour, are still the main cause for environmental changes in the marine system of Admiralty Bay, despite intense human activities (Sicinski et al. 2011). But studies that require

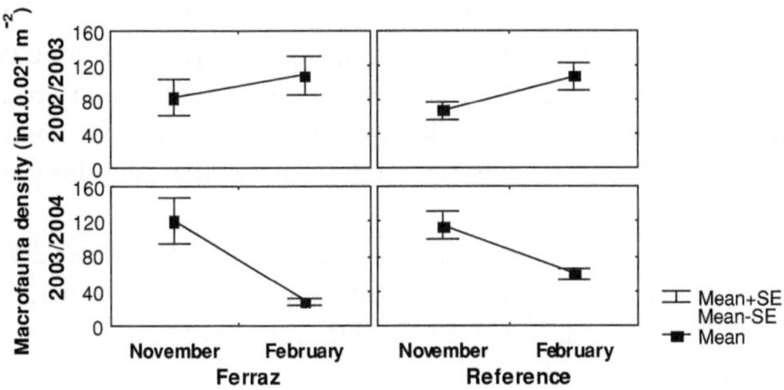

Fig. 8.7 Admiralty Bay macrofaunal spatio-temporal study near Ferraz Staion and other four reference areas (Austral Summers 2002/03 and 2003/04). Macrofaunal density variation "within year" and "between years" comparing two austral summers

Table 8.2 Nested ANOVA results of macrofauna near Ferraz to detect if "within year" variability was greater than "between years" variability; F-ratio based on density differences

	Years	Month (Year)	Area	Area * Year	Area * Month (Year)
20 m depth					
Polychaeta	1.70	3.05	1.54	**2.83***	0.91
Oligochaeta	2.26	**5.98****	**4.58****	1.14	**2.71***
Amphipoda	0.51	2.84	1.73	0.91	0.86
Cumacea	0.01	**7.51****	**2.63***	**2.61***	**2.85***
Bivalvia	0.07	**5.10****	**5.17****	2.5	0.73
Total number	2.88	**10.28*****	**2.56***	2.07	0.96
30 m depth					
Polychaeta	0.81	3.09	**9.40*****	**3.41****	1.03
Oligochaeta	0.01	**3.51***	3.17*	0.79	0.53
Amphipoda	0.24	**3.48***	2.21	**2.68***	1.06
Cumacea	3.04	**6.38****	**3.96****	**3.00***	**3.30****
Bivalvia	2.30	1.49	0.76	0.72	0.95
Total number	0.05	**5.82****	**3.28***	1.61	0.83
60 m depth					
Polychaeta	0.75	**3.32***	**5.84*****	0.97	1.82
Oligochaeta	1.86	**5.55****	**5.67*****	1.49	1.39
Amphipoda	0.76	**7.09****	**13.91*****	**7.81*****	**3.37*****
Cumacea	**8.61****	1.46	**3.54****	1.57	0.79
Bivalvia	1.11	1.04	**3.89****	0.59	0.89
Total number	0.41	**7.41****	**6.27*****	1.49	0.98

(*$p < 0.05$, **$p < 0.01$, ***$p < 0.001$)

time series should improve our understanding of the ecosystem processes within the bay. Evaluation on benthic communities during the austral summers 2002/03 and 2003/04 has shown that intra-annual differences in densities of benthic macrofauna could be more significant than spatial variation or inter-annual changes (Fig. 8.7, Table 8.2), and this was possibly related to short-time differences in water column production during summer. But inter-annual differences can also occur in some benthic groups, suggesting that benthic communities can also respond to inter-annual differences in the water-column production. The latter was higher in 2002/03 (the whole bay was frozen in previous winter) than in 2003/04 (Lange et al. 2007). Further studies in similar sampling stations supported this hypothesis.

The meiofauna was also investigated during the austral summer 2004/05 at two sites in Martel Inlet (Ferraz and Botany). Densities varied between 1,569 ± 928 to 16,245 ± 12,282 ind 10 cm^{-2} (mean ± SD) from 20 to 60 m, decreasing with increasing depth (Gheller 2007).

Intra-annual differences (decrease from Nov–Dec to Jan–Feb) were also observed at 20–30 m depth. The highest densities at the beginning of summer were probably due to intense deposition of sea-ice algae and phytoplankton blooms after ice break: an important food source for benthic organisms.

Often, the dominant megabenthic-community structure at 20–30 m has been described in relation to substrate types (see Sicinski et al. 2011). During IPY (2008 austral summer), the imaging of benthic megafauna was pursued in 14 sampling stations using a tow-camera (AVTECH Super-mini underwater TV camera, Model FM-1000) connected to an onboard computer (Fig. 8.1).

Three imaging transects were undertaken for five minutes in each station from 20 to 30 m. Transects were geo-referenced and films from each transect were later analyzed. Each frame was frozen and the occurrence of the identifiable megafauna recorded, but all frame records for each station were randomly sorted. Accounts of different organisms were arranged in the following trophic groups: producers (macroalgae); suspension feeders (Porifera, Cnidaria, Bivalvia, Crinoidea, Ascidiacea); deposit feeders and omnivores (Polychaeta, Echinoidea, Holothuroidea, Ophiuroidea); carnivores and scavengers [nemertean (*Parborlasia corrugatus*), gastropod (*Neobuccinum eatoni*), Crustacea, Asteroidea]. The frequency of occurrence and abundance of these trophic groups were used to compare with wave dynamics.

The analyses of 1,418 frames showed that some areas have similarities in the occurrence of suspension feeders such as sponges, ascidians and bivalves, but also macroalgae, mainly *Desmarestia* sp. and *Himantothalus* sp. and omnivores (Wanda and Dobrowolski Glaciers, Stenhouse, Yellow and Ullman Points). Primary producers such as macroalgae were found in all stations, but had higher frequency of occurrence in areas with higher wave turbulence (Plaza Point, Goetel Glacier, Botany, Hennequin and Ullman Point, and Wanda Glacier). Carnivores and scavengers (e.g. *Odontaster validus*, *Parborlasia corrugatus*) were more frequent at the inner parts of Mackelar and Martel Inlet, especially at Refuge 2 and Yellow Point respectively. Deposit feeders and omnivores such as polychaetes were more frequent in areas with medium–low wave intensity (e.g., Dobrowolski Glacier, Ullman and Botany Point, Krak, Goetel and Stenhouse glaciers). In contrast, ophiuroids (mainly *Ophionotus victoriae*) were more frequent at more exposed areas, where also suspension feeders appear to be dominant.

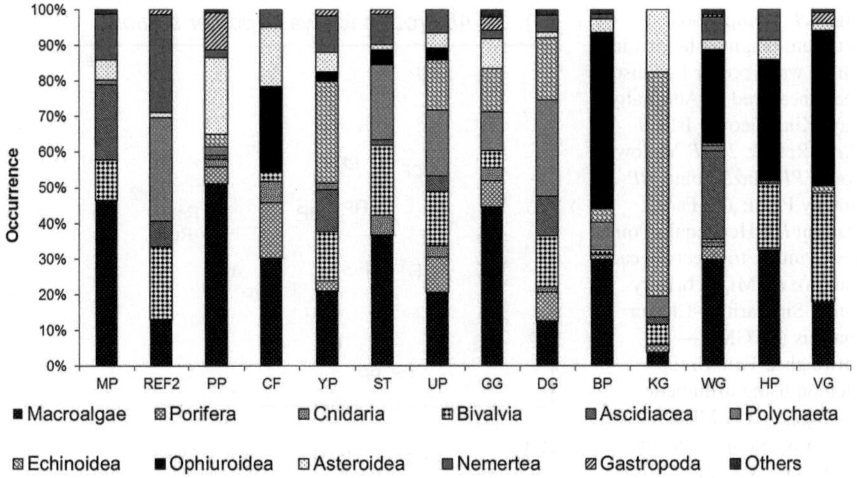

Fig. 8.8 Admiralty Bay, King George Island. Sampling stations used for taking images of the megafauna and macroalgae down to 30 m, frequency of occurrence of main taxa in each imaged sampling station. *MP* Machu Picchu; *REF2* Refuge 2; *PP* Plaza Point; *EACF* Ferraz Station; *YP* Yellow Point; *ST* Stenhouse Glacier; *UP* Ullman Point; *GG* Goetel Glacier; *DG* Dobrowolski Glacier; *BP* Botany Point; *KG* Krak Glacier; *WG* Wanda Glacier; *HP* Hennequin Point; *VG* Vieville Glacier

Specimens of the gastropod *Nacella concinna* were observed mainly in Plaza Point, but only a few organisms were observed in the subtidal zone. Conversely, this species is abundant in the tidal zone and it is an important food item for *Larus dominicanus* (Favero et al. 1997; Figueiredo & Lavrado 2011 Figueiredo 2012). The density of megabenthos near Stenhouse was low, possibly related to intense ice falling from the glacier. The images revealed a clear heterogeneity in the dominance of megabenthic organisms at different sampling stations (Fig. 8.8).

Stations with lower wave power (Yellow Point and Refuge 2) were clearly distinct from those with higher wave power (Plaza Point, and this from Ferraz Station, Hennequin and Botany Points). These differences are influenced not only by substrate types, as some stations have similar substrates (e.g. Ferraz and Yellow Point), but also dominant patterns of wave dynamics (Figs. 8.6, 8.9). Besides these factors, blocks of ice likewise considerably affect the benthic community structure through scour and anchor ice.

A short-term variation study near shore during the 2008/09 summer showed that some macrofaunal groups with swimming capabilities, e.g. Amphipoda (Crustacea) and other small organisms living in the uppermost sediment layer, e.g. *Mysella charcoti* (Bivalvia), exhibited significant changes along with grain size distribution in the same sampling station, correlated to storm and wind intensities causing turbulence in the coast in front of Ferraz in Martel Inlet (Monteiro 2010).

The Antarctic benthic fauna generally shows high diversity, high biomass and abundance of organisms, although species spatial distributions vary among distinct taxonomic groups (Clarke 2008). However, only a few studies have provided distribution information on a wide depth range (e.g. Brandt et al. 2004). Sicinski

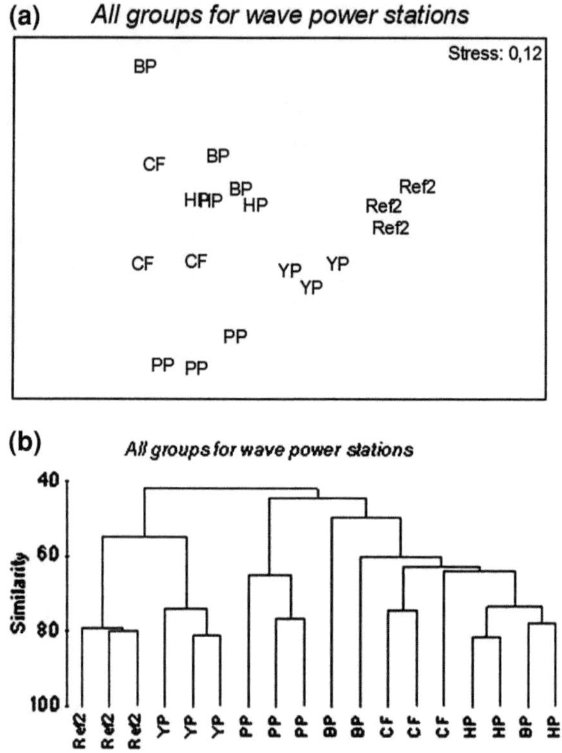

Fig. 8.9 Comparison between imaging stations in which wave power had also been measured at Admiralty Bay, King George Island (*Ref2* Refuge 2; *YP* Yellow Point; *PP* Plaza Point; *BP* Botany Point; *CF* Ferraz Station; *HP* Hennequin Point, n = 3 image transects in each station): **a** nMDS; **b** Bray-Curtis Similarity—Cluster analysis (UPGMA—Unweighed Pair-Group Method using arithmetic Averages— PRIMER V.6). ANOSIM Sample statistic (Global R): 0.843; Significance level of sample statistic: 0.1 %pp

et al. (2011) described the distribution found within the coast at Martel Inlet and mainly from Ezcurra Inlet to the central basin of the bay. We investigated if there was a clear zonation as well or not from Martel to the central basin.

The densities of meiofauna varied between 4,464 ± 1,972 and 6,682 ± 464 ind 10 cm^{-2}, and were high compared to other oceans (see Soltwedel 2000). Nematodes were more than 70 % of the organisms, followed by nauplii (9 %) and copepods (8 %). Echinodermata (28 %), Polychaeta (14 %), Ascidiacea (11 %), and Crustacea (10 %) were the most frequent megafauna groups in the deepest zones of the bay.

The density of meiofauna (Nematoda, Nauplii, Copepoda among others) and macrofauna (*stricto* sensu taking into account the main taxa, e.g. Crustacea, Mollusca and Annelida) did not show any differences along a depth gradient from Plaza Point to the central basin towards the opening of the bay. Also, the benthic megafauna did not show significant differences within the same depth range. Considering that the Antarctic continental shelf is deep (from about 450 m to more than 1000 m in some areas), some species have evolved tolerance to high depth, low and variable levels of food, and particularly low levels of energy for maintenance, as temperature is low (Clarke and Johnston 2003; Kaiser et al. 2011).

This may also explain our results, and the fact that sampling followed a transect over fine to very fine sediments (Figs. 8.1, 8.6), unlike zonation reported by Sicinski et al. (2011). These authors have provided an Antarctic benchmark study for Admiralty Bay, gathering 30 years of benthic data and recording over 1,300 marine species. But, even though their study showed indication of biotic temporal variation possibly related to climate change, here we have had indication that there is still much need for integrating available multidisciplinary information, including all trophic levels and marine realms, into ecological models that could appropriately predict the effect of changes on biodiversity within Admiralty Bay.

8.3 Concluding Remarks

The dominant basic energy flow in the Southern Ocean is the surface phytoplankton production followed by secondary consumption by the zooplankton (krill being its central player), tertiary consumption by pelagic organisms (e.g. squid, fish) and other predators such as seabirds and mammals (Hofmann et al. 2004, 2008). The decomposing organic matter flows to the seafloor and enter the benthic microbial loops and trophic web (Smith et al. 2006, 2008). At Admiralty Bay a few trophic observations have been done in the coastal zone (e.g. Corbisier et al. 2004, 2010). Usually, the most productive areas are within or associated to the zones of sea-ice formation, which cover most shallow areas around Antarctica and some deep oceanic areas each winter (Hofmann et al. 2008, 2011). This includes Admiralty Bay at King George Island. The Atmospheric Antarctic Oscillation (AAO) is tightly related to ice formation, sea-ice extent, ice melt, cloud cover (and incident light), wind speed and air temperature in the west of the Antarctic Peninsula (Liu et al. 2004), and these affect phytoplankton production, which can be considered the most immediate climate-change biological indicator in the bay.

The most non-linear process affecting waves in coastal waters is depth-induced breaking, which is poorly understood. It involves highly nonlinear hydrodynamics on a wide range of scales, from capillary to gravity waves, down to turbulence and wave-current interactions (Holthuijsen 2007). Despite these difficulties, we can predict that energy dissipation should be related to the power transmitted, and somehow this affects the sediment and nutrient resuspension at the bottom boundary layer. The circulation model and information on wave patterns may elucidate the physical influence on the biota distributions. Currents and waves partially explain the mobilization of nutrients through resuspension and consequently this may also influence the structuring of the benthic communities.

The environmental sensitivity of Admiralty Bay is particularly high, due to its location within the Antarctic Peninsula region, and has had potential for high risk of direct and indirect anthropogenic impacts besides climate change (Campos et al. 2012). Long-term multidisciplinary data are beginning to become available in the area and this will help in the understanding of cause-effect processes from global and human-induced changes in the local terrestrial and marine biodiversity. If integrated, these data may provide useful tools for modelling, detecting and predicting future changes, not only in Admiralty Bay, but possibly elsewhere in Antarctica, considering common features within the Southern Ocean.

Acknowledgments Data have been gathered under the CAML umbrella and in projects funded by CNPq IPY Project MABIREH (Process No. 52.0293/2006-1), the Brazilian National Institute of Science and Technology on Antarctic Environmental Research (INCT-APA) funded through CNPq (Process No. 574018/2008-5), FAPERJ (Process No. E-26170.023/2008), and CNPq Project Southern Ocean Benthic Observing Systems (SOBE, Process No. 557126/2009-6). Sampling and data integration have been supported by the Ministry of Science, Technology and Innovation (MCTI), Ministry of Environment (MMA), and Interministerial Secretariat for the Brazilian Marine Resources (SECIRM). We thank all the personnel involved in this work. The order of authors is alphabetical, except for the first author. Anonymous reviewers are thanked for their useful comments and contribution to this chapter. This is an EBA contribution under Workpackage 4.

References

Acevedo J, Rasmussen K, Felix F, Castro C, Llano M, Secchi E R, Saborio Mt, Aguayo-Lobo A, Haase B, Scheidat M, Dalla Rosa L, Olavarria C, Forestel P, Acuna P (2007) Migratory destinations of humpback whales from the Magellan Strait feeding ground, Southeast Pacific. Mar Mamm Sci 23:453–463

Aguayo-Lobo A, Acevedo Jr, Brito JL, Acuña PG, Bassoi M, Secchi ER, Dalla Rosa L (2011) Presence of the leopard seal, *Hydrurga leptonyx* (De Blainville, 1820), on the coast of Chile: an example of the Antarctica—South America connection in the marine environment. Oecol Aust 15(1):69–85. doi:10.4257/oeco.2011.1501.07

Albuquerque MA, Pellizari V, Schaefer CEGR, Luz AP, Corrêa DM (2004) Ecologia microbiana dos solos da Antártica Marítima. In: Schaefer CEGR, Francelino MR, Simas FNB, Filho MRA (eds) Ecossistemas costeiros e monitoramento ambiental da Antártica Marítima, Baía do Almirantado. Ilha Rei George, NEPUT, pp 73–89

Allcock AL, Barratt I, Eléaume M, Linse K, Norman MD, Smith PJ, Steinke D, Stevens DW, Strugnell JM (2011) Cryptic speciation and the circumpolarity debate: a case study on endemic Southern Ocean octopuses using the COI barcode of life. Deep-Sea Res II 58:242–249

Aronson RB, Thatje S, Clarke A, Peck LS, Blake DB, Wilga CD, Seibel BA (2007) Climate change and invasibility of the Antarctic benthos. Ann Rev Ecol Evol Systemat 38:129–154

Bageston JV, Wrasse CM, Gobbi D, Tahakashi H, Souza P (2009) Observation of mesospheric gravity waves at Comandante Ferraz Antarctic Station (62°S). Ann Geophys 27:2593–2598

Barker PF, Burrell J (1982) The influence on Southern Ocean circulation, sedimentation and climate of the opening of Drake Passage. In: Craddock C (ed) Antarctic Geoscience. University Wisconsin Press, Madison, pp 377–385

Barker PF, Thomas E (2004) Origin, signature and palaeoclimatic influence of the Antarctic circumpolar current. Earth Sci Rev 66:143–162

Basterretxca G, Arístegui J (1999) Phytoplankton biomass and production during austral spring (1991 and summer (1993) in the Bransfield Strait. Polar Biol 21:11–22

Brandini FP (1993) Phytoplankton biomass in an Antarctic coastal environment during stable water conditions—implications for the iron limitation theory. Mar Ecol Prog Ser 93:267–275

Brandini FP, Rebello J (1994) Wind field effect on hydrography and chlorophyll dynamics in the coastal pelagial of Admiralty Bay, King George Island Antarctica. Antarct Sci 6(4):433–442

Brandt A, De Broyer C, De Mesel I, Ellingsen KE, Gooday AJ, Hilbig B, Linse K, Thomson MRA, Tyler PA (2007) The biodiversity of the deep Southern Ocean benthos. Phil Trans R Soc B 362:39–66. doi:10.1098/rstb.2006.1952

Brandt A, De Broyer C, Gooday AJ, Hilbig B, Thomson MRA (2004) Introduction to ANDEEP (ANtarctic Benthic DEEP-sea biodiversity: colonization history and recent community patterns)—a tribute to Howard L Sanders. Deep-Sea Res II 51:1457–1465

Braun M, Rau F, Simões JC (2001) A GIS-based glacier inventory for the Antarctic Peninsula and the South Shetland Islands: a first case study on King George Island. Geo-spatial Inf Sci 4:15–24

Campos LS, Montone RC, Moura RB, Yoneshigue-Valentin Y, Kawall HG, Convey P (2012) Anthropogenic impacts on sub-Antarctic, Antarctic islands and the adjacent marine environment. In: Verde C, di Prisco G (eds) Adaptation and evolution in marine environments—the impacts of global change on biodiversity, Volume 2. Springer, Berlin

Catewicz Z, Kowalik Z (1983) Harmonic analysis of tides in Admiralty Bay. Oceanol 15:97–109

Chisholm SW (2000) Oceanography: stirring times in the Southern Ocean. Nature 407:685687. doi:10.1038/35037696

Clarke A (2008) Antarctic marine benthic diversity: patterns and processes. J Exp Mar Biol Ecol 366(1):48–55

Clarke A, Johnston NM (2003) Antarctic marine benthic diversity. Oceanogr Mar Biol 41:47–114

Clarke A, Barnes DKA, Hodgson DA (2005) How isolated is Antarctica? Trends Ecol Evol 20(1):1–3. doi:10.1016/j.tree.2004.10.004

Convey P (2010) Terrestrial biodiversity in Antartica: recent advances and future challenges. Polar Sci 4:135–147

Corbisier TN, Bromberg S, Petti MAV (2010) Admiralty bay trophic relationships: a summary of results from isotopic analysis. In: Yoneshigue-Valentin Y, Dalto AG, Lavrado HP, de Carvalho ALPS (eds) Annual activities report 2009—INCT-APA. Editora Cubo, São Carlos, pp 65–67

Corbisier TN, Petti MAV, Skowronski RSP, Brito TAS (2004) Trophic relationships in the nearshore zone of Martel Inlet (King George Island, Antarctica): $\delta^{13}C$ stable-isotope analysis. Polar Biol 27(2):75–82

Correia E (2011) Study of Antarctic-South America connectivity from ionospheric radio soundings. Oecol Aust 15(1):32–39. doi:10.4257/oeco.2011.1501.03

Correia E, Raulin JP, Kufmann P, Bertoni FCP, Moro J (2011) Monitoring of atmospheric changes related to Sun-Earth interactions. In: Yoneshigue-Valentin Y, Dalto AG, Lavrado HP (eds) Annual activities report 2010—INCT-APA. Editora Cubo, Rio de Janeiro, pp 20–26

Costa ES, Alves MAS (2007) Biologia reprodutiva e ecologia comportamental de skuas Antárticas *Catharacta maccormicki* e *C. lonnbergi*. Oecol Bras 11:78–94

Costa ES, Alves MAS (2008) The breeding birds of Hennequin point: an ice free area of Admiralty Bay (Antarctic Specially Managed Area), King George Island, Antarctica. Rev Bras Ornitol 16:137–141

Costa LM, Francelino MR, Schaefer CEGR, Dias LE, Borges Jr M, Mendonça ES (2004) Química das águas de degelo na Ilha Rei George, Antártica. In: Schaefer CEGR, Francelino MR, Simas FNB, Filho MRA (eds) Ecosistemas costeiros e monitoramento ambiental da Antártica Marítima, Baía do Almirantado, Ilha Rei George. NEPUT, Viçosa, pp 91–94

Dalla Rosa L, Secchi ER, Maia YG, Zerbini NA, Heide-Jørgensen MP (2008) Movements of satellite-monitored humpback whales on their feeding ground along the Antarctic Peninsula. Polar Biol 31:771–781. doi:10.1007/s00300-008-0415-2

Duarte RTD (2010) Micro-organismos em ambientes criogênicos: gelo glacial, solos expostos por recuo de geleiras, e permafrost polares. Dissertation, Universidade de São Paulo. http://www.teses.usp.br/teses/disponiveis/87/87131/tde-07102010-153139/

Favero M, Silva P, Ferreyra G (1997) Trophic relationships between the kelp gull and the Antarctic limpet at King George Island (South Shetland Islands, Antarctica) during the breeding season. Polar Biol 17:431–436

Figueiredo MIS (2012) Biologia populacional de *Nacella concinna* (Strebel, 1908) (Gastropoda: Nacellidae) na Baía do Almirantado, Ilha Rei George, Antártica. Dissertation, Museu Nacional da Universidade Federal do Rio de Janeiro 109 pp

Figueiredo MIS, Lavrado HP (2011) Aspects of the population structure of *Nacella concinna* (Strebel, 1908) (Gastropoda: Nacellidae) at Admiralty Bay, King George Island, Antarctica. In: Yoneshigue-Valentin Y, Dalto AG, Lavrado HP (eds) Annual activity report 2010—INCT-APA. Editora Cubo, Rio de Janeiro, pp 167–170

Gheller PF (2007) A meiofauna e os Nematoda da enseada Martel (Antártica) e seu uso em monitoramento ambiental. Dissertation, Universidade de São Paulo, São Paulo 103 pp

Gonçalves VN, Vaz ABM, Rosa CA, Rosa LH (2012) Diversity and distribution of fungal communities in lakes of Antarctica. FEMS Microbiol Ecol (In press). DOI 10.1111/j.1574-6941.2012.01424.x

Gordon AL, Nowlin WDJ (1978) The basin waters of the Bransfield Strait. J Phys Oceanogr 8:258–264

Griffiths HJ (2010) Antarctic marine biodiversity—what do we know about the distribution of life in the Southern Ocean? PLoS ONE 5(8):1–11

Griffiths HJ, Barnes DKA, Linse K (2009) Towards a generalised biogeography of the Southern Ocean benthos. J Biogeog 36:162–177. doi:10.1111/j.1365-2699.2008.01979.x

Griffiths HJ, Linse K, Barnes DKA (2008) Distribution of macrobenthic taxa across the Scotia Arc, Antarctica. Antarct Sci 20(3):213–226. doi:10.1017/S0954102008001168

Gruber NLS (1990) Contribuição ao estudo da sedimentação glacio-marinha da Baıa do Almirantado, ilha Rei George, Antártica. Dissertation, Universidade Federal do Rio Grande do Sul, Porto Alegre

Gutt J, Barratt I, Domack EC, D'acoz CU, Dimmler W, Gre′Mare A, Heilmayer O, Isla E, Janussen D, Jorgensen E, Kock K-H, Lehnert LS, López-Gonzales P, Langner S, Linse K, Manjón-Cabeza ME, Meißner M, Montiel A, Raes M, Robert H, Rose A, Schepisi ES, Saucède T, Scheidat M, Schenke H-W, Seiler J, Smith C (2011) Biodiversity change after climate-induced ice-shelf collapse in the Antarctic. Deep-Sea Res II 58(1–2):74–83

Hodgson DA, Convey P, Verleyen E, Vyverman W, McInnes SJ, Sands CJ, Fernández-Carazo R, Wilmotte A, De Wever A, Peeters C, Tavernier I, Willems A (2010) The limnology and biology of the Dufek Massif, Transantartic Mountais. Polar Sci 4:197–214

Hofmann EE, Wiebe PH, Costa DP, Torres JJ (2004) Integrated ecosystem studies of Western Antarctic Peninsula continental shelf waters and related Southern Ocean regions. Deep-Sea Res II 51(17–19):1921–2344

Hofmann EE, Wiebe PH, Costa DP, Torres JJ (2008) Dynamics of plankton, krill, and predators in relation to environmental features of the western Antarctic Peninsula and related areas: SO GLOBEC Part II. Deep-Sea Res II 55(3–4):269–558

Hofmann EE, Wiebe PH, Costa DP, Torres JJ (2011) Introduction to understanding the linkages between Antarctic food webs and the environment: a synthesis of Southern Ocean GLOBEC studies. Deep Sea Res II 58(13–16):1505–1507

Holthuijsen LH (2007) Waves in oceanic and coastal waters. Cambridge University Press, Cambridge

Kaiser S, Griffiths HJ, Barnes DKA, Brandão SN, Brandt A, O'Brien PE (2011) Is there a distinct continental slope fauna in the Antarctic? Deep-Sea Res II 58:91–104

Kopczynska EE (2008) Phytoplankton variability in Admiralty Bay, King George Island, South Shetland Islands: six years of monitoring. Polar Res (29)2:117–139

Krüger L, Petry MV (2011) On the relation of Antarctic and subantarctic seabirds with abiotic variables of South and Southeast Brazil. Oecol Aust 15(1):40–50

Kruszewski G (1999) Zlodzenie Zatoki Admiralicji w latach 1977–1996. Problemy Klimatologii Polarnej 9:173–191

Lange PK (2011) O fitoplâncton da baía do Almirantado (Antártica): variações temporais e espaciais entre os anos de 2002 e 2009. Dissertation, Universidade Federal do Rio Grande, 125 pp

Lange PK, Tenenbaum DR, Braga ES, Campos LS (2007) Microphytoplankton assemblages in shallow waters at Admiralty Bay (King George Island, Antarctica) during the summer 2002–2003. Polar Biol 30:1483–1492

Lemos LN, Suleiman AKA, Pereira AB, Roesch LFW (2011) Global patterns in soil bacterial community composition across a continental scale. In: Yoneshigue-Valentin Y, Dalto AG, Lavrado HP (eds) Annual activity report 2010—INCT-APA. Editora Cubo, Rio de Janeiro, pp 63–67

Ligowski R, Kopczÿnska EE (1993) Phytoplankton. In: Rakusa-Suszczewski S (ed) The maritime Antarctic coastal ecosystem of Admiralty Bay. Polskiej Akademii Nauk, Warsaw, pp 45–48

Linse K, Griffiths HJ, Barnes DKA, Clarke A (2006) Biodiversity and biogeography of Antarctic and sub-Antarctic Mollusca. Deep-Sea Res 53:985–1008

Lipski M (1987) Variations of physical conditions, nutrients and chlorophyll a contents in Admiralty Bay (King George Island, South Shetland Islands). Polar Res 8:307–332

Liu J, Curry JA, Martinson DG (2004) Interpretation of recent Antarctic sea ice variability. Geophys Res Lett 31:L02205. doi:10.1029/2003GL018732

Maciel MS, Rangel TP, Brito FP, Pimentel DC, Gobo RR, Gobo AAR, Campos LS, Lavrado HP, Rezende CE (2006) Concentração de metais pesados nos sedimentos costeiros na baía do Almirantado—Península Antártica. In: I Congresso Brasileiro de Biologia Marinha, 2006, Niterói. I Congresso Brasileiro de Biologia Marinha 2:163–164

Marani L, Alvalá PC (2011) Monitoring greenhouse gases in Comandante Ferraz Antarctic station, King George Island. In: Yoneshigue-Valentin Y, Dalto AG, Lavrado HP (eds) Annual activity report 2010—INCT-APA. Editora Cubo, Rio de Janeiro, pp 44–47

Mariano EVC, Leme NMP, Alvalá PC (2011) Atmospheric SO_2 measurements at the Brazilian Antarctic station. In: Yoneshigue-Valentin Y, Dalto AG, Lavrado HP (eds) Annual activity report 2010—INCT-APA. Editora Cubo, Rio de Janeiro, pp 38–43

Martins CC, Bícego MC, Rose NL, Taniguchi S, Lourenço RA, Figueira RC, Mahiques MM, Montone RC (2010) Historical record of polycyclic aromatic hydrocarbons (PAHs) and spheroidal carbonaceous particles (SCPs) in marine sediments from Admiralty Bay, King George Island Antarctica. Env Pollut 158(1):192–200

Martins CC, Montone RC, Gamba RC, Pellizari VH (2005) Sterols and fecal indicator microorganisms in sediments from Admiralty Bay, Antarctica. Bras J Oceanogr 53:1–12

Medlin LK, Priddle J (1990) Polar marine diatoms. British Antarctic Survey. Natural Environmental Council, Cambridge

Monteiro, GSC (2010) Variação temporal de pequena escala da macrofauna bentônica da zona costeira rasa da enseada Martel (Baía do Almirantado, Antártica), com ênfase em Annelida Polychaeta. Dissertation, Universidade de São Paulo

Montone RC, Campos LS, Alvarez CE, Ito RG, Lavrado HP, Bícego MC, Pellizari VH, Schaefer CEGR, Corbisier TN, Sander M, Pereira AB, Gomes V, Carelli RF, Braga ES, Ngan PV, Mahiques MM, Castro BM, Petti MAV, Nakayama CR, Martins CC, Francelino M, Weber RR (2012) Environmental assessment of Admiralty Bay, King George Island, Antarctica. In: Verde C, di Prisco G (eds) Adaptation and evolution in marine environments—the impacts of global change on biodiversity, Vol 2. Springer, Berlin

Montone RC, Martins CC, Bícego MC, Taniguchi S, Silva DAM, Campos LS, Weber RR (2010) Distribution of sewage input in marine sediments around a maritime Antarctic research station indicated by molecular geochemical indicators. Sci Total Env 408:4665–4671

Moura RB (2009) Estudo taxonômico dos Holothuroidea (Echinodermata) das Ilhas Shetland do Sul e do Estreito de Bransfield, Antártica. Dissertation, Universidade Federal do Rio de Janeiro

Nakayama CR, Kuhn E, Araújo ACV, Alvalá PC, Ferreira WJ, Vazoller RF, Pellizari VH (2011a) Revealing archaeal diversity patterns and methane fluxes in Admiralty Bay, King George Island, and their association to Brazilian Antarctic Station activities. Deep-Sea Res II 58:128–138

Nakayama CR, Ushimaru PI, Lima DV, Pellizari VH (2011b) Occurrence of microbial faecal pollution indicators in sediment and water samples at Admiralty Bay, King George Island, Antarctica. In: Annual activities report (2010) of National Institute of Science and Tecnology on Antarctic Environmental Research. São Carlos: Editora Cubo, pp 162–166

Pearse JA, Leat PT, Barker PF, Millar IL (2001) Geochemical tracing of Pacific-to-Atlantic upper-mantle flow through Drake passage. Nature 410:457–461

Peck LS, Convey P, Barnes DKA (2006) Environmental constraints on life histories in Antarctic ecosystems: tempos, timings and predictability. Biol Rev Camb Philos Soc 81(1):75–109

Pinheiro DK, Leme NP, Peres LV, Kall E (2011) Influence of the Antarctic ozone hole over the south Brazil in 2008 and 2009. In: Yoneshigue-Valentin Y, Dalto AG, Lavrado HP (eds) Annual activity report 2010—INCT-APA. Editora Cubo, Rio de Janeiro, pp 33–37

Pörtner HO, Peck L, Somero G (2007) Thermal limits and adaptation in marine Antarctic ectotherms: an integrative view. Phil Trans R Soc Lond B: Biol Sci 362:2233–2258
Pruszak Z (1980) Currents circulation in the waters of Admiralty Bay (region of Arctowski Station on King George Island). Polar Res 1:55–74
Pugh PJA, Convey P (2008) Surviving out in the cold: Antarctic endemic invertebrates and their refugia. J Biogeogr 35:2176–2186
Rakusa-Suszczewski S (1980) Environmental conditions and the functioning of Admiralty Bay (South Shetland Islands) as part of the near shore Antarctic ecosystem. Polar Res 1(1):11–27
Rakusa-Suszczewski S (1993) Matter transport and budgets in the Admiralty Bay area. In: Rakusa-Suszczewski S (ed) The Antarctic coastal ecosystem of Admiralty Bay. Polish Academy of Sciences, Warsaw, pp 199–212
Rakusa-Suszczewski S, Mietus M, Piasecki J (1993) Weather and climate. In: Rakusa-Suszczewski S (ed) The maritime Antarctic coastal ecosystem of Admiralty Bay, Department of Antarctic Biology, Polish Academy of Sciences, Warsaw, pp 19–25
Rintoul SR, Sparrow M, Meredith MP, Wadley V, Speer K, Hofmann E, Summerhayes C, Urban E, Bellerby R (2012) The Southern Ocean observing system: initial science and implementation strategy. ISBN 978-0-948277-27-6
Robakiewicz M, Rakusa-Suszczweski S (1999) Application of 3D circulation model to Admiralty Bay, King George Island Antarctica. Polar Res 20(1):43–58
Rodrigues M, Mahiques MM, Souza LAP, Faria EO, Tolentino CV (2003) Caracterização da superfície de fundo da Enseada Martel (Ilha Rei George, Antártica) por meio do Sonar de Varredura Lateral. In: Congresso Internacional da Sociedade Brasileira de Geofísica—SBGF, 8. Rio de Rede 2 Gerenciamento ambiental na baía do Almirantado, ilha Rei George, Antártica—239—Janeiro. Boletim de Resumos Expandidos CD-ROM
Rogers AD, Johnston NM, Murphy EJ, Clarke A (2012) Antarctic ecosystems: an extreme environment in a changing world. Phil Trans Royal Soc, Wiley (In press)
Rosa LH, Vaz ABM, Caligiorne RB, Campolina S, Rosa CA (2009) Endophytic fungi associated with the Antarctic grass *Deschampsia antarctica* Desv. (*Poaceae*). Polar Biol 32:161–167
Rosa LH, Vieira MLA, Santiago IF, Rosa CA (2010) Endophytic fungi community associated with the dicotyledonous plant *Colobanthus quitensis* (Kunth) Bartl. (*Caryophyllaceae*) in Antarctica. FEMS Microbiol Ecol 73:178–189
Rosado AS, Cury JC, Peixoto RS, Jesus HE, Schaefer CEGR, Bícego MC, Jurelevicius DA, Seldin L, Seabra PN, Greer CW (2011) Bioremediation, hydrocarbon depletion and microbial genetic diversity of Antarctic oil-polluted soil. In: Yoneshigue-Valentin Y, Dalto AG, Lavrado HP, de Carvalho ALPS (eds) Annual activity report 2009—INCT-APA. Editora Cubo, São Carlos, pp 207–210
Santiago IF, Alves TMA, Rabello A, Sales-Júnior PA, Romanha AJ, Zani CL, Rosa CA, Rosa LH (2012) Leishmanicidal and antitumoral activities of endophytic fungi associated with the Antarctic angiosperms *Deschampsia antarctica* Desv. and *Colobanthus quitensis* (Kunth) Bartl. Extremophiles 16:95–103
Santos IR, Favaro DI, Schaefer CE, Silva-Filho EV (2007) Sediment geochemistry in coastal maritime Antarctica (Admiralty Bay, King George Island): evidence from rare earths and other elements. Mar Chem 107:464–474
Santos IR, Silva-Filho EV, Schaefer CEGR, Albuquerque-Filho MR, Manoel R, Campos LS (2005) Heavy metal contamination in coastal sediments and soils near the Brazilian Antarctic station, King George Island. Mar Poll Bull, Holanda 50:185–194
Schaefer CEGR, Dias LE, Campos LS, Albuquerque Filho MR, Costa LM, Borges Júnior M (2004) Monitoramento ambiental em sedimentos costeiros da Baía do Almirantado: granulometria, teores de macronutrientes e metais biodisponíveis. In: Schaefer CEGR, Francelino MR, Simas FNB, Albuquerque Filho MR (eds) Ecossistemas costeiros e monitoramento ambiental da Antártica Marítima, Baía do Almirantado, Ilha Rei George. NEPUT Universidade Federal de Viçosa, Viçosa, pp 119–129

Schiaparelli S, Danis B, Wadley V, Stoddart DM (2012) The Census of Antarctic Marine Life (CAML): the first available baseline for Antarctic marine biodiversity. In: Verde C, di Prisco G (eds) Adaptation and evolution in marine environments—the impacts of global change on biodiversity, Vol 2. Springer, Berlin

Setzer AW, Oliveira MO, Francelino MR, Schaefer CEGR, Costa LM, Bremer UF (2004) Regime climático na Baía do Almirantado: relações com o ecossistema terrestre. In: Schaefer CEGR, Francelino MR, Simas FNB, Filho MRA (eds) Ecossistemas costeiros e monitoramento ambiental da Antártica Marítima, Baía do Almirantado, Ilha Rei George. Viçosa, NEPUT, pp 1–6

Setzer A, Villela FNJ, Dechiche AGP (2010) Antarctic meteorology. In: Yoneshigue-Valentin Y, Dalto AG, Lavrado HP, de Carvalho ALPS (eds) Annual activity report 2009—INCT-APA. Editora Cubo, São Carlos, pp 20–21

Siciński J, Jażdżewski K, De Broyer C, Ligowski R, Presler P, Nonato EF, Corbisier TN, Petti MAV, Brito TAS, Lavrado HP, Błażewicz-Paszkowycz M, Pabis K, Jażdżewska A, Campos LS (2011) Admiralty Bay Benthos diversity: a long-term census. Census of Antarctic marine life special volume. Deep-Sea Res II 58:30–48

Simões JC, Bremer UF, Aquino FE, Ferron FA (1999) Morphology and variations of glacial drainage basins in the King George Island ice field, Antarctica. Ann Glaciol 29:220–224

Smith CR, Mincks S, Demaster DJ (2006) A synthesis of bentho-pelagic coupling on the Antarctic shelf: Food banks, ecosystem inertia and global climate change. Deep-Sea Res II 53:875–894

Smith CR, Mincks S, Demaster DJ (2008) The FOODBANCS project: Introduction and sinking fluxes of organic carbon, chlorophyll-a and phytodetritus on the western Antarctic Peninsula continental shelf. Deep-Sea Res II 55:2404–2414

Sodré ED (2011) Impacto sobre a atmosfera e sobre a biota terrestre devido às emissões antrópicas na Baía do Almirantado/Ilha Rei George—Antártica. Dissertation, Universidade Estadual do Rio de Janeiro

Sodré ED, Evangelista H, Brito L, Corrêa SM (2011) Considering new parameters in the study of atmospheric impacts at Admiralty Bay. In: Yoneshigue-Valentin Y, Dalto AG, Lavrado HP (eds) Annual activity report 2010—INCT-APA. Editora Cubo, Rio de Janeiro, pp 48–53

Soltwedel T (2000) Metazoan meiobenthos along continental margins: a review. Prog Oceanogr 46:59–84

Taniguchi S, Montone RC, Bícego MC, Colabuono FI, Weber RR, Sericano JL (2009) Chlorinated pesticides, polychlorinated biphenyls and polycyclic aromatic hydrocarbons in the fat tissue of seabirds from King George Island Antarctica. Mar Pollut Bull 58(1):129–133

Teixeira LCRS, Peixoto RS, Cury JC, Sul WJ, Pellizari VH, Tiedje J, Rosado AS (2010) Bacterial diversity in rhizosphere soil from Antarctic vascular plants of Admiralty Bay, maritime Antarctica. ISME J 4:989–1001

Tenenbaum DR, Barrera-Alba JJ, Duarte RB, Tenório MMB (2011) Plankton structure in a shallow coastal zone at Admiralty Bay, King George Island, West Antarctic Peninsula: pico-, nano-, microphytoplankton, and chlorophyll biomass. In: Yoneshigue-Valentin Y, Dalto AG, Lavrado HP (eds) Ann activity report 2010—INCT-APA. Editora Cubo, Rio de Janeiro, pp 108–114

Tenório MMB, Duarte RB, Barrera-Alba JJ, Tenenbaum DR (2011) Plankton structure in a shallow coastal zone at Admiralty Bay, King George Island, West Antarctic Peninsula: chlorophyll biomass and size-fractioned chlorophyll during the austral summer 2009/2010. In: Yoneshigue-Valentin Y, Dalto AG, Lavrado HP (eds) Annual activity report 2010—INCT-APA. Editora Cubo, Rio de Janeiro, pp 115–120

Thomson MRA (2004) Geological and paleoenvironmental history of the Scotia Sea region as a basis for biological interpretation. Deep-Sea Res II 51:1467–1487

Tokarczyk R (1987) Classification of water masses in the Bransfield Strait and southern part of the Drake passage using a method of statistical multi-dimensional analysis. Polar Res 8:333–366

Trivelpiece WZ, Trievelpiece SG, Volkman NJ (1987) Ecological segregation of Adélie, Gentoo, and chinstrap penguins at King George Island. Ecol 68(2):351–361

Turner J, Bindschadler R, Convey P, di Prisco G, Fahrbach E, Gutt J, Hodgson D, Mayewski P, Summerhayes C (2009) Antarctic Climate Change and the Environment, a Contribution to the International Polar Year 2007–2008. SCAR, Cambridge 526 pp

Vaz ABM, Rosa LH, Vieira MLA, de Garcia V, Brandão LR, Teixeira LCRS, Moliné M, Libkind D, van Broock M, Rosa CA (2011) The diversity, extracellular enzymatic activities and photoprotective compounds of yeasts isolated in Antarctica. Braz J Microbiol 42:937–947

Yoneshigue-Valentin Y, Dalto AG, Lavrado HP (2011) Annual activity report 2010. National Institute of Science and Technology on Antarctic Environmental Research/Instituto Nacional de Ciência e Tecnologia Antártico de Pesquisas Ambientais (INCT-APA). Editora Cubo, Rio de Janeiro 240 pp

Chapter 9
Environmental Assessment of Admiralty Bay, King George Island, Antarctica

Rosalinda C. Montone, Cristina E. Alvarez, Márcia C. Bícego, Elisabete S. Braga, Tania A. S. Brito, Lúcia S. Campos, Roberto F. C. Fontes, Belmiro M. Castro, Thaïs N. Corbisier, Heitor Evangelista, Marcio Francelino, Vicente Gomes, Rosane G. Ito, Helena P. Lavrado, Neusa Paes Leme, Michel M. Mahiques, César C. Martins, Cristina R. Nakayama, Phan V. Ngan, Vivian H. Pellizari, Antonio B. Pereira, Monica A. V. Petti, Martin Sander, Carlos E. G. R. Schaefer and Rolf R. Weber

9.1 Introduction

Monitoring has been developed by Antarctic Programmes such as Antarctic New Zealand (www.antarcticanz.govt.nz) and the US Antarctic Programme (USAP) (http://www.usap.gov) considering expansion of human activities and inevitable environmental impacts (review in Tin et al. 2009). The expansion of human

R. C. Montone (✉) · M. C. Bícego · E. S. Braga · B. M. Castro · T. N. Corbisier · V. Gomes · R. G. Ito · M. M. Mahiques · C. C. Martins · P. V. Ngan · V. H. Pellizari · M. A. V. Petti · R. R. Weber
Instituto Oceanográfico, Universidade de São Paulo, Praça do Oceanográfico 191, São Paulo, São Paulo 05508-120, Brazil
e-mail: rmontone@usp.br

M. C. Bícego
e-mail: marciabicego@usp.br

E. S. Braga
e-mail: edsbraga@usp.br

B. M. Castro
e-mail: bmcastro@usp.br

T. N. Corbisier
e-mail: tncorbis@usp.br

V. Gomes
e-mail: vicgomes@usp.br

C. Verde and G. di Prisco (eds.), *Adaptation and Evolution in Marine Environments, Volume 2*, From Pole to Pole, DOI: 10.1007/978-3-642-27349-0_9,

activities in Antarctica highlights the need for efficient implementation of measures that ensure protection of the environment. These include environmental impact assessments, long-term monitoring, mitigation of introduced species, ecosystem-based management of living resources, and increased regulation of National Antarctic Programmes and tourism activities (Tin et al. 2009).

R. G. Ito
e-mail: rgito@usp.br

M. M. Mahiques
e-mail: mahiques@usp.br

C. C. Martins
e-mail: ccmart@ufpr.br

P. V. Ngan
e-mail: phanvn@usp.br

V. H. Pellizari
e-mail: vivianp@usp.br

M. A. V. Petti
e-mail: mapetti@usp.br

R. R. Weber
e-mail: rweber@usp.br

C. E. Alvarez
Universidade Federal do Espírito Santo, Av Fernando Ferrari 514,
Vitória ES, Espírito Santo 29073-910, Brazil
e-mail: cristina.engel@ufes.br

T. A. S. Brito
Fundação UNESCO-HidroEX, Av Mário Palmério 1000,
Frutal, Minas Gerais 38200-000, Brazil
e-mail: tania.brito@hidroex.mg.gov.br

L. S. Campos · H. P. Lavrado
Instituto de Biologia, Universidade Federal do Rio de Janeiro, Av Carlos Chagas Filho 373,
Rio de Janeiro, Rio de Janeiro 21941-902, Brazil
e-mail: campos-lucia@biologia.ufrj.br

H. P. Lavrado
e-mail: hpasseri@biologia.ufrj.br

R. F. C. Fontes
Campus Experimental do Litoral Paulista, Universidade Estadual Paulista, Praça Infante
Dom Henrique s/n°, São Vicente, São Paulo 11330-900, Brazil
e-mail: rcfontes@clp.unesp.br

H. Evangelista
Universidade do Estado do Rio de Janeiro, Rua São Francisco Xavier 524, Rio de Janeiro,
Rio de Janeiro 20550-013, Brazil
e-mail: heitor@uerj.br

M. Francelino
Universidade Federal Rural do Rio de Janeiro, Br 465 km 7, Seropédica,
Rio de Janeiro 23890-000, Brazil
e-mail: marciorocha@ufrrj.br

Admiralty Bay was designated Antarctic Specially Managed Area (ASMA no. 1) by the Antarctic Treaty System Consultative Meeting in 1996 (ATCM XXVIII 2005) to manage the impact of the activities of the nations in the area. The Scientific Committee for Antarctic Research (SCAR) recommends studies to evaluate the impact caused by scientists, tourists, station staff, infrastructure and logistics.

Admiralty Bay is a region of great environmental, historical, scientific and aesthetic value. Human occupation has increased in recent years, creating a need for environmental monitoring to help management for preserving the region. Adequate forecasting is necessary to assist decision-making bodies within the scope of the Antarctic Treaty System. Better understanding of ecosystem functioning in the context of the effect of human activities and climate change is critical for terrestrial and marine biodiversity. During IPY, the main results from a Brazilian multidisciplinary study at the ASMA were assembled to allow precise evaluation of the implications of future changes on life. The work generated baseline strategies in support of adequate environmental management of the ASMA and planning of future activities. Thus, here we illustrate the main features of this region in the context of local and global environmental impacts.

N. P. Leme
Instituto Nacional de Pesquisas Espaciais, Av dos Astronautas 1.758, São José dos Campos, São Paulo 12227-010, Brazil
e-mail: nleme@crn.inpe.br

C. C. Martins
Centro de Estudos do Mar, Universidade Federal do Paraná, Av Beira-mar s/n, Pontal do Paraná, Paraná 83255-971, Brazil

C. R. Nakayama
Campus Diadema, Universidade Federal de São Paulo, Rua Artur Riedel 275, Diadema, São Paulo 09972-270, Brazil
e-mail: crnakayama@unifesp.br

A. B. Pereira
Universidade Federal do Pampa, Av Antônio Trilha, São Gabriel, Rio Grande do Sul 97300-000, Brazil
e-mail: antoniopereira@unipampa.edu.br

M. Sander
Universidade do Vale do Rio dos Sinos, Av Unisinos 950, São Leopoldo, Rio Grande do Sul 93022-000, Brazil
e-mail: sander@bios.unisinos.br

C. E. G. R. Schaefer
Universidade Federal de Viçosa, Av Peter Henry Rolfs s/n, Viçosa, Minas Gerais 36570-000, Brazil
e-mail: carlos.schaefer@ufv.br

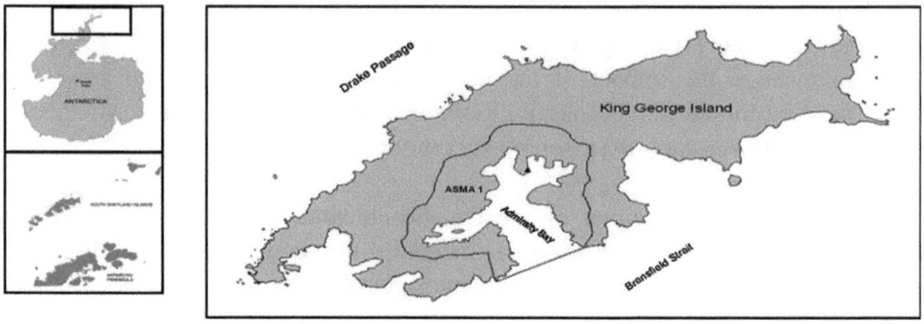

Fig. 9.1 Location of Admiralty Bay Antarctic Specially Managed Area (ASMA 1)

9.2 Study Area

The ASMA is located in the central sector of King George Island, South Shetland Islands, 120 km north of the Antarctic Peninsula (Fig. 9.1). It is an area of environmental and scientific interest, as it is subject to large seasonal fluctuations in sea-ice coverage and is known to respond to climate change (Santos et al. 2006).

The ASMA covers an area of 362 km^2, divided in sectors covered by ice, permanently ice free, and the bay itself. 195 km^2 (54 %) are occupied by 36 glaciers that drain the glacial mass of the island, which reaches an altitude of 700 m at the center of the ice field. The waters of the bay occupy 138 km^2 (38 %) and depths reach more than 550 m in the bay fjord and connects it to the Bransfield Strait (Rakuza-Suszczewski 1993). Only 29 km^2 (8 %) are ice free and located along the coast, including inland rocky outcrops such as Keller Peninsula, Hennequin Point, Demay Point and Thomas Point (the area surrounding the Polish station Henryk Arctowski).

The geological substrate of the ASMA is composed of a stratigraphic sequence of sedimentary rocks at its base covered by extensive volcanic rocks of Jurassic to Cenozoic ages (Birkenmajer 1982). The geological fault planes are oriented east-northeast to south-southwest and control the fjord pattern of the bay. The Island has three types of glaciers: (1) mild glaciers truncated in ice cliffs; (2) steep-slope glaciers incised by short and straight valleys and ending with cascading ice; (3) glaciers that end on land and exhibit a frontal moraine deposit (Bremer 1998). These deposits, as well as the rocky outcrops, may host dense moss and algae communities characteristic of the early stages of plant successions (Pereira et al. 2007). In these areas, there are dense populations of many species of birds and mammals that breed and rest (Arigony 2001).

The meteorological conditions of the ASMA, and of the entire archipelago, are controlled by cyclonic systems that originate in the Bellingshausen Sea and move east bringing moisture and heat, causing fog and maintaining mild temperatures (Rakuza-Suszczewski et al. 1993; Setzer and Hungria 1994). The annual mean air temperature is −2.8 °C, varying from 0.9 °C in summer to −7 °C in winter

(Ferron et al. 2004). The mean relative moisture is greater than 80 % and the annual precipitation is 500 mm, exceeding 1,000 mm in the upper parts of the glacial caps (Rakuza-Suszczewski et al. 1993).

Setzer et al. (2008) studied air temperatures for the last 60 years and showed that the highest temperature occurred in 1989 and that there has been a systematic increase over the last 20 years (1986–2006). Trends in 1958–2002 (Chapman and Walsh 2004) suggest modest warming over much of the region from 60° to 90°S. All seasons showed warming, winter trends being the largest at +0.172 °C decade^{-1}, while summer rates are only +0.045 °C decade^{-1}, contributing to approximately 12 % glacier retreat in 1956–2000 (Arigony Neto 2001).

Studies of biodiversity are mostly related to flora assessments, census of marine-bird and mammal populations, and diversity and ecology of macroalgae, invertebrates and fish in coastal shallow areas (Zadróżny 1996; Oliveira et al. 2009; Sicinski et al. 2011). In the adjoining ice-free areas, the distribution of plant communities is closely related to stable landforms and nesting birds and ornithogenic soils (Simas et al. 2006; Schaefer et al. 2007). Where edaphic conditions are favourable, moss carpets occur associated with lichens and fungi (Pereira et al. 2007). Other lichened mycobiota is on rock fragments and surfaces, sometimes associated with bird colonies (Francelino 2004). The coastal areas are the most densely covered, mainly by moss carpet formations in waterlogged soils (Simas et al. 2008). Near the Brazilian Station two of these areas are almost 300 m in length. As elevations increase rocky outcrops become covered with fruticose/crustose lichens and mosses (Pereira et al. 2007).

Within the ASMA, 13 species of birds breed. Three somatically breeding Pygoscelid penguins make up 91 % of the number and 95 % of the biomass of the breeding community. Other breeding seabirds are southern giant petrel; blue-eyed shag; brown and south polar skua, Wilson's storm petrel; black-bellied storm petrel, cape petrel, kelp gull, Antarctic tern and American sheathbill. ASPA N° 128, Cape Vauréal, Chabrier Island and surroundings, is one of the most important bird breeding area. Areas around Vaureal Cape harbour nests for all of the species mentioned above, including the blue-eyed shags, and 50 % of the giant petrels (Rakuza-Suszczewski 1993).

All six Antarctic pinniped species occur in the ASMA. The most frequent during winter is the crabeater seal (*Lobodon carcinophagus*). During summer, elephant (*Mirounga leonina*) and fur seals (*Arctocephalus gazella*) are most abundant. Fur seals, once relatively rare, have increased in number in recent years. Elephant and Weddell seals (*Leptonychotes weddelli*) breed in the area. Leopard seals (*Hydrurga leptonyx*) are found throughout the year in varying numbers; Ross seals (*Ommatophoca rossii*) rarely occur (Salwicka and Rakuza-Suszczewski 2002). Humpback whales (*Megaptera novaeangliae)* are the most frequent cetacean visitors during summer (Rakuza-Suszczewski 1993).

The shallow bottom marine community is characterized by macroalgae, mainly Phaeophyta and Rhodophyta, down to 50–60-m depth, and giant kelp *Himantothallus grandifolius* beds. In intertidal zones, epifauna is usually small and cryptic, with the exception of *Nacella concinna*, a common food item for the kelp gull

Larus domini-canus (Favero et al. 1997). The vagile benthos is abundant with a high variety and density of Amphipoda. Below 4–5 m, substrata are typically sandy and dominated by Isopoda, particularly *Paraserolis polita*. With increasing depth, vagile species such as the urchin *Sterechinus neumayeri,* the gastropod *Neobuccinum eatoni* and the nemertean *Parborlasia corrugatus* dominate. Below 25 m depth, on a muddy and more stable substratum, sessile forms include sponges, anemones and tunicates, with the dominance of the sedentary bivalve *Laternula elliptica* in shallower zones. High densities of echinoderms, e.g. *Amphioplus acutus, Ophionotus victoriae* and *Odontaster validus*, have been reported (Nonato et al. 2000; see Sicinski et al. 2011). Species are largely the same as observed on similar substrata at other sites in the region. Fishes comprise fifteen Nototheniidae, mainly the genus Notothenia and Trematomus, and two Channichthydae species, Hapagiferidae and Zoarcidae (Zadróżny 1996; Sicinski et al. 2011).

The ASMA hosts dense research and logistic activities. There are five research facilities from Brazil, Poland, Peru, Ecuador and the US. Two of them, the Brazilian Station "Comandante Ferraz" (EACF) and the Polish Station Henryk Arctowski, operate during the entire year, the other three (the Peruvian Station Machu Picchu, a permanent facility, the frequently occupied American Pieter J. Lenie Copacabana Refuge and the Equatorian Refuge) only during the summer. Due to relatively easy access, it is frequently visited by tourists.

9.3 Environmental Assessment

An environmental evaluation of the status of the ASMA began in 2002 with a Brazilian multi-institutional and multidisciplinary approach. This provided the basis for further studies during IPY, and for the establishment of a National Science and Technology Institute on Antarctic Environmental Research (INCT-APA). The latter has continued long-term environmental monitoring and investigations on human activities in the climate-change context.

The programme was implemented in two stages: first, assessment of environmental conditions from past data; and second, integration of data collected over time in order to develop a long-term monitoring strategy. Data from 2002 to 2006 included (1) *atmosphere*: monitoring of solar radiation, particularly the impact of UV on the environment, greenhouse gases, occurrence of metals and persistent organic pollutants (POPs), aerosol transport, long-range atmospheric inputs to King George Island; (2) *terrestrial*: research on plants, birds and microbial diversity, soil contamination, residue production by Ferraz activities; (3) *marine*: planktonic and benthic diversity, carbonate system, metal contamination in sediment, indicators of chemical and microbiological pollution, biomarker bioassays that detect human impacts. A synthesis of the main results is presented in this chapter.

9.3.1 Atmospheric Environment

Studies of the atmosphere during 25 years produced a database with climate information on King George Island and the Peninsula, and 15 years of measurements of the ozone layer and UV radiation in the sub-Antarctic and South America. Only long-term measurements can provide responses to changes and input to numerical models for weather and climate (Justino et al. 2010). A UV impact on the amphipod *Gondogeneia antarctica* was observed (Gomes et al. 2009). Atmospheric studies also focused on occurrence of POPs, aerosol transport and long-range input of metals the island. The occurrence of atmospheric PCBs in the vicinity of the Brazilian Station (62°05′S, 58°23′W) was studied by Montone et al. (2003) during the summer of 1995–1996. The low concentrations of atmospheric PCBs (12.1–92.6 pg m^{-3}) were similar to the preliminary study of 1993–1994 (Montone et al. 2001a) and in other Antarctic islands (Larsson et al. 1992; Kallenborn et al. 1998). Levels of PCBs were correlated with meteorological conditions and the highest levels may be associated with the passage of the frontal systems from South America. Evidence of transport of PCBs and organochlorine pesticides were provided by Weber and Montone (1990) and Montone et al. (2005) in the transect from Southwest Atlantic Ocean to its Antarctic sector. The presence of PCBs in algae and sediment from the bay (Montone et al. 2001a, b) also confirms that long-distance atmospheric transport is the major source of PCBs. Recently, Martins et al. (2010) confirmed the long-range transport of anthropogenic emissions of spheroidal carbonaceous particles (SCPs), an indicator of industrial fossil fuel combustion, from several regions of the globe, particularly South American countries. Leal et al. (2008) have studied local emission levels against the component derived from global pollution in Admiralty Bay, showing significant correlation between concentration of atmospheric aerosol and freshly deposited particles in the snow, and permitted an estimate of the snow deposition factor for Na, Mg, K, Cl, Mn, Cu, Zn, Fe, Br, Si, Pb, Al, and Ti. Long-term aerosol data suggest that besides the local aerosol sources, the long-range continental atmospheric transport of airborne particles is not significantly affected by the particles produced by local human impacts at King George Island.

9.3.2 Terrestrial Environment

King George Island has been one of the most visited and densely populated areas of Antarctica since it was discovered by William Smith in 1819. Local plant communities were deteriorating and it was suspected that it was due to construction and use of station buildings. Plant communities in decaying state were observed in the vicinity of the stations at Keller Peninsula (Ferraz Station), Crepin Point (Machu Picchu) and Thomas Point (Arctowski). In contrast, higher diversity was reported at Ullmann Point in an area unaffected by human activities along Martel Inlet. (Pereira et al. 2007).

In the last decades, a nesting area of birds decreased by 7 % in the ice-free areas of Admiralty Bay (Sander et al. 2007). This reduction was most pronounced for birds that are most sensitive to human disturbance, such as the giant petrel (Petry and Krünger 2011). On the other hand, populations of opportunist birds and predators (skuas and kelp gulls) increased in numbers and reproductive areas (Sander et al. 2007; Costa and Alves 2008; Carneiro et al. 2010).

The soils in Keller Peninsula where Ferraz is located have high amount of coarse fragments, low degree of chemical weathering and carbon content (periglacial erosion), with little plant development (Simas et al. 2007, 2008).

Around Ferraz, soils have undergone diesel contamination especially beneath the fuel tanks (Oliveira 2005; Luz et al. 2006). Aromatic-hydrocarbon-degrading bacteria have been reported in these fuel-affected areas (Luz et al. 2006).

Activities at Ferraz have generated solid residues such as plastics ($\sim$1,200 kg), metal ($\sim$1,300 kg), paper ($\sim$1,200 kg) and glass ($\sim$300 kg); waste production followed a seasonal pattern, being highest during summers. These are returned to Brazil. Maintenance and aging of Ferraz infrastructure have released contaminants, and these have been monitored over the years, together with the effect of noise and operations with helicopters and motorcycles. These are known to cause stress on birds and pinnipeds (Price 2008; Albores-Barajas et al. 2009).

9.3.3 Marine Environment

The marine environment is rich and diverse, and can be highly influenced by seasonal changes in ice formation and melt (Lange et al. 2007; Sicinski et al. 2011).

The carbonic acid system indicates that the ocean surface is a source of CO_2 to the atmosphere of Admiralty Bay. The sea-air CO_2 flux (mean value $\pm$ standard deviation) ranged between 2.3 ± 1.0 and 5.2 ± 3 mmol $m^{-2}d^{-1}$ in 1999, 2001, 2002 and 2003, in accordance to Wanninkhof (1992) (Ito et al. in preparation).

Relatively high metal concentrations were found in sediments, but bioavailability is low due to the reducing conditions of the sediment (Santos et al. 2005), suggesting low risk to the biota. In front of Ferraz, sediment features, e.g. anoxia in the first centimetres, abundance of pirite and calcopirite (Schaefer et al. 2004), and higher organic matter, can be responsible for metal enrichment.

Monitoring has shown that the major sources for potential chronic pollution in the marine environment are by-products of fossil-fuel combustion and discharge of sewage from stations (see Bícego et al. 2009; Martins et al. 2002; Rosado et al. 2011). Sewage discharge feeds organic compounds that are oxidized in the ocean, increasing consumption of dissolved oxygen (DO). Although DO shows slight under-saturation at the discharge area, this can be caused by local topography and physical mixing (e.g. Campos et al. 2012).

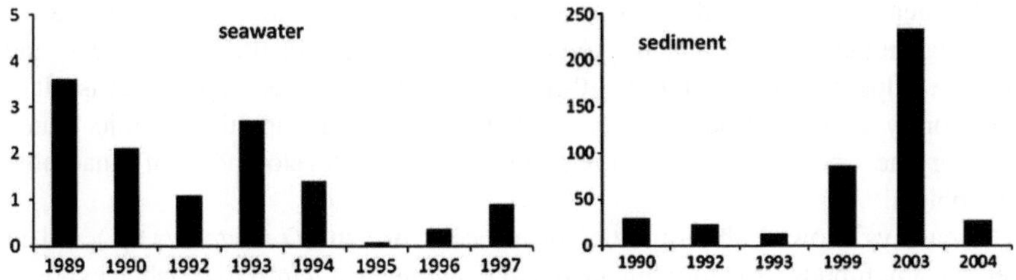

Fig. 9.2 PAH concentrations in front of EACF: seawater (μg L^{-1}); sediment (ng g^{-1})

The introduction of hydrocarbons into the bay and at EACF-adjacent area has been studied since 1989 by the Brazilian Antarctic Programme. Polycyclic aromatic hydrocarbons (PAHs) and aliphatic hydrocarbons (AHs) were detected in seawater and sediment (Bícego et al. 1996, 1998, 2003, 2009; Martins et al. 2004, 2010). The most contaminated sites are near the Brazilian and Polish Stations, mostly due to summer activities. Seawater concentrations of PAHs near EACF have decreased (3.50 to 1.0 μg L^{-1}) from 1989 to 1997 (Fig. 9.2). In addition, the sediments revealed hydrocarbons, suggesting a mix of direct input of oil and combustion-generated hydrocarbons. But a considerable increase of PAHs in the sediment in front of EACF was detected during 2000 and 2002, which may be related to increase in the number of people from 1994. A decrease was observed from 2004 after improvement of sewage treatment at the station, indicating that contribution of human wastes may be significant (Bícego et al. 2009).

Recently, Martins et al. (2010) described the temporal distribution of higher-molecular-weight PAHs in short sediment cores (<20 cm) collected in three different inlets of Admiralty Bay. The maximum concentration occurred at 3.5 cm (454.9 ng g^{-1}) in the sediments close to Ferraz in 1995–1997. Relatively high concentrations were also detected in the lower sections as 1969–1972 (106.3 ng g^{-1}) and 1957–1966 (46.9 ng g^{-1}). The highest levels were detected in the layers of sediment cores relating to the last 30 years, reflecting increase in human occupation, resulting in more fossil-fuel consumption, combustion of organic matter and petroleum derivatives, and input of sewage effluent.

The sewage contribution, especially at Martel Inlet, has been studied since 1997 (Martins et al. 2002, 2005; Montone et al. 2010). Chemical indicators (faecal sterols) near EACF (10–20-m depth) increased during the last years, reflecting intensification of human activities and, consequently, increase in sewage discharge. However, the influence of sewage was limited to 200 m from the sewage outlet in the water column, and of 700 m in the sediment. The dispersion of the sewage effluent in the shallow coastal zone of Martel Inlet is favoured by hydrodynamics, especially tides. Nonetheless, assessment of microbial indicators, as *Escherichia coli*, indicated that contamination in Admiralty Bay is punctual and restricted to the proximities of the EACF wastewater outlet. Increased availability of easily degradable organic matter from sewage can also stimulate methanogenesis by autochthonous archaea (see Nakayama et al. 2011).

The density of the benthic macrofauna did not show variations between areas of potential human impact and the reference ones, except in the area close to the sewage effluent of the Station, at 20-m depth, 100–200-m far from the coast. It is noteworthy that ice blocks the passage, which may disrupt the benthic fauna, making the sediment more anoxic and enabling development of anaerobic microbiota.

Bioassays showed abnormalities in red cells of fish (*Trematomus* sp.) and in benthic amphipods (*Bovallia gigantea* and *Gondogeneia antarctica*) when exposed to seawater collected in front of fuel tanks and sewage outlet, in in vitro and in situ experiments (Ngan et al. 2007). However, significant changes in the benthic fauna were not detected, indicating that human impacts at Ferraz are of small magnitude and impact in the benthic system.

9.4 Monitoring Strategy Proposal

A set of biotic and abiotic variables (Table 9.1), that have potential significance as environmental indicators of impact or could explain the ecosystem functioning of the ASMA, were selected after considering the main results from this study. In order to better understand the Antarctic ecological processes, long-term data series are required, because natural impacts may often have strong effects on biodiversity, but they should be distinguished from those caused by human impacts.

The report for SCAR and the Council of Managers of National Antarctic Programmes (COMNAP) (Kennicutt et al. 1996) and the handbook from the Committee for Environmental Protection (CEP 2006) were also used as guidelines for planning the Brazilian monitoring strategy. In this context, a number of parameters were chosen to address the objectives, in compliance with logistic and operational restrictions imposed by the environment.

9.4.1 Terrestrial Environment Indicators

The terrestrial environment is generally subjected to human impact and any change in the ecosystem is reflected on the atmospheric and marine environments. Since the local impact is restricted to areas of intense use, near the stations of Admiralty Bay, it was given emphasis to the surroundings of EACF.

Natural processes such as ice freezing and thawing, are significant agents of environmental changes. For the terrestrial environment, the parameters chosen as indicators were based on solid-residue production, wastewater discharge, fuel handling, acoustic impact, modification of landscape/vegetation, glacier retreat, soil quality, bird population dynamics.

Monitoring solid residue and relationship with the resident population makes it possible to identify the main sources as a function of the activities and, from

Table 9.1 Indicators proposed for management at Admiralty Bay (ASMA 1)

Environment	Indicators	Sampling frequency
Atmospheric		
Solar radiation	Global, UV, radiation balance	Annual
Black carbon, heavy metals	Aerosols	Annual
Dispersion of pollutants	Aerosols, winds	Annual
Meteorological variables	Wind, temperature, humidity, pressure	Annual
Greenhouse gases	CO_2, CO, O_3, NO_2, CH_4	Annual
Marine		
Sea-water quality	Biochemical, cellular, histopathological biomarkers	Annual
	T, S, DO, pH, N, P, silicate, chlorophyll-*a*	Biannual
Sediment quality	Methane flux/biological methane balance	Annual
	Clostridium, microbial molecular structure	Biannual/ quinquennial
	Benthic fauna, trophic web	Biannual/ quinquennial
	Hydrocarbons, faecal sterols, LABs, heavy metals	Triennial
Terrestrial		
Solid waste	Type, vol, weight, *per capita* production	Annual
Waste water	Volume, faecal coliforms	Annual
Fuel handling	Consumption control	Annual
Acoustic impact	Noise levels during routine and station maintenance	Annual
Landscape	Temporal analyses of images from the stations areas	Biannual
Glacier retraction	Variations in the extension of glaciers	Quinquennial
Soil quality	Geomorphological features	Quinquennial
	Heterotrophic bacteria, microbial community molecular structure	Annual/ quinquennial
Biological parameters	Vegetation coverage, retraction, community biodiversity	Quinquennial
	Population size, distribution and breeding success of birds	Annual/ quinquennial

T temperature, *S* salinity, *N* nitrate, *P* phosphate, *DO* dissolved oxygen, *LABs* linear alkyl benzenes

annual results, the look for actions aiming at residue reduction. Microbiological indicators can be used to monitor sewage treatment and assess soil contamination. Efficiency of the system and contamination of the area adjacent to the wastewater outlet can be monitored through counts of faecal indicators (total coliforms, *Escherichia coli* and *Clostridium perfringens*). Soil contamination can also be assessed through counts of heterotrophic bacteria, since variation in soil bacterial densities related to introduction of contaminants was determined in previous studies with samples collected around the Station. Also, microbial-community analysis provides important baseline data and allows for long- term impact studies from the introduction of contaminants (Luz et al. 2006).

The level of noise will be monitored mainly during routine and maintenance activities. The cleaning procedures of corroded parts has caused considerable acoustic impact and generated residues, with an estimated 10 % of the particles scattered by the local winds. The fuel consumption has a direct relationship with specific needs, i.e. quality of the fuel, equipment efficiency, storage conditions and appropriate use. The control of consumption in all instances allows recognition of flaws and use of mitigating measurements. This is not a proper indicator, but an auxiliary variable in understanding observed effects in the environment.

Monitoring of the aesthetic landscape can be carried out through analyses of images, adopting specific methodologies for their attainment. Although landscape evaluation must consider the dynamic scenario, parameters and methods that can be applied to this situation were not established. This tool is more than an indicator, since it allows verifying the natural plant coverage variations and possible expansion of EACF. Images can also be used to monitor and measure plant communities. These terrestrial communities are the first to be affected by degradation. The increase in the thawing area over the last years, mainly at Hennequin Point, is an important factor for monitoring the plant-community evolution, due to a larger visual exposure of vegetated areas selected for monitoring. Variations in the extension of glaciers can be verified through remote sensing. Geomorphological features (extension or retraction) related to climate change can be observed in aerial photographs, allowing comparative analyses of periodical variation.

Birds often occupy the top position of the food web. By evaluating fluctuations and trends of bird population in Admiralty Bay it is possible to evaluate its environmental quality. Because birds show large population fluctuations, annual assessment in long-term programmes is recommended. For decades, the bay has served as reproduction site for many birds, which nest in the same place, with few exceptions (Sander et al. 2007). Comparison of data along a time series allows the detection of reductions in the number of reproductive sites, which could indicate a loss in the ability of the population to maintain itself. It can also indicate the occupation of other areas without interfering with the overall stock population.

9.4.2 Marine Environment Indicators

Monitoring the marine environment include the pelagic and benthic systems. In the pelagic system, five sites of water sampling are proposed (Fig. 9.3): Martel Inlet (in front of the sewage outlet of EACF) and Botany Point (as a reference area); Mackelar Inlet (in front of the Peruvian station); Ezcurra Inlet (Thomaz Point) and Shag Point (in front of the Polish station).

Some basic parameters were chosen as seawater quality indicators to be used in a biannual time scale: (1) temperature and salinity are necessary for verifying the physical–chemical conditions of the marine environment, freshwater discharge and

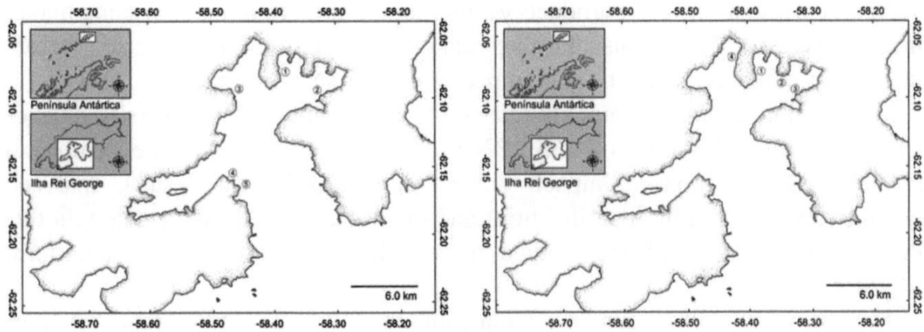

Pelagic system

Benthic system

Fig. 9.3 Suggested sampling strategy for Admiralty Bay: Pelagic system: (1) EACF, (2) Botany Point, (3) Peruvian Station, (4) Thomaz Point, (5) Sahg Point. Benthic system: (1) EACF, (2) Ullmann Point, (3) Botany Point, (4) Refuge 2

calculating other indicators; (2) DO: sewage discharge or organic compounds are oxidized, generating consumption of DO; (3) pH: during organic-matter oxidation CO_2 is produced and pH lowers; (4) the nutrient indicators (nitrate, phosphate and silicate) allow identification of the water-mass features, and are fundamental for understanding the biogeochemical cycles within the bay; (5) chlorophyll-*a*: it indicates phytoplankton biomass, allowing rapid evaluation of eutrophization or planktonic biota. The seawater quality will be monitored through the cellular biomarkers in an annual time scale. The toxic potential of the effluents may be identified by their effects on selected vital processes in bio-indicator organisms.

Regarding the benthic system, the sampling strategy has involved at least four areas for long-term monitoring (Fig. 9.3). One is close to Ferraz (Martel Inlet), and three have been considered as reference areas: Ullmann Point, Botany Point (Martel Inlet) and Refuge 2 (Mackellar Inlet).

In each area two distinct sites were chosen, 200-m far from each other; at Ferraz the two sites are potential pollution sources (in front of the oil tank and close to the sewage outfall). It will be possible to evaluate the natural variation in small scale and to distinguish influence of pollution sources in stations. The biota will be studied at 20–30-m depth, where previous results showed large benthic diversity and low influence of the natural impact, making it easier to detect human effects.

Considering the preliminary evaluation of the possible effects of human activity at EACF on the marine biota and which parameters are best suited to discriminate these effects, the following additional indicators were suggested for future monitoring of the Brazilian station.

Clostridium is a faecal pollution indicator that may be used in long-term monitoring of contamination of remote populations. This bacterium produces spores and is more resistant to extreme environmental conditions (Hughes and Thompson 2004; Martins et al. 2005).

The presence of methane, although non-exclusive, may be related to larger availability of degradable organic matter. Methane contributes to the greenhouse effect and is one of the final products of microbial transformation of organic matter under anaerobic conditions. Further data accumulation is necessary before identifying methane production as impact indicator (Nakayama et al. 2011).

Microbial communities quickly respond to disturbances. The diversity indexes applied to clone libraries of 16S rRNA can be useful to evaluate ecological dynamics and disturbance impacts (Zhang et al. 2008). The results, analyzed qualitatively (taxonomic identification) and quantitatively, allow to ascertain differences in diversity of the studied areas. They are considered a long-term indicator.

The sediment benthic fauna is a biota component widely used in impact studies, especially in coastal areas. Although many evaluations are based on structural changes of the community, this organization also involves analyses of functional changes. Although individual in vitro responses to contaminants are important in environmental evaluations, analyses of populations and communities in their natural milieu allow a more realistic view on the health of the environment.

Along with remote samplers, such as the mini-box corer for macrofauna, non-destructive methods have been proposed to monitor coastal megabenthic communities, e.g. camera systems and small remotely operated vehicles for visual census and to record natural physical disturbances (e.g. scouring), or anthropogenic effects. The ideal sampling frequency is annual, because of the large Antarctic inter-annual variability. Because of logistic difficulties, a 2-year interval has been considered as minimum to evaluate responses from the environment; a larger interval would not allow accurate assessment, and it would be difficult to distinguish between natural and anthropogenic impacts.

Isotope ratios $\delta^{13}C$ and $\delta^{15}N$ as trophic web tracers (Michener and Schell 1994) have also been used as an important tool in monitoring sewage pollution and eutrophication, since this contamination changes the isotopic signature in living organisms (Corbisier et al. 2004; Conlan et al. 2006).

The chemical indicators used in studies of faecal pollution and domestic effluents in the sediment are faecal sterols and linear alkylbenzenes (LABs). These are resistant to degradation, therefore less susceptible to environmental change, being thus useful indicators of sewage pollution.

Human activities require the fossil fuels as energy source, thereby placing the region at risk of hydrocarbon contamination. Due to continuous use of fossil fuel in the bay, the contribution of petroleum hydrocarbons (alkanes and PAHs) will be monitored in a triennial scale. The chosen metals are Cu, Zn, Pb and Fe, often associated with human activity, although in certain areas Cu can be naturally high.

The long-term time series of these parameters will give a perspective of environmental changes, as well as generate the dynamic baseline necessary to establish natural environmental variability.

9.5 Final Considerations

Past and current data available in the Admiralty Bay ASMA supply a general overview of the environmental conditions of the ecosystem.

In the terrestrial environment, local impact in areas under intensive human pressure around the stations at the bay was identified, with noticeable interference in diversity and abundance of the fauna and flora. Natural processes such as freezing and thawing are also significant agents; permafrost shows instability due to warming. The most important aspects to be monitored are dynamic soil and vegetation responses to human disturbances, following publication of reliable geochemical background data (Albuquerque Filho 2005; Santos et al. 2006, 2007).

In the marine environment, a local effect was identified on the benthic fauna and in the surrounding areas of Ferraz, with little space dispersion. Natural processes such as thawing, anchor ice and ice scour nowadays appear much more significant agents of environmental changes in the ASMA than human activity.

In the atmosphere, an external influence due to long-range transport was observed, as well as indications of local human influence (Justino et al. 2010).

The diagnosis and the comparative multidisciplinary study led to proposing environmental quality parameters, aiming to establish a monitoring programme. Coherent and efficient monitoring depends on long-term series of abiotic and biotic variables. Beside supplying an invaluable perspective of changes, they allow establishing a dynamic baseline for the natural variability of the environment.

The integrated knowledge of the environment also produced an organization and usage plan of the Brazilian Station (within its support capacity), coherent with the needs, logistics, and the Madrid Protocol.

Critical result evaluation identified the variables for environmental-impact assessment of Admiralty Bay. The bay is a continuously growing venue. Better planning and coordination of human activities will help to reduce and/or avoid the interference risk and minimize environmental impacts, providing an effective mechanism for conservation of the valuable features that characterize the area.

Acknowledgments Financial support was obtained from the Antarctic Brazilian Program (PROANTAR) by a grant from the Ministry of Environment (MMA) and National Research Council (CNPq) with logistical support from the Secretariat of the Interministerial Commission for Sea Resources (SECIRM). Studies were within the Brazilian Antarctic Environment Research Network (CNPq processes: 550345/02-7, 550347/02-0, 550348/02-6, 550349/02-2, 550350/02-0, 550351/02-7, 550354/02-6, 550356/02-9, 550357/02-5, 550359/02-8, 550365/02-8, 550368/02-7, 550371/02-8, 550373/02-0). This work is in the frame of IPY and the Brazilian National Science and Technology Institute on Antarctic Environmental Research (INCT-APA).

References

Albores-Barajas YV, Soldatini C, Furness RW (2009) Are burrow nesting seabird chicks affected by human disturbance? Waterbirds 32(4):572–578

Albuquerque Filho MR (2005) Geoquímica dos solos da Península Keller, Ilha Rei George, Antártica Marítima, como subsídio ao monitoramento ambiental. Ph.D. thesis. Universidade Federal de Viçosa. 217 pp

Arigony Neto J (2001) Determinação e interpretação de características glaciológicas e geográficas com sistema de informações geográficas na Área Antártica Especialmente Gerenciada Baía do Almirantado, Ilha Rei George, Antártica. Dissertation UFRGS-CEPSRM, Porto Alegre. 84 pp

ATCM XXVIII (2005) Review of the Admiralty Bay Antarctic specially managed area management plan (ASMA n 1), 31 pp

Bícego MC, Weber RR, Ito RG (1996) Aromatic hydrocarbons on surface waters of Aldmiralty Bay, King George Island, Antarctica. Mar Pollut Bull 32:549–553

Bícego MC, Zanardi E, Ito RG, Weber RR (1998) Hydrocarbons in surface sediments of Admiralty Bay, King George Island, Antarctica, Peninsula. Pesq Antartct Bras 3:15–21

Bícego MC, Zanardi-Lamardo E, Weber RR (2003) Four-year of dissolved/dispersed petroleum hydrocarbons on surface waters of Admiralty Bay, King George Island, Antarctica. Rev Bras Oceanogr 51:33–38

Bícego MC, Zanardi-Lamardo E, Taniguchi S, Martins CC, Silva DAM, Sassaki ST, Albergaria-Barbosa ACR, Paolo FS, Weber RR, Montone RC (2009) Results from a 15-year study on hydrocarbon concentrations in water and sediment from Admiralty Bay, King George Island, Antarctica. Antarct Sci 21:209–220

Birkenmajer K (1982) Pliocene tillite-bearing succession of King George Island (South Shetland Islands, Antarctica). Studia Geol Pol 74:7–72

Bremer UF (1998) Morfologia e bacias de drenagem da cobertura de gelo da Ilha Rei George, Antártica. Dissertation. Universidade Federal do Rio Grande do Sul, UFRGS, 119 pp

Campos LS, Barboza CAM, Bassoi M, Bernardes M, Bromberg S, Corbisier TN, Fontes RFC, Gheller PF, Hajdu E, Kawall HG, Lange PK, Lanna AM, Lavrado HP, Monteiro GCS, Montone R, Morales T, Moura RB, Nakayama CR, Oackes T, Paranhos R, Passos FD, Petti MAV, Pellizari VH, Rezende CE, Rodrigues M, Rosa LH, Secchi E, Tenenbaum DR, Yoneshigue-Valentin Y (2012) Environmental processes, biodiversity and changes in Admiralty Bay, King George Island, Antarctica. In: Verde C, di Prisco G (eds) Adaptation and evolution in marine environments—the impacts of global change on biodiversity, vol 2. Springer, Heidelberg

Carneiro APB, Polito MJ, Sander M, Trivelpiece WZ (2010) Abundance and spatial distribution of sympatrically breeding *Catharacta* spp. (skuas) in Admiralty Bay, King George Island, Antarctica. Polar Biol 33(5):673–682

CEP (2006) Committee for Environmental Protection Handbook, 185 p

Chapman WI, Walsh JE (2004) Observed climate change in the Arctic. Recent variations of sea ice and air temperature in high latitudes. In: Arctic climate impact assessment, ACIA overview report. Cambridge University Press, Cambridge, 140 pp

Conlan KE, Rau GH, Kvitek RG (2006) Delta C-13 and delta N-15 shifts in benthic invertebrates exposed to sewage from McMurdo Station, Antarctica. Mar Pollut Bull 52:1695–1707

Corbisier TN, Petti MAV, Skowronski RSP, Brito TAS (2004) Trophic relationships in the nearshore zone of Martel Inlet (King George Island, Antarctica): $\delta^{13}C$ stable isotope analysis. Polar Biol 27(2):75–82

Costa ES, Alves MAS (2008) The breeding birds of Hennequin Point: an ice-free area of Admiralty Bay (Antarctic Specially Managed Area), King George Island, Antarctica. Rev Bras Ornitol 16:137–141

Favero M, Silva P, Ferreyra G (1997) Trophic relationships between the kelp gull and the Antarctic limpet at King George Island (South Shetland Islands, Antarctica) during the breeding season. Polar Biol 17:431–436

Ferron FA, Simões JC, Aquino FE, Setzer AW (2004) Air temperature time series for King George Island, Antarctica. Pesq Antart Bras 4:155–169

Francelino MR (2004) Geoprocessamento aplicado ao monitoramento ambiental da Antártica Marítima: geomorfologia, solos e cobertura vegetal da Península Keller, PhD thesis. Universidade Federal de Viçosa, 243 pp

Gomes V, Passos MJACR, Paes Leme NM, Santos TCA, Campos DYF, Hasue FM, Phan VN (2009) Photo-induced toxicity of anthracene in the Antarctic shallow water amphipod, *Gondogeneia antarctica*. Polar Biol 32:1009–1021

Hughes KA, Thompson A (2004) Distribution os sewage pollution around a marine Antarctic research station indicated by faecal coliforms, *Clostridium perfrigens* and faecal sterol markers. Env Pollut 127:315–321

Justino F, Setzer A, Bracegirdle TJ, Mendes D, Grimm A, Dechiche G, Schaefer CEGR (2010) Harmonic analysis of climatological temperature over Antarctica: present day and greenhouse warming. Int J Climatol (Published online in Wiley InterScience). (www.interscience.wiley.com) doi:10.1002/joc.2090

Kallenborn R, Oehme M, Wynn-Williams DD, Schlabach M, Harris J (1998) Ambient air levels and atmospheric long-range transport of persistent organochlorines to Signy Island, Antarctica. Sci Total Environ 220:167–180

Kennicutt MC, Sayers JCA, Walton S, Wratt G (1996) Monitoring of environmental impacts from science and operations in Antarctica. A report for SCAR and COMNAP

Lange PK, Tenenbaum DR, Braga ES, Campos LS (2007) Microphytoplankton assemblages in shallow waters at Admiralty Bay (King George Island, Antarctica) during the summer 2002–2003. Polar Biol 30:1432–2056

Larsson P, Jarnmark C, Sodergren A (1992) PCBs and chlorinated pesticides in the atmosphere and aquatic organisms of Ross Island, Antarctica. Mar Pollut Bull 25:281–287

Leal MA, Joppert M, Licínio MV, Evangelista H, Maldonado J, Dalia KC, Lima C, Barros Leite CV, Correa SM, Medeiros G, Cunha KD (2008) Atmospheric impacts due to Anthropogenic activities in remote areas: the case study of Admiralty Bay/King George Island/Antarctic Peninsula. Water Air Soil Pollut 188(1–4):67–80

Luz AP, Ciapina EMP, Gamba RC, Lauretto MS, Farias EWC, Bícego MC, Taniguchi S, Montone RC, Pellizari VH (2006) Potential for bioremediation of hydrocarbon polluted soils in the maritime Antarctic. Antarct Sci 18(2):335–343

Martins CC, Venkatesan MI, Montone RC (2002) Sterols and linear alkylbenzenes in marine sediments from Admiralty Bay, Antarctica. Antarct Sci 14(3):244–252

Martins CC, Bícego MC, Taniguchi S, Montone RC (2004) Aliphatic (AHs) and aromatic hydrocarbons (PAHs) in surface sediments in Admiralty Bay, King George Island, Antarctica. Antarct Sci 16:117–122

Martins CC, Montone RC, Gamba RC, Pellizari VH (2005) Sterols and fecal indicator microorganisms in sediments from Admiralty Bay, Antarctica. Braz J Oceanogr 53:1–12

Martins CC, Bícego MC, Rose NL, Taniguchi S, Lourenço RA, Figueira RCL, Mahiques MM, Montone RC (2010) Historical record of polycyclic aromatic hydrocarbons (PAHs) and spheroidal carbonaceous particles (SCPs) in marine sediment cores from Admiralty Bay, King George Island, Antarctica. Environ Poll 158:192–200

Michener RH, Schell DM (1994) Stable isotopes ratios as tracers in marine aquatic foodwebs. In: Lajtha K, Michener RH (eds) Stable isotopes in ecology and environmental sciences. Blackwell, Oxford, pp 138–157

Montone RC, Taniguchi S, Sericano JL, Weber RR, Lara WH (2001a) Determination of polychlorinated biphenyls in Antarctic macroalgae *Desmarestia* sp. Sci Total Environ 277:181–186

Montone RC, Taniguchi S, Weber RR (2001b) Polychlorinated biphenyls in marine sediments of Admiralty Bay, King George Island, Antarctica. Mar Pollut Bull 42:611–614

Montone RC, Taniguchi S, Weber RR (2003) PCBs in the atmosphere of King George Island, Antarctica. Sci Total Environ 308:167–173

Montone RC, Taniguchi S, Boian C, Weber RR (2005) PCBs and chlorinated pesticides (DDTs, HCHs and HCB) in the atmosphere of the southwest Atlantic and Antarctic Oceans. Mar Pollut Bull 50(7):778–782

Montone RC, Martins CC, Bícego MC, Taniguchi S, Silva DAM, Campos LS, Weber RR (2010) Distribution of sewage input in marine sediments around a maritime Antarctic research station indicated by molecular geochemical indicators. Sci Total Environ 408:4665–4671

Nakayama CR, Kuhn E, Araújo ACV, Alvalá PC, Ferreira WJ, Vazoller RF, Pellizari VH (2011) Revealing archaeal diversity patterns and methane fluxes in Admiralty Bay, King George Island, and their association to Brazilian Antarctic Station activities. Deep Sea Res II 58(1–2):128–138

Ngan PV, Gomes V, Passos MJACR, Ussami KA, Campos DYF, Rocha AJS, Pereira BA (2007) Biomonitoring of the genotoxic potential (micronucleus and erythrocyte nuclear abnormalities assay) of the Admiralty Bay water surrounding the Brazilian Antarctic Research Station "Comandante Ferraz", King George Island. Polar Biol 30:209–217

Nonato EF, Brito TAS, Paiva PC, Petti MAV, Corbisier TN (2000) Benthic megafauna of the nearshore zone of Martel Inlet (King George Island, South Shetland Islands, Antarctica): depth zonation and underwater observations. Polar Biol 23:580–588

Oliveira LM (2005) Hidrocarbonetos em solos da área da estação antártica brasileira Comandante Ferraz. Viçosa: Dissertation. Universidade Federal de Viçosa, 132 pp

Oliveira EC, Absher TM, Pellizzari FM, Oliveira MC (2009) The seaweed flora of Admiralty Bay, King George Island. Antarct Polar Biol 32:1639–1647

Pereira AB, Spielmann AA, Martins MFN, Francelino MR (2007) Plant communities from ice-free areas of Keller Peninsula, King George Island. Antarct Oecol Brasil 11(1):14–22

Petry VM, Krünger L (2011) Monitoring of the dynamic and spatial distribution of Antarctic seabird populations in the South Shetland Islands. In: INCT-APA annual activity report 2010. Editora Cubo, São Carlos, pp 27–30

Price C (2008) The impact of human disturbance on birds: a selective review. Aust Zool 34(special issue):163–196

Rakuza-Suszczewski S (ed) (1993) The maritime Antarctic coastal ecosystem of Admiralty Bay. Polish Academy of Sciences (Polskiej Akademii Nauk Warsaw), 216 pp

Rakuza-Suszczewski S, Mietus M, Piasecki J (1993) Weather and climate. In: Rakuza-Suszczewski S (ed) The maritime Antarctic coastal ecosystem of Admiralty Bay. Polskiej Akademii Nauk, Warsaw, pp 19–25

Rosado AS, Cury JC, Peixoto RS, Jesus HE, Schaefer CEGR, Bícego MC, Jurelevicius DA, Seldin L, Seabra PN, Greer CW (2011) Bioremediation, hydrocarbon depletion and mi-crobial genetic diversity of Antarctic oil-polluted soil. In: INCT-APA annual activity report 2010. Editora Cubo, São Carlos, pp 207–210

Salwicka K, Rakuza-Suszczewski S (2002) Long-term monitoring of Antarctic pinnipeds in Admiralty Bay (South Shetland Islands). Acta Thériol 47:443–457

Sander M, Balbão TC, Polito MJ, Costa ES, Carneiro APB (2007) Recent decrease in chinstrap penguin (*Pygoscelis antarctica*) populations at two of Admiralty Bay's inlets on King George Island, South Shetland Islands, Antarctica. Polar Biol 30:659–661

Santos IR, Silva Filho EV, Schaefer CEGR, Albuquerque Filho MR, Campos LS (2005) Heavy metals contamination in coastal sediments and soils near the Brazilian Antarctic Station, King George Island. Mar Pollut Bull 50:185–194

Santos IR, Silva Filho EV, Schaefer CEGR, Sella S, Silva CA, Gomes V, Passos MJA, Ngan PV (2006) Baseline mercury and zinc concentration in terrestrial and coastal organisms of Admiralty Bay, Antarctica. Environ Pollut 140:304–311

Santos IR, Schaefer CEGR, Silva Filho EV, Fávaro DIT (2007) Sediment geochemistry in coastal maritime Antarctica (Admiralty Bay, King George Island): evidence from rare earths and other elements. Mar Chem 107:1–12

Schaefer CEGR, Dias LE, Campos LS, Albuquerque Filho MR, Costa LM, Borges Júnior M (2004) Monitoramento ambiental em sedimentos costeiros da Baía do Almirantado: granulometria, teores de macronutrientes e metais biodisponíveis. In: Schaefer CEGR, Francelino MR, Simas FNB, Albuquerque Filho MR (eds) Ecossistemas costeiros e monitoramento ambiental da

Antártica Marítima, Baía do Almirantado, Ilha Rei George. NEPUT Universidade Federal de Viçosa, Viçosa, 192 pp
Schaefer CEGR, Santana RM, Simas FNB, Francelino MR, Fernandes Filho EI, Albuquerque Filho MR, Calijuri ML (2007) Geoenvironments from the vicinity of Arctowski station, Admiralty Bay, King George Island, Antarctica: vulnerability and valuation assessment. US Geol Surv 1047:15–19
Setzer AW, Hungria CS (1994) Meteorologia na Península Antártica—Alguns aspectos práticos. Technical report. Instituto Nacional de Pesquisas Espaciais, Brazil, 101 pp
Setzer AW, Romão M, Aquino FE (2008) Antártica: Relação Climática com a América do Sul. Climanálise (São José dos Campos) 23:1–12
Siciński J, Jażdżewski K, De Broyer C, Ligowski R, Presler P, Nonato EF, Corbisier TN, Petti MAV, Brito TAS, Lavrado HP, Błażewicz- Paszkowycz M, Pabis K, Jażdżewska A, Campos LS (2011) Admiralty Bay Benthos diversity: a longterm census. Census of Antarctic Marine Life special volume. Deep-Sea Res II 58:30–48
Simas FNB, Schaefer CEGR, Melo VF, Guerra MBB, Saunders M, Gilkes RJ (2006) Clay-sized minerals in ornithogenic cryosols from Admiralty Bay, King George Island, Antarctica. Clays Clay Miner 54:721–736
Simas FNB, Schaefer CEGR, Melo VF, Albuquerque Filho MR, Michel RFM, Pereira VV, Gomes MRM, Costa LM (2007) Ornithogenic cryosols from maritime Antarctica: phosphatization as soil forming process. Geoderma 138:191–203
Simas FNB, Schaefer CEGR, Albuquerque Filho MR, Francelino MR, Costa LM (2008) Genesis, properties and classification of cryosols from Admiralty Bay, maritime Antarctica. Geoderma 144:242–248
Tin T, Fleming ZL, Hughes KA, Ainley DG, Convey P, Moreno CA, Pfeiffer S, Scott J, Snape I (2009) Impacts of local human activities on the Antarctic environment. Antarct Sci 21(1):3–33
Wanninkhof R (1992) Relationship between wind speed and gas exchange over the ocean. J Geophys Res 97:7373–7382
Weber RR, Montone RC (1990) Distribution of organochlorines in the atmosphere of the South Atlantic and Antarctic Oceans. In: Kurtz DA (ed) Long range transport of pesticides. Lewis Publishers, Chelsea, pp 185–197
Zadróżny T (1996) Fishes of Admiralty Bay caught in 1994 and 1995 (King George Is-land, South Shetland Islands, Antarctica). Pol Arch Hydrobiol 43(3):347–354
Zhang W, Ki J, Qian P (2008) Microbial diversity in polluted harbour sediments I: bacterial community assessment based on four clone libraries of 16S rDNA. Estuar Coast Shelf Sci 76:668–681

Chapter 10
Anthropogenic Impacts on Sub-Antarctic and Antarctic Islands and the Adjacent Marine Environments

Lúcia S. Campos, Rosalinda C. Montone, Rafael B. Moura, Yocie Yoneshigue-Valentin, Helena G. Kawall and Peter Convey

10.1 Introduction

The impacts of human activities are more apparent at present in the sub-Antarctic than in the Antarctic. The sub-Antarctic islands, despite their physical isolation in the Southern Ocean, are generally more accessible, have less challenging climates, and a longer history of human presence than the more southern latitudes of the Peninsula and continent (Frenot et al. 2005; Selkirk 2007; Convey and Lebouvier 2009; Trathan and Reid 2009). Parts of the sub-Antarctic and Antarctica (in particular,

L. S. Campos (✉) · R. B. Moura · Y. Yoneshigue-Valentin
Instituto de Biologia, Universidade Federal do Rio de Janeiro, Av. Carlos Chagas Filho, 373, Ilha do Fundão, Rio de Janeiro, RJ 21941-902, Brazil
e-mail: campos-lucia@biologia.ufrj.br

R. B. Moura
e-mail: lytechinusvariegatus@gmail.com

Y. Yoneshigue-Valentin
e-mail: yocie@biologia.ufrj.br

R. C. Montone · H. G. Kawall
Instituto Oceanográfico, Universidade de São Paulo, Praça do Instituto Oceanográfico, 191, Cidade Universitária, São Paulo, SP 05508-120, Brazil
e-mail: rmontone@usp.br

H. G. Kawall
e-mail: helena.kawall@gmail.com

R. B. Moura
Universidade Federal de Pernambuco, Av. Prof. Moraes Rego, 1235, Cidade Universitária, Recife, PE 50670-420, Brazil

P. Convey
British Antarctic Survey, High Cross, Madingley Road, Cambridge, CB3 0ET, UK
e-mail: pcon@bas.ac.uk

C. Verde and G. di Prisco (eds.), *Adaptation and Evolution in Marine Environments, Volume 2*, From Pole to Pole, DOI: 10.1007/978-3-642-27349-0_10,

the Peninsula and Scotia arc, including the South Shetland Islands) have experienced the most rapid rates of climate change in the Southern Hemisphere over recent decades (Convey et al. 2009a; Turner et al. 2009), and are receiving the highest contemporary rates of human visitation, associated with the fisheries industry, military, tourism, and scientific activities (Frenot et al. 2005; Lynch et al. 2010).

Anthropogenic impacts in the sub-Antarctic and Antarctic are direct and indirect. Direct impacts are related to changes in an environment component that result from direct cause–effect consequences of interactions between an environment and a human action, for example, the decrease of a limpet population due to an oil spill. Indirect impacts are related to changes in a component that result from interaction between the environment and other impacts (direct or indirect). Alteration in a seagull population because of decrease in limpet population affected by the oil spill is an example of indirect impact. Independent of the type of impact, consequences to organisms can include (1) population decrease or increase; (2) changes in distribution patterns; (3) changes in diversity patterns; (4) threat of extinction.

Many environmental impacts are more easily detectable in terrestrial rather than marine environments. Indeed much sub-Antarctic research has revealed the cause–effect relationships of different anthropogenic impacts in the terrestrial environment at local and regional scales across the Southern Ocean Islands, including studies on terrestrial plants and insects, and also on marine birds and mammals (Woehler and Croxall 1997; Woehler et al. 2001; Eriksson and Burton 2003; Hooker and Gerber 2004; Chown et al. 2005; Frenot et al. 2005; Convey and Lebouvier 2009). A particular focus in recent years has been on the establishment and invasion of non-indigenous species and their potential impacts on native biodiversity (Frenot et al. 2005; Greenslade and Convey 2012; Chown et al. 2012; Hughes and Convey 2012).

There is also increasing concern over human activities and their impacts in the marine environment itself, which are much more complex to measure and distinguish from natural sources. Most marine impact assessments in these regions relate to the near-shore environment, as this is where most human activities concentrate, except offshore fisheries. The rate and intensity of contamination largely depend on the intuitively sensible factors of (a) human settlement period, (b) human population size and frequency of visitation, and (c) environmental protection through effective monitoring and management plans (Conlan et al. 2004).

South of latitude 60°, Antarctic governance is primarily provided through consensus agreement in the Antarctic Treaty System. North of this latitude, the sub-Antarctic islands and adjacent marine areas fall under national governance. However, most regulation of the marine environment of the Antarctic and the greater part of the sub-Antarctic that lies beyond national jurisdictions falls under the umbrella of the Convention on the Conservation of Antarctic Marine Living Resources (CCAMLR) (Agnew 1997).

In the Antarctic Peninsula region, King George Island, one of the South Shetland Islands (at around 62°S), is one of the most visited parts of the Antarctic, and is closest (c. 800 km) to another continent (South America). It lies near the northern limit of the maritime Antarctic and its boundary with the sub-Antarctic. This island hosts ten permanent research stations, and is also used as a stepping stone for many scientists and tourists travelling to and from other parts of Antarctica. The first Antarctic Specially Managed Area (ASMA No. 1), Admiralty Bay, is located on this island, and was designated by the Antarctic Treaty Consultative Meeting XX in 1996 (ATCM XXVIII document ATCM 2005) due to its outstanding environmental, historical, aesthetic and scientific values. It is characterized by remarkable glaciated mountain landscapes, varied geological features, rich seabird and mammal breeding grounds, abundant terrestrial plant communities and highly diverse marine habitats (Sicinski et al. 2011). Admiralty Bay was a reference site under the SCAR EASIZ (1994–2004) programme, and it has also been considered a case study within the Census of Antarctic Marine Life during the 2007–2008 International Polar Year (Sicinski et al. 2011). A long-term evaluation of the marine environment taking into account physical and biological characteristics of the area has been established through the Brazilian Antarctic Programme in order to assess local environmental impacts. This takes into account past information, as well as applying a multidisciplinary approach using integrated data sampling.

In this paper, information on environmental assessment for monitoring and management of Admiralty Bay at King George Island is provided in the context of a general overview on impacts from human activities in the sub-Antarctic and Antarctic islands (Fig. 10.1), and their adjacent marine environments.

10.2 Southern Ocean: Anthropogenic Pressures

In the Southern Ocean marine environment, direct human impacts include fisheries and associated bycatch, ship strikes (higher predators), and pollution events. Indirect threats include the food-web ramifications of overexploitation at any trophic level (although mainly applying to krill), as well as those of habitat degradation through acoustic and chemical pollution, marine debris and physical habitat destruction (e.g. trawling methods causing long-lasting damage to the sea floor) (Hooker and Gerber 2004).

The Southern Ocean islands are best known for their diversity and abundance of seabirds and mammals, many of them being breeding platforms for petrels, albatrosses, penguins and seals (Bergstrom and Chown 1999). The terminology applied to these islands is varied (Selkirk 2007). For instance, Shirihai (2002) classified the islands into four biogeographical zones: (1) pack ice/Antarctic maritime; (2) Antarctic waters/peri-Antarctic; (3) cool sub-Antarctic; (4) temperate sub-Antarctic. They can also be divided into 'older' islands of continental origin, whose biota is derived from the nearby continents (e.g. Falkland Islands, South Georgia, Auckland and Campbell Islands, and the younger islands (some volcanic, others emergent seafloor) (Chown et al. 1998; Bergstrom and Chown 1999; Bergstrom et al. 2006). In general, each isolated Southern Ocean island

Fig. 10.1 Map showing the main sub-Antarctic and Antarctic islands (modified from Boyer and Haywood 2006). Land areas south of 60°S fall under the provisions of the Antarctic treaty system (ATS)

shows different characteristics depending on the position and evolutionary history, and more recently has suffered from varying anthropogenic pressures as summarized in Table 10.1.

The continental shelves in the Southern Ocean, including those surrounding the different island groups, support a remarkably diverse fauna (Clarke et al. 2005; Jacob et al. 2006). The most important physical environmental factors controlling distribution patterns and biodiversity in the marine realm are temperature, ice cover, oxygen, light, ultra-violet radiation, and wind, factors which vary over a range of temporal scales (hours, daily, through seasonal to interannual) and spatial scales (Knox 1994). As well as natural variability, anthropogenic changes currently affecting the Southern Ocean are linked with global processes, including warming and ozone-hole-related increased levels of UV-B radiation (Convey et al. 2009a; Turner et al. 2009), as well as direct threats from industrial, research and other visitor pressure causing pollution and introduction of alien species (Bergstrom and Chown 1999; Chown and Gaston 2002; Gaston et al. 2003; Chown et al. 2005; Frenot et al. 2005, Clarke et al. 2005). One of the greatest challenges today

Table 10.1 Summary of marine ecological information, human activities and impacts to marine organisms and the environment across the Southern Ocean Islands

Area	Southern ocean Islands	Marine ecological notes	Human activities/impacts
Continental Antarctic	Balleny	Adélie and Chinstrap penguin breeding area; 4th specially protected area under ATS; Weddell, crabeater, and southern elephant seals on some beaches	Never inhabited, few tourist landings
	Scott island	Breeding populations of giant fulmars, Antarctic prion and petrel	Uninhabited. Rarely visited
Maritime Antarctic	Peter I Øy	Small breeding populations of Adélie and chinstrap penguins, south polar skuas and southern fulmars. Large number of seals, particularly crabeater and leopard seals	Uninhabited. Occasional tourist ships but no regular landings
	South Orkney Islands	Large populations of Adélie and chinstrap, smaller numbers of gentoo and macaroni penguins; southern giant petrels and many smaller petrel species; few Antarctic fur seals breed (but large summer moulting numbers; Weddell and crabeater seals common, and leopard seals frequent around penguin colonies). Orcadas station, occupied since 1903; longest continuous climate record in the Antarctic. Signy island: c. 30 year runs of year-round marine, terrestrial and limnic environmental and biological data	Inhabited. Scottish, French and Argentine sealing, and scientific expeditions. Whaling station on Signy island in early Twentieth Century. Main activities related to research stations, Orcadas (Laurie Island) and Base H (Signy Island, UK)
	South Sandwich Islands	Large populations of chinstrap penguins in the archipelago, Adélie, gentoo, macaroni also present	Now uninhabited, Argentinian station on Southern Thule abandoned in 1982. South Africa operates and maintains unmanned meteorological stations. Very few tourist visits
	South Shetland Islands	Several penguin species breed on the Islands (Adélie, chinstrap, gentoo, macaroni). Antarctic fur, southern elephant and Weddell common, crabeater and leopard seals also found. Many species of whales offshore during summer. Some of the longest time series on bird populations, meteorological and benthic data	Inhabited. Busiest Antarctic region, nine stations in King George island, many visits by tourist cruises. National operator and tourist air operations Admiralty Bay in King George island is an ASMA, with four research stations (Poland, Brazil, USA, Peru) and a small Ecuadorian refuge

(continued)

Table 10.1 (continued)

Area	Southern oceanIslands	Marine ecological notes	Human activities/impacts
Sub-Antarctic	Heard and McDonald Islands	Pristine marine biological communities; important breeding location for seabirds and mammals (e.g. king, gentoo, macaroni, rockhopper penguins; southern elephant, Antarctic fur and sub-Antarctic fur seals); World Heritage Site in 1971. These active volcanic islands are surface exposures of the surrounding submarine Kerguelen Plateau, where the confluence of major east-moving oceanic currents and small-scale upwelling, eddies and gyres caused by submarine topography create benthic and pelagic habitats. Surface sediments with siliceous diatom mud or ooze, calcareous substrata and foraminifera in some areas. Echinoderms are the most prevalent benthic invertebrates, by-catch of benthic trawls	Uninhabited. Minimal human visitation over last 100 years: temporary research station and field camps. Nineteenth century elephant seal fishery for oil. Heard island occasionally visited by national scientific parties, adventurers and a few tourists. Mining prohibited. Commercial fishing controlled and occurs within the HIMI EEZ outside a 1 nm buffer around the 12 nm territorial sea: main targets: *Champsocephalus gunnari* and *Dissostichus eleginoides*. Marine pollution associated with shipping activities (fuel or oil spills, sewage or waste water discharge and anthropic debris. Human waste and grey water may temporarily change local nutrient regime and ecosystem functioning
	South Georgia	Large seabird populations; large numbers of Antarctic fur and southern elephant seals; many whales found offshore, shag breeding population in Shag Rocks	Inhabited. Cooper island, Specially Protected Area; Bird and Annenkov Islands, Sites of Special Scientific Interest. South Georgia, centre for the land-based South Atlantic whaling industry until 1965. BAS operates a station on Bird island. It manages an applied fisheries research station at King Edward Point on behalf of South Georgia Government. Popular destination for tourist ships. Shag Rocks uninhabited and rarely visited. Marine area, fishing ground managed under CCAMLR
	Îles Kerguelen	Populations of seabirds, abundant king, macaroni, rockhopper and gentoo penguins; increased populations of Antarctic fur and southern elephant seals	Inhabited. Past whaling and sealing destroyed much of native fauna. Port aux Français station continuously occupied by scientists since 1949
	Macquarie Island	Hosts seabirds and mammals (king, gentoo, royal and rockhopper penguins, fur and elephant seals); World Heritage Area because of unique geological values; Macquarie island Marine Park, one of the largest protected areas in the world	Inhabited. Sealing throughout the nineteenth and twentieth centuries, penguins also exploited for oil. During the Twentieth century, visited by scientific parties. Macquarie Station established in 1948

(continued)

Table 10.1 (continued)

Area	Southern oceanIslands	Marine ecological notes	Human activities/impacts
	Prince Edward and Marion Islands	Surrounded by huge kelp beds. Large seabird colonies (e.g. king, macaroni and rockhopper penguins); breeding site for southern elephant, sub-Antarctic and Antarctic fur seals; both Special Nature Reserves; intention of declaring a large Marine Protected Area around Prince Edward islands	Inhabited. Sealing was a very intense activity in late Nineteenth century. Tourism prohibited and scientific research strictly controlled. Marion island's weather station and later scientific research base continuously staffed since 1947
	Îles Crozet	Breeding populations of c. 25 million seabirds, including king, macaroni, rockhopper and gentoo penguins; except for Île de la Possession, these Islands are designated as Specially Protected Areas under Territoire des Terres Australes and Antarctiques Françaises (TAAF)	Alfred-Faure research station on Île de la Possession permanently occupied. The islands were serviced sealing and whaling industries towards the end of the nineteenth Century. French National Park since 1938
New Zealand shelf Islands	Campbell	Breeding site to yellow-eyed and rockhopper penguins and other seabirds; World Heritage Area	Now uninhabited. Human presence restricted and tourist landings regulated, but it has many introduced species that contributed to depletion of seabirds and mammals that breed on land. Sealing and whaling in late nineteenth and early twentieth centuries
	Chatham Islands	Breeding site for New Zealand;fur seal	Inhabited. Fishing and farming main source of income
	Antipodes	Breeding site to many seabirds, World Heritage Site	Uninhabited. Tourists ships with no landing
	Auckland	Breeding area of Hooker's sea lion and southern right whale in the Southwest Pacific; World Heritage Area	Uninhabited. Restricted tourist landings
	The Snares	Uninhabited. Nature Reserve. Regular tourist ships with no landing. Landing by special research permit only	Breeding site to New Zealand fur seal and Hooker's sea lion; World Heritage Site
	Bounty Islands	Erect-crested penguin breeding area; designated World Heritage Area	Uninhabited. Occasional tourists ships with no landing. Since the Island's discovery fur seals nearly wiped out

(continued)

Table 10.1 (continued)

Area	Southern oceanIslands	Marine ecological notes	Human activities/impacts
Cool temperate oceanic Islands	Falkland Islands/ Islas Malvinas	Rockhopper, magellanic, gentoo, king and macaroni penguins, black-browed albatross, southern elephant seal, South American sea lion and South American fur seal. 16 whale species and 7 dolphin species recorded off the archipelago	Inhabited since the 17th Century. Main marine activities, fishing industry (mainly squid) and more recently oil exploration, to date limited and in tandem with detailed environmental assessment. Ecosystems exten-sively altered following colonization by destruction of seal and penguin communities, and introduction of terrestrial grazing mammals
	Islas Diego Ramírez	Populations of macaroni, rockhopper and magellenic penguins, black- browed and grey-headed albatrosses	Inhabited. Occupied by an Armada de Chile detachment
	Amsterdam & Saint-Paul	Abundant rock lobsters	Inhabited. Ship-based fishery. Terrestrial habitats heavily impacted by introduced cattle grazing
	Gough	Northern-most breeding ground for southern elephant and sub-Antarctic fur seals; World Heritage Area	Inhabited. Visited by sealers, whalers and penguin hunters from the Nineteenth century. British meteorological and radio station during World War II, now maintained by the South African Weather Bureau. Staffed year-round. Least disrupted cool temperate Island and marine ecosystems
	Tristan da Cunha	Many breeding seabirds (albatross, northern rockhopper penguin); breeding colonies of sub-Antarctic fur seals	Inhabited. Seals drastically reduced by the beginning of the Eighteenth century. Current population of around 300 Islanders

(Based on information from Ainley et al. 1984; Armstrong 1994; Barnes 2006; Boyer and Haywood 2006; Chown and Gaston 2002; Kirkman et al. 2001; Gutt et al. 2006; Hopkins and Olleveant 2002; Jacob et al. 2006; Lescroël and Bost 2006; http://www.austlii.edu.au/au/legis/cth/consol_act/epabca1999588/access April 2, 2012). See Selkirk 2007; Selkirk et al. 2007; Convey and Lebouvier 2009; Trathan and Reid 2009 for descriptions of many of the islands/groups and their history of human exploitation and impact. The 'area' definitions used are taken from those widely adopted in the terrestrial research community in this region, recognising that there is no one universally applied definition of the 'sub-Antarctic' (see Smith 1984; Convey 2007; Selkirk 2007)

is to determine how changes to these various drivers may affect ecosystems and biodiversity in the Southern Ocean (Clarke et al. 2005).

10.2.1 Climate Change

Climate influences many ecological processes, and acts through local conditions such as temperature, wind, precipitation, oceanic currents and their interactions (Georges and Le Maho 2003). Global effects such as those caused by climate change, which have consequences across marine ecosystems and for all trophic levels, require mitigation at a global level (Hooker and Gerber 2004). Local environmental changes are often, but not necessarily, associated with these large-scale phenomena, again driving changes in energy and nutrient flows, biomass, growth, recruitment and movement of species (Georges and Le Maho 2003).

The climate history of the Southern Ocean has influenced different levels of biological organization in the terrestrial and marine realms (Chown et al. 1998; Turner et al. 2009). Continental scale ice-sheet formation was an important factor that provided the context for evolution in the Southern Ocean over the last 30 million years (Crame 1999; Lawver and Gahagan 2003). The combination of geographic and climate change has led to a biota rich in endemic taxa (Chown et al. 2000; Griffiths et al. 2009; Convey et al. 2012a), with diverse marine benthic systems on continental shelves (Gutt and Starmans 1998) and in the deep-sea (Brandt et al. 2004; Gage 2004), contrasting with depauperate but nevertheless unique and often ancient systems on land (Convey 2007; Convey et al. 2009b).

The position of the Southern Ocean islands in relation to the movement of the Polar Frontal Zone over time has greatly influenced their glaciation: those lying south of this zone and/or larger and higher were glaciated more heavily compared to those lying to the north of it (Hall 1990), which has some important implications in understanding biogeographic patterns and possible extinctions associated with past glaciation events (Clarke et al. 2005; Thatje et al. 2005, 2008). Increased iceberg calving in more southern regions may have led to extensive scouring of continental shelves resulting in temporal/local extinction of communities and generating processes of re-colonisation and succession, producing distinct diversity patterns (Thatje et al. 2008; Gage 2004). Periodic extensions and contractions of the continental icesheets may have influenced speciation on the continental shelves by limiting gene flow between isolated populations in the Southern Ocean (Rogers et al. 2007; González-Wevar et al. 2011a, b).

Human activities are very strongly implicated as underlying contemporary global climate trends. Since the 1950s, the Antarctic climate system has experienced marked warming on the western side of the Peninsula both in air and ocean temperatures, with the extent of seasonal sea ice formation drastically decreasing along with the rapid and widespread retreat of glaciers and multiple ice-shelf collapses (Meredith and King 2005; Turner et al. 2009). Although overall sea ice is currently increasing (as a result of the protection of the bulk of East Antarctica and associated regions of the Southern Ocean from global warming trends by the formation of the ozone hole), sea ice is predicted to decrease over the coming century as the ozone hole recovers and this protection decreases (Turner et al. 2009).

Climate change may affect marine organisms directly (through influencing their physiology) or indirectly via other elements of the food web (Georges and Le Maho 2003). Furthermore, changing climate and environmental conditions are likely to enable new taxa to establish populations, particularly in the sub-Antarctic and adjacent Antarctic regions, leading to significant changes in Southern Ocean ecosystems and their food webs (Clarke et al. 2005).

10.2.2 Human Activity in Antarctica

Human visitation and exploitation around the Southern Ocean islands commenced in the late Eighteenth Century (Headland 1984; Trathan and Reid 2009). The impacts of anthropogenic exploitation have been dramatic in many areas of the Southern Ocean, especially on vertebrate top predators such as whales and seals (Croxall 1987; Agnew and Nicol 1996; Kock 2007). As seabirds and seals must return to land or to pack-ice to reproduce, they are likely to be affected by events that influence the terrestrial, sea- and fast-ice environments (Croxall 1987; Woehler and Croxall 1997; Woehler et al. 2001). One of the most visible legacies of human activities in the sub-Antarctic has been the introduction of non-native vertebrate grazers and predators that have caused often drastic and irreversible effects on native vegetation and ground nesting and burrowing marine birds (Croxall 1987; Bergstrom and Chown 1999; Frenot et al. 2005).

Human presence is generally associated with a range of impacts to the marine environment, associated with production of sewage and grey water that generally outflows to the sea, litter, oil spills, and increased noise levels, which may affect marine biota (Hooker and Gerber 2004). Fisheries, tourism and other activities (e.g. oil exploration) are more likely to be associated with or lead to development of human settlements and facilities and, unless strictly controlled, anthropogenic impacts tend to increase significantly in such areas (Tin et al. 2009).

Environmental controls relevant to the sub-Antarctic islands are applied by the relevant sovereign governments, while those areas lying within the Antarctic Treaty governance area (in particular, in the context of this work, archipelagos such as the maritime Antarctic South Shetland Islands) fall under the Antarctic Treaty System and subsidiary conventions, such as CCAMLR (Table 10.1).

Even areas under formal environmental protection (MPAs, EPAs, ASPAs, ASMAs) are not isolated from introduced alien species and pests, nor from larger-scale environmental trends (Edgar et al. 2000; Lewis et al. 2003, 2004; Clarke et al. 2005; Hughes and Convey 2010). Pathways for both marine and terrestrial introductions must be better understood if the integrity of Southern Ocean and Antarctic biodiversity is to be protected to the greatest extent possible (Lewis et al.

2005; Chown et al. 2012). Robust baseline survey data are needed to establish the baseline states of these ecosystems and to ensure that alien species are recognised as such, and long-term monitoring programmes should be established, so that appropriate management measures can be put in place (Wall et al. 2011; Hughes and Convey 2012).

10.2.3 Sealing, Whaling and Fisheries

Major living resources that have been exploited in the Southern Ocean include whales, seals, penguins, fish, krill, stone crab and squid, with in several cases the history of exploitation extending over two centuries (Croxall 1987; Rodhouse et al. 1994, 1995; Agnew and Nicol 1996; Kock 2007; Trathan and Reid 2009). Even in some of the most isolated regions of the sub-Antarctic and Antarctic, uncontrolled exploitation of seals (especially fur seals), whales and penguins, for fur, oil, food and fuel, reduced some populations and species to the verge of extinction. Exploitation ceased only when these fisheries became unviable because of diminished financial returns (Croxall 1987; Kock 1994).

Commercial harvesting has recently been directed mainly at fish, primarily *Dissostichus eleginoides* and *D. mawsoni* and the icefish *Champsocephalus gunnari*, and Antarctic krill *Euphausia superba* (Nicol and de la Mare 1993; Kock 2007; CCAMLR 2011). The most important active fisheries are around South Georgia, the Ob and Lena Seamounts, the Crozet archipelago and the Kerguelen Plateau (Knox 1994; Agnew and Phegan 1995; Agnew 1997).

With these fisheries being regulated under CCAMLR, although accepting a possibly large component of Illegal, Unregulated and Unreported (IUU) fishery in some areas (Fallon and Kriwoken 2004; Hooker and Gerber 2004), the main direct impact from this activity relates to bycatch of non-target species (Kock 1994, 2001; Hooker and Gerber 2004), the production of various types of litter, some of which (such as packing bands, net debris, and ingestible plastic) can have serious impacts on biota (Lewis et al. 2003, 2005; Ivar do Sul et al. 2011). Clearly, without appropriate and effective management control, krill and fish fisheries may impact the entire ecosystem as they target keystone species that are the source of energy to many top predators in the Southern Ocean food web. Decreases in top-predator populations have been already described for many species (Nicol and de la Mare 1993; Ainley and Blight 2008; Trivelpiece et al. 2011).

The main fishing techniques employed—trawling or longlining—may also have other direct impacts on marine biota. For instance, longline fisheries on the shelf areas surrounding the main breeding centres of sub-Antarctic avifauna may lead to considerable mortality and contribute to the decline of the global population of many seabird species, with particular prominence being given to the impacts on albatrosses and petrels in recent years (Croxall 1987; Moloney et al. 1994; Croxall and Prince 1996; Weimerskirch et al. 1997; Nel et al. 2000; Kock 2001; Ainley and Blight 2008).

The Scientific Committee of CCAMLR is responsible for the development of a rational scientific basis for the management of commercial fisheries over an extensive portion of sub-Antarctic and Antarctic waters (Agnew 1997; Croxall and Nicol 2004). Current levels of knowledge of the quantitative dynamics of predator–prey interactions in these regions, while still imperfect, are far superior to that in any other pelagic ecosystem (Croxall 1987; Hempel 1987; Agnew 1997; Hooker and Gerber 2004). Even though regulating these large and complex marine ecosystems is a task for which managers currently have neither sufficient knowledge nor adequate tools, CCAMLR's approach is to regulate human activities proactively, with the intention of preventing substantial changes in both the target stocks and the ecosystem as a whole (Hooker and Gerber 2004).

10.2.4 Tourism

Tourism is a permitted activity in the Antarctic under the Antarctic Treaty. This activity was initiated with the first tourist flight in 1956. Tourism has increased rapidly in recent years (Frenot et al. 2005; Naveen et al. 2001; Lynch et al. 2010), as also observed in the Falkland Islands (Ingham and Summers 2002). For instance, in 1992–1993 there were less than 9,000 Antarctic tourists in total which, by the 2010–2011 season, had increased to 53,650 (www.iaato.org/tourism-statistics). Most tourist activities are ship-based with the large majority taking place in the Antarctic Peninsula and Scotia Arc region (including South Georgia), and smaller elements of the industry active in the Ross Sea and the other sub-Antarctic islands. Smaller air-supported tourist operations take place from Punta Arenas to the inland Ellsworth Mountains and, increasingly, to King George Island.

Expedition vessels visiting the Antarctic and sub-Antarctic are operated to high environmental standards as required by membership of the International Association of Antarctic Tour Operators (IAATO). However, vessels carrying more than 400 passengers may not become members of IAATO, and are therefore not subject to the same self-regulating guidelines (Ingham and Summers 2002), and they are not accepted as visitors to Antarctic stations by national operators.

The nature and severity of possible cumulative effects of repeated visits by ship-based tourists may differ from site to site depending on their characteristics and variables such as the frequency of visitation (Tin et al. 2009). Repeated anchoring of ships, for instance, while passengers are transported to sites on shore may damage benthic substrata and generally disturb community composition. Also, fuel leaks and illegal dumping of sewage and litter may have cumulative effects on nearshore benthic communities close to sites repeatedly visited (Barnes and Conlan 2007; Ainley and Blight 2008). Added to these pollutants, noise from ship and small boat operations, and repeated approaches to whales, seals, and penguins could interfere with biologically important activities such as feeding, breeding, nursing, and over time, cause animals to avoid areas traditionally used for such purposes (Hooker and Gerber 2004; Tin et al. 2009).

Sustainable tourism depends primarily on the maintenance of a pristine natural environment. Perhaps one of the most valuable mechanisms used globally to promote and manage successful wildlife ecotourism is the establishment of marine protected areas (MPA) providing a powerful and convincing method for marketing the marine environment. The entire area of an MPA should not necessarily be open to tourism. However, few MPAs and oceanic islands have management plans that include strategies for sustainable tourism (Hoyt 2005), and the MPAs currently existing in Southern Ocean locations (South Orkney Islands, South Georgia, Heard Island) were not created with this in mind.

An alternative means of minimising effects from tourism is to incentivize ecotourism, however many 'ecotourism' enterprises fail to achieve the minimum standards required to qualify in the marine tourism sector (Hall and McArthur 1993; Hoyt 2005). The control of tourist landings on islands is relatively straightforward to establish at locations under national sovereignty (Bergstrom and Chown 1999), but could be less easy to achieve at locations such as the South Shetland Islands, administered under the ATS (Hall and McArthur 1993). Unregulated landings may cause stress to seabirds and marine mammals that breed on land. Also, associated with the climate changes occurring in certain parts of the sub-Antarctic and maritime Antarctic, increased number of tourist (and other) vessels may increase the risk of introducing alien species through larval transport in ballast waters, or organisms encrusted on hulls across the Polar Front (Lewis et al. 2003; Barnes 2005; Clarke et al. 2005; Barnes and Conlan 2007). The banning of antifouling compounds could, in particular, threaten the ice-free sub-Antarctic islands used as an anchorage by vessels en-route to Antarctica (Lewis et al. 2004).

10.2.5 Invasive Species

The impact caused by invasive organisms to native communities is a primary threat to global biodiversity even in the most remote regions of the Earth (Frenot et al. 2005; Lewis et al. 2005). The translocation of species by humans is considered one of the major threats to native communities in the Antarctic and Southern Ocean (Frenot et al. 2005; Lee and Chown 2009). Despite international agreements and restrictions through the Protocol on Environmental Protection to the Antarctic Treaty (agreed in 1991, entered into force in 1998) and CCAMLR, and also specific regulations and practices adopted by individual national programmes, vessel traffic and cargo and human transit between the Southern Ocean and other regions continues to create opportunities for unintended introductions (Frenot et al. 2005; Hughes et al. 2010; Chown et al. 2012; Hughes and Convey 2012).

Around 200 established alien species have been identified to date on land in sub-Antarctic and more than ten in the Antarctic (Frenot et al. 2005; Tin et al. 2009; Hughes and Convey 2012; Greenslade et al. 2012), and this figure is likely to increase because of the current lack of baseline survey and monitoring data in many locations and groups, and the underlying trends of global change and expanding human interest in these regions (Frenot et al. 2005; Lee and Chown 2009). Virtually all records of alien species to date are from the terrestrial environment (Frenot et al. 2005). There is also a small number of records of non-indigenous species in the marine environment (Clarke et al. 2005, Frenot et al. 2005; Tin et al. 2009), although the current status of any of these is unknown. For instance, the North Atlantic spider crab *Hyas araneus* has been reported once from the Bransfield Straight between the South Shetland Islands and the Antarctic Peninsula (Tavares and Melo Tavares and De Melo 2004). A deep-water Southern Ocean king crab has been collected on the Antarctic continental slope at 1,100 m depth (Thatje et al. 2008), however it is unknown whether this actually represents a change in local distribution already occurring adjacent to the slope/shelf region. Even if this occurrence is related to a change in deep-water temperatures (warming) allowing movement up slope and onto the shelf, it is not at all clear whether this change is linked to any human influence on 'warming', rather than being part of a natural cycle.

The hulls of ships and flotsam such as plastic debris provide important anthropogenic surfaces for biological settlement in oceanic habitats (Barnes 2002; Lewis et al. 2005). As potential marine invasive species of concern are almost exclusively from shallow coastal habitats and only a sub-set of the marine debris in the ocean will move from one shallow coastal area to another, at least in the sub-Antarctic and Antarctic regions, fouling on vessel hulls may be a more likely mechanism of transferring biological communities across the biogeographical barriers of the Southern Ocean than marine debris (Lewis et al. 2005). However, Barnes and Fraser (2003) have demonstrated multi-year survival of encrusting communities on marine debris in the Southern Ocean, while Convey et al. (2002) reported accumulation of anthropogenic and natural debris on the remote Southern Ocean South Sandwich Islands, at locations that had never previously been visited by humans. Thus, the latter mechanisms cannot be discounted.

10.2.6 Offshore Exploration, Military and Scientific Activities

The Antarctic Treaty forbids military, commercial exploration and exploitation activities in the entire region under control of the ATS, other than fisheries activities approved under CCAMLR. As yet, no oil or other mineral exploration activities have taken place or been proposed around sub-Antarctic or wider Southern Ocean islands, other than the cold temperate Falkland Islands and linked areas of the South American shelf. As these are all administered under various national sovereignties, regulation of any such activities would take place at national level.

Apart from mechanical impacts to the seafloor as a result of drilling, or chemicals being released into the water column, noise pollution could be another cause of concern. There are limited data available about effects of sound on marine organisms such as cetaceans (Popper et al. 2000; McCauley et al. 2000) and fish but they may show behavioural and physiological effects including permanent loss of hearing (Popper 2003), swim bladder rupture, eye and ear damage (Gausland 2000; Hirst and Rodhouse 2000). Krill may also be affected (Popper 2003). Tissue damage resulting from high-intensity sounds produced by underwater blasting is expected to be similar for most species (Cooper 1982; Brown and Adams 1983). Other than these impacts, noise levels seem to be more important to marine mammals and birds that use the land for breeding. Significant responses from penguin chicks, petrels and skuas to overflights have been reported (Giese and Riddle 1999; Pfeiffer 2005; Hughes et al. 2008).

Light pollution can cause disorientation in birds, especially on vessels where impacts of these animals are common (Cherel et al. 1996; Black 2005; Helena Kawall, personal observation). The Southern Ocean squid-fishing industry, which relies on the use of powerful lights to attract the squid, is a particular concern here, but ‘bird strikes’ are frequently observed on the decks of ships displaying lights at night, and similarly at research stations on land.

Research stations generate similar sources of anthropic impact on the adjacent marine environment, i.e., sewage, litter and oil spills, although to a lesser extent when compared to villages or towns, which are present on some of the more peripheral Southern Ocean islands (Table 10.1). In general, environmental awareness and adequate management procedures are available in the station context. However, even the scientific work itself and its logistic support have increasingly become a cause of concern, especially in the Antarctic Treaty area (Bargagli 2005). Tin et al. (2009) provide a review of the impacts of local human activities on the Antarctic environment, including information on sewage outfalls measured around Antarctic stations. We revisit this in Table 10.2, where additional data are provided with respect to the impact of stations in Admiralty Bay.

Other than the exploitation of marine resources (even if for scientific purposes), construction of infrastructure, increased marine and air traffic, possible accidents and leaks certainly leave imprints on the Antarctic environment, some of which are irreversible (Tin et al. 2009; Convey et al. 2012b). Examples of some of these accidents are provided by recent events at the Brazilian Ferraz Station, including a major station fire, and the sinking and recovery of a fuel barge in the 2011-2012 season. Actions are now being taken in order to evaluate the extent of the impacts of these events. However, some local environmental assessments and monitoring baselines exist, with data being obtained in the area over at least the past ten years (Campos et al. 2012; Montone et al. 2012).

Undoubtedly, a combination of effective environmental impact assessments, long-term monitoring, mitigation measures for non-indigenous species, ecosystem-based management of living resources, and increased regulation of national

Table 10.2 Examples of extent of pollution from Antarctic station sewage outfalls, using chemical and microbial indicators

	McMurdo, US	Rothera, UK	Dumont d'Urville, France	Ferraz, Brazil	Casey, Australia
Pollution indicator	(all data before STP installed in 2003)	(STP installed in 2003)	(no STP)	(STP operational prior to fire, 2012)	(STP operational)
Faecal microorganisms (seawater)	<600 m (~1 km along shoreline and >300 m offshore)	<500 m (no STP), 100 m (STP operational)	<2000 m	<200 m along shoreline and >100 m offshore 10–20 m depth	~200 m
Clostridium sp. (sediment)	<3000 m (most within 200 m)	300 m (no STP)			
Clostridium sp. (marine invertebrates)	>822 m < 3000 m				
Faecal sterols	<594 m	<200 m (no STP)		~50 m 10–20 m depth	
Hydrocarbons				~10 m 10–20 m depth	
C and N isotope ratios	<612 m				
Metals Cd, Pb, Cu, Zn				20-40 m	
References	Howington et al. 1992	Hughes 2003	Delille and Delille 2000	Martins et al. 2002, 2010; Montone et al. 2010, this volume Ribeiro et al. 2011	Morris et al. 2000

(Derived from Tin et al. 2009), with additional recent data from Ferraz Station). Distance from the outfall to sites where pollution indicators are close to background levels

Antarctic programmes and tourism activities are amongst the measures that should be implemented in order to safeguard future protection of the Antarctic environment (Tin et al. 2009; Convey et al. 2012a).

10.3 Marine Environmental Management

Environmental assessment for monitoring and management has been applied on several sub-Antarctic islands, but there are fewer marine impact assessment studies than those relating to the terrestrial environment or the atmosphere.

Current human impacts in the sub-Antarctic islands off New Zealand derive from research and management teams, fishing and tourism. Tourists are allowed ashore only on Campbell, Enderby and Auckland Islands and are confined to parts of those islands (West 2005). All islands may be viewed from the sea, including from small boats, and the latter are used regularly off the Snares Islands. The islands are managed by the Southland Conservancy of the Department of Conservation, whose activities include pest eradication, weed control, threatened-species recovery, archaeological survey and facilities maintenance.

Future impacts in these islands relate to exploration for hydrocarbons in the Great South Basin; the islands may come under increased pressure as bases for shipping, perhaps even oil terminals, and the risk of pollution may increase dramatically (West 2005).

A process to protect the marine environment around the New Zealand sub-Antarctic islands has been established by the Department of Conservation and the Ministry of Fisheries since 2006 and currently the final recommendations are being prepared to be submitted to the New Zealand government for consideration and implementation (http://www.biodiversityvt.nz, accessed April 2, 2012).

Heard Island and McDonald Islands (HIMI) located in the Indian Ocean sector of the Southern Ocean have included a marine protected area since 2002 (http://www.austlii.edu.au/au/legis/cth/consol_act/epabca1999588/, accessed April 2, 2012). Recent land-based studies have focused on vegetation mapping, glaciology, cultural heritage, terrestrial ecology, seabird populations, and the marine ecosystem, while marine research has addressed physical and biological oceanography, and marine geology of the HIMI region. (http://www.heardisland.aq/research, accessed April 2, 2012). The Australian Antarctic Division coordinates research programmes in this region to understand the environment and how it is changing and to provide information to guide the sustainable management of the HIMI fishery and to supply to CCAMLR (http://www.heardisland.aq/research/management-research, accessed April 2, 2012). The islands and surrounding waters show very little evidence of anthropogenic influence, displaying features of relatively undisturbed environmental condition. The AAD has created a Catalogue of Australian Antarctic and Sub-Antarctic Metadata in which 208 metadata records are available in the areas of environmental impacts, habitat changes and natural hazards (http://data.aad.gov.au/, accessed on April, 2 2012).

According to Boyd and Murray (2001), the changing status of the marine ecosystem has been studied at South Georgia using up to 27 variables measured over 22 years from three upper trophic level predators (macaroni penguin, gentoo penguin and Antarctic fur seal) that specialise in foraging upon krill (*Euphausia superba*). These variables include population size, breeding performance, offspring growth rate, foraging behaviour and diet of the three predators. These studies have demonstrated that the marine ecosystem at South Georgia experiences acute but transient changes that are reflected in the predator populations. However, there is less certainty that trends in populations are resultant from shifts in the degree to which the ecosystem can support krill-feeding seals and penguins.

The island archipelagos near the Antarctic Peninsula, although considered within the maritime Antarctic, share many features of their climate with parts of the sub-Antarctic, and have experienced more intensive use and human activities than other parts of the continent since the first landings of sealers in the early Nineteenth Century (Kock 2007; Tin et al. 2009). One of the most intensively used islands within the South Shetlands archipelago is King George Island. This hosts ten stations as well as temporary refuges and field camps, and functions as a stepping stone to other regions of the continent, particularly the Antarctic Peninsula. One bay, Admiralty Bay is occupied by four permanent stations operated by different nations: Henryk Arctowski Station at Thomaz Point (Poland); Comandante Ferraz Antarctic Station (EACF) at Keller Peninsula (Brazil); Machu Picchu at Crepin Point (Peru); Copacabana at Llano Point (USA). Ecuador also operates a refuge at Hennequin Point. Also within the bay there are several small removable and permanent installations at Mackelar and Martel Inlets (ATCM XXVIII ATCM 2005).

Despite the Treaty requirement to monitor and report human activities, few nations do so (Hughes 2010; Hughes and Convey 2012). Kennicutt et al. (2010) documented long-term trends in various pollution sources at McMurdo Station in the Ross Sea. A long term evaluation of the atmospheric, terrestrial and marine environments taking into account physical and biological characteristics of the Admiralty Bay area has been established by Brazil in order to assess local and global environmental impacts (Weber and Montone 2006; INCT-APA 2010, 2011; Montone et al. 2012). A multidisciplinary approach using integrated data sampling has taken into account past information and gathered new data in order to generate models that may assist the management of the bay (Fig. 10.2). This study has helped select potentially significant biotic and abiotic environmental indicators (Montone et al. 2012).

Anthropogenic impacts are currently restricted to the proximities of stations in the area. Detection of biogenic methane has indicated enrichment of methanogenic communities near Ferraz (Nakayama et al. 2011). Intra- and inter-annual differences in densities of benthic macrofauna have been greater than their spatial variation (Lavrado et al. unpublished). It is clear that benthic faunal abundances change in relation to water column production (Lavrado and Rodrigues 2011), although carbon and nitrogen stable isotopic analysis has not indicated differences between areas related to organic enrichment in the food web (Corbisier et al. 2011).

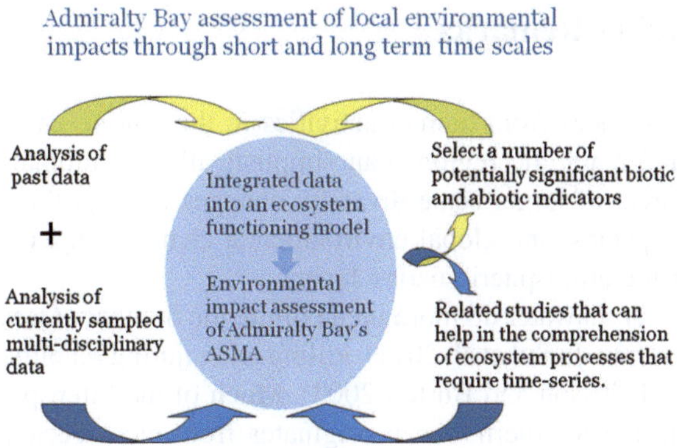

Fig. 10.2 Admiralty Bay Case Study: Brazilian assessment of local environmental impacts through short- and long-term time scales. Multidisciplinary studies were established to test strategies for environmental monitoring and provide support information for the Management of Admiralty Bay ASMA#1

The Brazilian environmental studies have shown that a suitable temporal replication within each year in sampling surveys is necessary in order to avoid confounding effects in long-term trend assessments within Antarctic environmental programmes. In the marine environment, natural rather than anthropogenic processes appear to be the main cause of sedimentary changes. The effects of contamination sources have been chronic and are so far very restricted in extent (maximum 200 m from the shoreline adjacent to Ferraz) (Montone et al. 2012). There has, however, been a decline in breeding areas of some birds, especially that of the southern giant petrel (Petry and Krüger 2010). Conversely, the populations of opportunistic scavenging and predatory species, such as skuas and kelp gulls, have increased considerably (Costa and Alves 2007; Costa 2008).

In general, natural (such as anchor-ice formation and ice scour) rather than anthropogenic processes, remain the main drivers of environmental changes in the marine system in Admiralty Bay. A number of potentially significant biotic and abiotic indicators related to the pelagic and benthic realms have been selected to improve marine environmental impact assessment, including the quality of marine water, state of the water column biota (plankton and nekton), quality of the marine sediment and state of the benthic biota.

A synthesis of benthic biodiversity knowledge against the background of habitat heterogeneity has shown that Admiralty Bay has high species richness and assemblage diversity (Sicinski et al. 2011). Less studied zones were sampled during IPY. The availability of marine-biodiversity data has increased, and analysis of the benthic community in the context of oceanic processes has shown that it could be influenced by circulation and turbulence as well as substratum type (Campos et al. 2012). However, the available ecological information still includes many gaps relating to the actual human impacts on terrestrial and marine diversity in the area (Sicinski et al. 2011; INCT-APA 2011; Campos et al. 2012).

10.4 Concluding Remarks

The main consequences from human activities in the marine environment in the sub-Antarctic and Antarctic regions can simplistically be summarised as marine pollution, the threat of larger-scale diversity changes through the introduction of non-indigenous species, and global environmental changes (e.g. climate warming and thinning of the atmospheric ozone layer).

Marine debris in seawater do more than pollute the Southern Ocean and visually degrade beaches (Ivar do Sul et al. 2011), killing and injuring seabirds and mammals (Nel et al. 2000; Eriksson and Burton 2003). Much of the litter, plastic and metal debris present in the Southern Ocean originates from meso-oceanic regions, primarily from fishing industry activities (Lewis et al. 2005, Ivar do Sul et al. 2011). Many seabirds are prone to the accumulation of large amounts of man-made litter leading to a decrease in body condition and possibly starvation (Nel et al. 2000 and references therein). Also, some of this debris (e.g. floating plastics) can provide substrata to cosmopolitan pelagic species, while other elements may sink to the deep sea (e.g. glass and metal bottles and containers) (Ivar do Sul et al. 2011). The impacts of pollution such as fuel oil spills or leaking storage tanks may not become clear until many years later in the sub-Antarctic and Antarctic regions. Even today evidence of wastes from old whaling stations may be detected from the deepest sediment layers sampled, such as in King Edward Cove, South Georgia (Mackie et al. 1978). In the continent, chemical and biological signatures of human pollution over several decades are apparent at the US McMurdo Station (Kennicutt et al. 2010), in a rare example of a published monitoring study of human impacts of Antarctic operations (see also Hughes 2010).

As the Southern Ocean includes some of the most productive and unique marine ecosystems on Earth, which have been heavily exploited in the past, sustainable management of marine resources requires the ability to distinguish the effects of human exploitation from those related to climate variability and change (Rintoul et al. 2012). Hooker and Gerber (2004) suggest that pelagic marine reserves are increasingly needed, and in this context marine predators may be most effectively used as indicator species of underlying prey distribution and ecosystem processes. For instance, a reduction in sea ice off the Antarctic Peninsula has been linked to a decline in krill abundance resulting in community-structure and associated food-web changes (Atkinson et al. 2004; Murphy et al. 2007; Hofmann et al. 2011). Establishing target areas and boundaries for ecosystem reserves can greatly benefit marine biota, and such areas can then also be used to monitor larger-scale environmental changes such as the effects of global changes.

The Southern Ocean may be subdivided into different biogeographical zones (Griffiths et al. 2009; Griffiths 2010), but human impacts are similar in many sub-Antarctic and Antarctic locations. Some more isolated islands remain in relatively pristine state, but risks increase with increasing human visitation to Antarctica.

Terrestrial and marine biodiversity may be affected by natural processes and local and global anthropic impacts. Understanding the types of human activities and their consequences to the environment have become increasingly important in the sub-Antarctic and Antarctic regions. More than ever, integrated, multidisciplinary studies, careful analysis of impact origins, and long-term monitoring are required in order to assess the effects of human activities on Antarctic environments and their biodiversity (Convey 2011; Wall et al. 2011). The outcomes from such studies provide robust and appropriate tools for environmental management, especially when taking into account areas considered as biodiversity hotspots.

Acknowledgments Part of this review was stimulated by the Census of Antarctic Marine Life during IPY, and especially Drs. Michael Stoddart and Victoria Wadley. The Brazilian assessment of Admiralty Bay was contributed by the Brazilian Antarctic Environment Research Network (CNPq processes: 550345/02-7, 550347/02-0, 550348/02-6, 550349/02-2, 550350/02-0, 550351/02-7, 550354/02-6, 550356/02-9, 550357/02-5, 550359/02-8, 550365/02-8, 550368/02-7, 550371/02-8, 550373/02-0), the IPY project MABIREH (Process No. 52.0293/2006-1), and continues through the Brazilian National Institute of Science and Technology on Antarctic Environmental Research (INCT-APA) financed by CNPq (Process No. 574018/2008-5), FAPERJ (Process No. E-26170.023/2008). We are thankful to the Ministry of Science, Technology and Innovation (MCTI), Ministry of Environment (MMA), and Interministerial Secretariat for the Brazilian Marine Resources (SECIRM). G. di Prisco and C. Verde are thanked for useful comments and contribution to this chapter.

References

Agnew DJ (1997) Review: the CCAMLR ecosystem monitoring programme. Antarct Sci 9(3):235–242

Agnew DJ, Nicol S (1996) Marine disturbances: commercial fishing. Foundation for ecological research west of the Antarctic Peninsula. Antarct Res Ser 70:417–435

Agnew DJ, Phegan G (1995) A fine-scale model of the overlap between penguin foraging demands and the krill fishery in the South Shetland Islands and Antarctic Peninsula. CCAMLR Sci 2:99–110

Ainley DG, Blight LK (2008) Ecological repercussions of historical fish extraction from the Southern Ocean. Fish Fish 9:1–26

Ainley DG, O'Connor EF, Boekelheide RJ (1984) The marine ecology of birds in the Ross Sea, Antarctica. Ornithol Monogr 32:1–97

Armstrong PH (1994) Human impact on the Falkland Islands environment. Environmentalist 14(3):1573–2991

ATCM XXVIII (2005) Review of the admiralty bay antarctic specially managed area management plan (ASMA No. 1). ATCM XXVIII, Stockholm, pp 1–31

Atkinson A, Siegel V, Pakhomov E, Rothery P (2004) Long-term decline in krill stock and increase in salps within the Southern Ocean. Nature 432:100–103

Bargagli R (2005) Antarctic ecosystems: environmental contamination, climate change, and human impact. Springer, Berlin

Barnes DKA (2002) Invasions by marine life on plastic debris. Nature 416:808–809

Barnes DKA (2005) Changing chain: past, present and future of the Scotia Arc's and Antarctica's shallow benthic communities. Atlantic 69:65–89

Barnes DKA (2006) A most isolated benthos: coastal bryozoans of Bouvet Island. Polar Biol 29:114–119

Barnes DKA, Conlan KE (2007) Disturbance, colonization and development of Antarctic benthic communities. Phil Trans R Soc Lond B Biol Sci 362:11–38

Barnes DKA, Fraser KPP (2003) Rafting by five phyla on man-made flotsam in the Southern Ocean. Mar Ecol Prog Ser 262:289–291

Bergstrom DM, Chown SL (1999) Life at the front: history, ecology and change on the southern ocean islands. Tree 14(12):472–477

Bergstrom D, Hodgson DA, Convey P (2006) The physical setting of the Antarctic. In: Bergstrom DM, Convey P, Huiskes AHL (eds) Trends in Antarctic terrestrial and limnetic ecosystems: Antarctica as a global indicator. Springer, Dordrecht, pp 15–33

Black A (2005) Light induced seabird mortality on vessels operating in the Southern Ocean: incidents and mitigation measures. Antarctic Sci 17(1):67–68

Boyd IL, Murray AWA (2001) Monitoring a marine ecosystem using responses of upper trophic level predators. J Anim Ecol 70(5):747–760

Boyer P, Haywood E (2006) Islands of the Southern Ocean. Information for delegates to the International Forum of the Sub-Antarctic. Hobart. 27 pp

Brandt A, De Broyer C, Gooday AJ, Hilbig B, Thomson MRA (2004) Introduction to ANDEEP (ANtarctic benthic DEEP-sea biodiversity: colonization history and recent community patterns)—a tribute to HL Sanders. Deep Sea Res II 51:1457–1465

Brown CR, Adams NJ (1983) The effect of underwater explosions on Rockhopper Penguins *Eudyptes chrysocome*. Cormorant 11:68

Campos LS, Barboza CAM, Bassoi M, Bernardes M, Bromberg S, Corbisier TN, Fontes RC, Gheller PF, Hajdu E, Kawall HG, Lange PK, Lanna AM, Lavrado HP, Monteiro GCS, Montone RC, Morales T, Moura RB, Nakayama CR, Oackes T, Paranhos R, Passos FD, Petti MAV, Pellizari VH, Rezende CE, Rodrigues M, Rosa LH, Secchi E, Tenenbaum DR, Yoneshigue-Valentin Y (2012) Environmental processes, biodiversity and changes in Admiralty Bay, King George Island, Antarctica. In: Verde C, di Prisco G (eds) Adaptation and evolution in marine environments—the impacts of global change on biodiversity, vol 2. Springer, Heidelberg

CCAMLR (2011) Statistical Bulletin, vol 23 (Database Version). www.ccamlr.org. Accessed 23 Mar 2012

Cherel Y, Weimerskirch H, Duhamel G (1996) Interactions between longline vessels and seabirds in Kerguelen waters and a method to reduce seabird mortality. Biol Conserv 75:63–70

Chown S, Gaston KJ (2002) Island-hopping invaders hitch a ride with tourists in South Georgia. Nature 408:637

Chown SL, Gaston KJ, Gremmen NJM (2000) Including the Antarctic: insights for ecologists everywhere. In: Davison W, Howard-Williams C, Broady P (eds) Antarctic ecosystems: models for wider ecological understanding. N Z Nat Sci, Christchurch, pp 1–15

Chown SL, Gremmen NJM, Gaston KJ (1998) Ecological biogeography of Southern Ocean islands: species-area relationships, human impacts, and conservation. Am Nat 152(4):562–575

Chown SL, Huiskes AHL, Gremmen NJM, Lee JE, Terauds A, Crosbie K, Frenot Y, Hughes KA, Imura S, Kiefer K, Lebouvier M, Raymond B, Tsujimotoi M, Ware C, Van de Vijver B, Bergstrom DM (2012) Continent-wide risk assessment for the establishment of non-indigenous species in Antarctica. Proc Nat Acad Sci. doi:10.1073/pnas1119787109

Chown SL, Hull B, Gaston KJ (2005) Human impacts, energy availability ad invasion across Southern Ocean Islands. Global Ecol Biogeogr 14:521–528

Clarke A, Barnes DKA, Hodgson DA (2005) How isolated is Antarctica? Trends Ecol Evol 20(1):1–3

Conlan KE, Kim SL, Lenihan HS, Oliver JS (2004) Benthic changes during 10 years of organic enrichment by McMurdo station, Antarctica. Mar Pollut Bull 49:43–46

Convey P (2007) Antarctic ecosystems. Encyclopedia of biodiversity, Levin SA. Elsevier, San Diego

Convey P (2011) Antarctic terrestrial biodiversity in a changing world. Polar Biol 11:1629–1641

Convey P, Lebouvier M (2009) Environmental change and human impacts on terrestrial ecosystems of the sub-Antarctic islands between their discovery and the mid-Twentieth Century. Pap Proc R Soc Tasman 143:33–44

Convey P, Barnes DKA, Griffiths H, Grant S, Linse K, Thomas DN (2012a) Biogeography and regional classifications of Antarctica. In: Rogers AD, Johnston NM, Murphy E, Clarke A (eds)

Chapter 15—Antarctica: an extreme environment in a changing world. Blackwell, Oxford, pp 469–491

Convey P, Barnes DKA, Morton A (2002) Artefact accumulation on Antarctic oceanic island shores. Polar Biol 25:612–617

Convey P, Bindschadler RA, di Prisco G, Fahrbach E, Gutt J, Hodgson DA, Mayewski P, Summerhayes CP, Turner J (2009a) Antarctic climate change and the environment. Antarctic Sci 21:541–563

Convey P, Hughes KA, Tin T (2012) Continental governance and environmental management mechanisms under the Antarctic Treaty System: sufficient for the biodiversity challenges of the next century? Biodiversity (in press)

Convey P, Stevens MI, Hodgson DA, Smellie JL, Hillenbrand CD, Barnes DKA, Clarke A, Pugh PJA, Linse K, Cary SC (2009b) Exploring biological constraints on the glacial history of Antarctica. Quat Sci Rev 28:3035–3048

Cooper J (1982) Methods of reducing mortality of seabirds caused by underwater blasting. Cormorant 10:109–113

Corbisier TN, Bromberg S, Gheller PF, Piera FE, Petti MAV (2011) Monitoring the impact of human activities in Admiralty Bay, King George Island, Antarctica: isotopic analysis of C and N in the summer 2005/2006. In: INCT-APA Annual Activity Report 2010, Editora Cubo, São Carlos, pp 182–187

Costa ES (2008) Aves de Ponta Hennequin (Península Antártica) com ênfase na Dis-tribuição, Abundância e Biologia Reprodutiva de skuas (Aves: Stercorariidae). Dissertation, Universidade Federal do Rio de Janeiro

Costa ES, Alves MAS (2007) Biologia reprodutiva e ecologia comportamental de skuas antárticas *Catharacta maccormicki* e *C. lonnbergi*. Oecol Brasil 11:78–94

Crame JA (1999) An evolutionary perspective on marine faunal connections between southernmost South America and Antarctica. Sci Mar 63(1):1–14

Croxall JP (1987) The status and conservation of Antarctic seals and seabirds: a review. Environ Int 13:55–70

Croxall JP, Nicol S (2004) Management of southern ocean fisheries: global forces and future sustainability. Antarctic Sci 16(4):569–584

Croxall JP, Prince PA (1996) Potential interaction between wandering albatrosses and longline fisheries for Patagonia toothfish at South Georgia. CCAMLR Sci 3:101–110

Delille D, Delille E (2000) Distribution of enteric bacteria in Antarctic seawater surrounding the Dumont d'Urville permanent station (Adélie Land). Mar Pollut Bull 40:869–872

Edgar GJ, Barrett NS, Graddon DJ, Last PR (2000) The conservation significance of estuaries: a classification of Tasmanian estuaries using ecological, physical and demographic attributes as a case study. Biol Conserv 92:383–397

Eriksson C, Burton H (2003) Origins and biological accumulation of small plastic particles in fur-seal scats from Macquarie Island. Ambio 32(6):380–384

Fallon LD, Kriwoken LK (2004) International influence of an Australian nongovernment organization in the protection of patagonian toothfish. Ocean Dev Int L 35:221–266

Frenot Y, Chown SL, Whinam J, Selkirk PM, Convey P, Skotnicki M, Bergstrom DM (2005) Biological invasions in the Antarctic: extent, impacts and implications. Biol Rev 80:45–72

Gage JD (2004) Diversity in deep-sea benthic macrofauna: the importance of local ecology, the larger scale, history and the Antarctic. Deep Sea Res II 51:1689–1708

Gaston KJ, Jones AG, Häenel C, Chown SL (2003) Rates of species introduction to a remote oceanic island. Proc Biol Sci 270:1091–1098

Gausland I (2000) Impact of seismic surveys on marine life. Lead Edge 19:903–905

Georges JY, Le Maho Y (2003) Réponses des ecosystems marins et insulaires aux changements climatiques. Geoscience 335:551–560

Giese M, Riddle M (1999) Disturbance of emperor penguin *Aptenodytes forsteri* chicks by helicopters. Polar Biol 22:366–371

González-Wevar CA, David B, Poulin E (2011a) Phylogeography and demographic inference in *Nacella* (Patinigera) *concinna* (Strebel, 1908) in the western Antarctic Peninsula. Deep Sea Res II 58:220–229
González-Wevar CA, Nakano T, Cañete JI, Poulin E (2011b) Concerted genetic, morphological and ecological diversification in *Nacella* limpets in the Magellanic Province. Mol Ecol 20(9):1936–1951
Greenslade P, Convey P (2012) Exotic *Collembola* on subantarctic islands: pathways, origins and biology. Biol Invasions 14:405–417
Greenslade P, Potapov M, Russel D, Convey P (2012) Lessons from history—global Collembola on Deception Island. J Insect Sci (in press)
Griffiths HJ (2010) Antarctic marine biodiversity—what do we know about the distribution of life in the Southern Ocean? PLoS One 5(8):1–11
Griffiths HJ, Barnes DKA, Linse K (2009) Towards a generalised biogeography of the Southern Ocean benthos. J Biogeogr 36:162–177
Gutt J, Starmans A (1998) Megabenthic structure and biodiversity in the Weddell and Lazarev Seas (Antarctica): ecological role of physical parameters and biological interactions. Polar Biol 20:229–247
Gutt J, Fricke A, Teixidó N, Potthoff M, Arntz W (2006) Mega-epibenthos at Bouvet Island (South Atlantic): a spatially isolated biodiversity hot spot on a tiny geological spot. Polar Biol 29:97–105
Hall KJ (1990) Quaternary glaciations in the southern ocean: Sector 0° long.−180° long. Q Sci Rev 9:217–228
Hall CM, McArthur S (1993) Case study: ecotourism in Antarctica and adjacent sub-Antarctic islands: development, impacts, management and prospects for the future. Tour Manag 14:117–122
Headland R (1984) The island of South Georgia. Cambridge University Press, New York
Hempel G (1987) The krill-dominated pelagic system of the Southern Ocean. Environ Int 13:33–36
Hirst AG, Rodhouse PG (2000) Impacts of geophysical seismic surveying on fishing success. Rev Fish Biol Fish 10:113–118
Hofmann EE, Wiebe PH, Costa DP, Torres JJ (2011) Introduction to understanding the linkages between Antarctic food webs and the environment: a synthesis of Southern Ocean GLOBEC studies. Deep Sea Res II 58(13–16):1505–1507
Hooker SK, Gerber LR (2004) Marine reserves as a tool for ecosystem-based management: the potential importance of megafauna. Bioscience 54(1):27–39
Hopkins L, Olleveant A (2002) A collaborative approach to the environmental assessment process prior to oil exploration activities offshore the Falkland Islands. Aquat Conserv Mar Freshw Ecosyst 12(1):137–144
Howington JP, McFeters GA, Barry JP, Smith JJ (1992) Distribution of the McMurdo station sewage plume. Mar Pollut Bull 25:324–327
Hoyt E (2005) Sustainable ecotourism on Atlantic Islands, with special reference to whale watching, marine protected areas and sanctuaries for cetaceans. Biol Environ Proc R Ir Acad 105(3):141–154
Hughes KA (2003) Influence of seasonal environmental variables on the distribution of presumptive fecal coliforms around an Antarctic research station. Appl Env Microbiol 69:4884–4891
Hughes KA (2010) How committed are we to monitoring human impact in Antarctica? Env Res Lett 5(4):1–3. 041001. doi:10.1088/1748-9326/5/4/041002
Hughes KA, Convey P (2010) The protection of Antarctic terrestrial ecosystems from inter and intra-continental transfer of non-indigenous species by human activities: a review of current systems and practices. Glob Environ Chang Hum Policy Dimens 20:96–112
Hughes KA, Convey P (2012) Determining the native/non-native status of newly discovered terrestrial and freshwater species in Antarctica—current knowledge, methodology and management action. J Env Manag 93:52–66

Hughes KA, Convey P, Maslen NR, Smith RIL (2010) Accidental transfer of non-native soil organisms into Antarctica on construction vehicles. Biol Invasions 12:875–891

Hughes KA, Waluda CM, Stone RE, Ridout MS, Shears JR (2008) Short-term responses of king penguins *Aptenodytes patagonicus* to helicopter disturbance at South Georgia. Polar Biol 31:1521–1530

Ingham RJ, Summers D (2002) Falkland Islands cruise ship tourism: an overview of the 1999–2000 season and the way forward. Aq Cons Mar Freshw Ecosyst 12:145–152

INCT-APA (2010) Annual Activity Report 2009—National Institute of Science and Technology—Antarctic Environmental Research. São Carlos, Editora Cubo, 102 pp

INCT-APA (2011) Annual Activity Report 2010—National Institute of Science and Technology—Antarctic Environmental Research. São Carlos, Editora Cubo, 240 pp

Ivar do Sul JA, Barnes DKA, Costa MF, Convey P, Costa ES, Campos LS (2011) Plastics in the Antarctic environment: are we looking at the tip of the iceberg? Oecol Austral 15:150–170

Jacob U, Brey T, Fetzer I, Kaehler S, Mintenbeck K, Dunton K, Beyer K, Struck U, Pakhomov EA, Arnzt WE (2006) Towards the trophic structure of the Bouvet Island marine ecosystem. Polar Biol 29:106–113

Kennicutt MC II, Klein A, Montagna P, Sweet S, Wade T, Palmer T, Sericano J, Denoux G (2010) Temporal and spatial patterns of anthropogenic disturbance at McMurdo Station, Antarctica. Env Res Lett 5:034010

Kirkman SP, Hofmeyr GJG, Bester MN, Isaksen K (2001) Counts of southern elephant seals, Mirounga leonine at Bouvet Island. Polar Biol 24:62–65

Knox GA (1994) The biology of the Southern Ocean. Cambridge University Press, Cambridge

Kock KH (1994) Fishing and conservation in southern waters. Polar Rec 30:3–22

Kock KH (2001) The direct influence of fishing and fishery-related activities on non-target species in the Southern Ocean with particular emphasis on longline fishing and its impact on albatrosses and petrels: a review. Rev Fish Biol Fish 11:31–56

Kock KH (2007) Antarctic marine living resources—exploitation and its management in the Southern Ocean. Antarct Sci 19:231–238

Lavrado HP, Rodrigues E (2011) Thematic area 3—impact of human activities on the antarctic marine environment. In: INCT-APA annual activity report 2010. Editora Cubo, São Carlos, pp 100–106

Lawver LA, Gahagan LM (2003) Evolution of Cenozoic seaways in the circum-Antarctic region. Palaeogeogr Palaeoclimatol Palaeoecol 198:11–37

Lee JE, Chown SL (2009) Breaching the dispersal barrier to invasion: quantification and management. Ecol Appl 19:1944–1959

Lescroël A, Bost CA (2006) Recent decrease in gentoo penguin populations at Iles Kerguelen. Antarct Sci 18:171–174

Lewis PN, Hewitt CL, Riddle MJ, McMinn A (2003) Marine introductions in the Southern Ocean: an unrecognized hazard to biodiversity. Mar Pollut Bull 46:213–223

Lewis PN, Riddle MJ, Hewitt CL (2004) Management of exogenous threats to Antarctica and the sub-Antarctic islands: balancing risks from TBT and non-indigenous marine organisms. Mar Pollut Bull 49:999–1005

Lewis PN, Riddle MJ, Smith SDA (2005) Assisted passage or passive drift: a comparison of alternative transport mechanisms for non-indigenous coastal species into the Southern Ocean. Antarct Sci 17(2):183–191

Lynch HJ, Crosbie K, Fagan WF, Naveen R (2010) Spatial patterns of tour ship traffic in the Antarctic Peninsula region. Antarct Sci 22:123–130

Mackie PR, Platt HM, Hardy R (1978) Hydrocarbons in the marine environment II. Distribution of n-alkanes in the fauna and environment of the sub-Antarctic island of South Georgia. Estuar Coast Shelf Sci 6:301–313

Martins CC, Bícego MC, Rose NL, Taniguchi S, Lourenço RA, Figueira RCL, Mahiques MM, Montone RC (2010) Historical record of polycyclic aromatic hydrocarbons (PAHs) and spheroidal carbonaceous particles (SCPs) in marine sediment cores from Admiralty Bay, King George Island, Antarctica. Env Pollut 158:192–200

Martins CC, Venkatesan MI, Montone RC (2002) Sterols and linear alkylbenzenes in marine sediments from Admiralty Bay, King George Island, South Shetland Islands. Antarct Sci 14:244–252

McCauley RD, Fewtrell J, Duncan AJ, Jenner C, Jenner MN, Penrose JD, Prince RIT, Adhitya A, Murdoch J, McCabe K (2000) Marine seismic surveys—a study of environmental implications. APPEA J 2000:692–708

Meredith MP, King JC (2005) Rapid climate change in the ocean west of the Antarctic Peninsula during the second half of the 20th century. Geophys Res Lett 32(19):1–5

Moloney CL, Cooper J, Ryan PG, Siegfried WR (1994) Use of a population model to assess the impact of longline fishing on wandering albatross *Diomedea exulans* populations. Biol Conserv 70:195–204

Montone RC, Campos LS, Alvarez CE, Ito RG, Lavrado HP, Bícego MC, Pellizari VH, Schaefer CEGR, Corbisier TN, Sander M, Pereira AB, Gomes V, Carelli RF, Braga ES, Ngan PV, Mahiques MM, Castro BM, Petti MAV, Nakayama CR, Martins CC, Francelino M, Weber RR (2012) Environmental Assessment of Admiralty Bay, King George Island, Antarctica. In: Verde C, di Prisco G (eds) Adaptation and evolution in marine environments—the impacts of global change on biodiversity, vol 2. Springer, Heidelberg

Montone RC, Martins CC, Bícego MC, Taniguchi S, Silva DAM, Campos LS, Weber RR (2010) Distribution of sewage input in marine sediments around a maritime Antarctic research station indicated by molecular geochemical indicators. Sci Total Environ 408:4665–4671

Morris CE, George J, Tate, PM, Cathers B (2000) Impacts of wastewater discharge to the Antarctic marine environment. In: Hugston T, Ruckstuhl C (eds) Proceedings of the sixth international symposium on cold region development, Hobart, pp 155–158

Murphy EJ, Watkins JL, Trathan PN, Reid K, Meredith MP, Thorpe SE, Johnston NM, Clarke A, Tarling GA, Collins MA, Forcada J, Sreeve RS, Atkinson A, Korb R, Whitehouse MJ, Ward P, Rodhouse PG, Enderlein P, Hirst AG, Martin AR, Hill SL, Staniland IJ, Pond DW, Briggs DR, Cunningham NJ, Fleming AH (2007) Spatial and temporal operation of the Scotia Sea ecosystem: a review of large-scale links in a krill centered food web. Proc R Soc B 362:113–148

Nakayama CR, Kuhn E, Araújo ACV, Alvalá PC, Ferreira WJ, Vazoller RF, Pellizari VH (2011) Revealing archaeal diversity patterns and methane fluxes in Admiralty Bay, King George Island, and their association to Brazilian Antarctic Station activities. Deep Sea Res II 58(1–2):128–138

Naveen R, Forrest SC, Dagit RG, Blight LK, Trivelpiece WZ, Trivelpiece SG (2001) Zodiac landings by tourist ships in the Antarctic peninsula region, 189–99. Polar Rec 37:121–132

Nel DC, Nel JL, Rya PG, Klages NTW, Wilson RP, Robertson G (2000) Foraging ecology of grey-headed mollymawks at Marion Island, southern Indian Ocean, in relation to longline fishing activity. Biol Conserv 96:219–231

Nicol S, de la Mare WK (1993) Ecosystem management and the Antarctic krill. Am Sci 81:36–47

Petry VM, Krüger L (2010) Monitoring of the dynamic and spatial distribution of Antarctic seabird populations in the South Shetland Islands. In: INCT-APA annual activity report 2009. Editora Cubo, São Carlos, pp 27–30

Pfeiffer S (2005). Effects of human activities on southern giant petrels and skuas in the Antarctic. PhD thesis, Ecology Institute, University of Jena. Available at: www.db-thueringen.de/servlets/derivateservlet/derivate-6345/Pfeiffer.pdf

Popper AN (2003) Effects of anthropogenic sounds on fishes. Fisheries 28(10):24–31

Popper AN, Deferrari HA, Dolphin WF, Edds-Walton PL, Greve GM, McFadden D, Rhines PB, Ridgway SH, Seyfarth RM, Smith SL, Tyack PL (2000) Marine mammals and low-frequency noise: progress since 1994. Nat Acad Press, Washington, DC

Ribeiro AP, Figueira RCL, Martins CC, Silva CRA, França EJ, Bícego MC, Mahiques MM, Montone RC (2011) Arsenic and trace metal contents in sediment profiles from the Admiralty Bay, King George Island, Antarctica. Mar Pollut Bull 62:192–196

Rintoul SR, Sparrow M, Meredith MP, Wadley V, Speer K, Hofmann E, Summerhayes C, Urban E, Bellerby R (2012) The Southern Ocean observing system: initial science and implementation strategy. Scientific Committee on Antarctic Research, Cambridge, 74 pp

Rodhouse PG, Barton J, Hatfield EMC, Symon C (1995) *Illex argentius*: life cycle, population structure and fishery. ICES Mar Fish Symp 199:425–432

Rodhouse PG, Robinson K, Gajdatsy SB, Daly HI, Asmore MJS (1994) Growth, age structure and environmental history in the cephalopod *Martialia hyadesi* (Teuthoidea: Ommastrephidae) at the Antarctic Polar Frontal Zone and on the Patagonian Shelf Edge. Antarct Sci 6:259–267

Rogers AD, Murphy EJ, Johnston NM, Clarke A (2007) Introduction. Antarctic ecology: from genes to ecosystems. Part 2. Evolution, diversity and functional ecology. Phil Trans R Soc B 362:2187–2189

Selkirk PM (2007) The nature and importance of the sub-Antarctic. Pap Proc R Soc Tasman 141(1):1–6

Selkirk PM, Quilty PG, Davies M (eds) (2007) The sub-Antarctic, papers and proceedings of the Royal Society of Tasmania, vol 141, part 1

Shirihai H (2002) A complete guide to Antarctic wildlife. Alua Press, Degerby

Siciński J, Jażdżewski K, De Broyer C, Ligowski R, Presler P, Nonato EF, Corbisier TN, Petti MAV, Brito TAS, Lavrado HP, Błażewicz-Paszkowycz M, Pabis K, Jażdżewska A, Campos LS (2011) Admiralty Bay Benthos diversity: a long-term census. Deep Sea Res II 58(1–2):30–48

Smith RIL (1984) Terrestrial plant biology of the sub-Antarctic and Antarctic. In: Laws RM (ed) Antarctic ecology. Academic Press, London, pp 61–162

Tavares M, De Melo GAS (2004) Discovery of the first known benthic invasive species in the Southern Ocean: the North Atlantic spider crab *Hyas araneus* found in the Antarctic Peninsula. Antarct Sci 16(2):129–131

Thatje S, Hillerbrand C-D, Larter R (2005) On the origin of Antarctic marine benthic structure.Trends Ecol Evo 20(10):534–540

Thatje S, Hall S, Hauton C, Held C, Tyler P (2008) Encounter of lithodid crab *Paralomis birsteini* on the continental slope off Antarctica, sampled by ROV. Polar Biol 31:1143–1148

Tin T, Fleming ZL, Hughes KA, Ainley DG, Convey P, Moreno CA, Pfeiffer S, Scott J, Snape I (2009) Impacts of local human activities on the Antarctic environment. Antarct Sci 21(1):1–31

Trathan PN, Reid K (2009) Exploitation of the marine ecosystem in the sub-Antarctic: historical impacts and current consequences. Pap Proc R Soc Tasman 143:9–14

Trivelpiece WZ, Hinke JT, Miller AK, Reiss CS, Trivelpiece SG, Watters GM (2011) Variability in krill biomass links harvesting and climate warming to penguin population changes in Antarctica. Proc Nat Acad Sci, April 11

Turner J, Bindschadler R, Convey P, di Prisco G, Fahrbach E, Gutt J, Hodgson D, Mayewski P, Summerhayes C (2009) Antarctic climate change and the environment, a contribution to the international polar year 2007–2008. SCAR, Cambridge 526 pp

Wall DH, Lyons WB, Convey P, Howard-Williams C, Quesada A, Vincent WF (2011) Long term ecosystem networks to record change: an international imperative. Antarct Sci 23:209

Weber RR, Montone RC (2006) Rede 2—Gerenciamento ambiental na Baía do Almirantado, Ilha Rei George, Antártica. Technical Report, Universidade de São Paulo, 252 pp

Weimerskirch H, Brothers N, Jouventin P (1997) Population dynamics of wandering albatross *Diomedea exulans* and Amsterdam albatross *D. amsterdamensis* in the Indian Ocean and their relationships with long-line fisheries: conservation implications. Conserv Biol 79:257–270

West C (2005) New Zealand subantarctic islands research strategy. Southland conservancy, Invercargill

Woehler EJ, Croxall JP (1997) The status and trends of Antarctic and Subantarctic seabirds. Mar Ornithol 25:43–66

Woehler EJ, Cooper J, Croxall JP, Fraser WR, Kooyman GL, Miller GD, Nel DC, Patterson DL, Peter HU, Ribic CA, Salwicka K, Trivelpiece WZ, Weimerskirch H (2001) A statistical assessment of the status and trends of Antarctic and sub-Antarctic seabirds. Report on SCAR BBS Workshop on Southern Ocean seabird populations, 43 pp

Chapter 11
Polar Monitoring: Seabirds as Sentinels of Marine Ecosystems

Céline Le Bohec, Jason D. Whittington and Yvon Le Maho

11.1 The Global Importance of Polar Monitoring

The Intergovernmental Panel on Climate Change (IPCC 2007) has highlighted an urgent need to assess how ecosystems respond to climate change. This has placed a large Earth and Life Sciences focus on polar regions, as these areas are so far experiencing the strongest and the most rapid global environmental changes (Hansen et al. 2006; Bindoff et al. 2007; Parry et al. 2007; Serreze et al. 2007; Graversen et al. 2008; Hoegh-Guldberg and Bruno 2010). However, while some oceanographic parameters can be continuously monitored by satellites, monitoring the productivity of the higher trophic levels of the oceans requires larger, challenging and costly campaigns. The monitoring of Southern Ocean ecosystems is especially logistically difficult due to the very large dimension of this ocean, its heterogeneity, and the severity of its climate (storms, low temperatures, sea ice cover). This frozen ocean moreover has a complex geopolitical status with an

C. Le Bohec (✉) · J. D. Whittington · Y. Le Maho
European Associated Laboratory 647 'BioSensib' Centre Scientifique de Monaco & Centre National de la Recherche Scientifique, 8 quai Antoine 1er, MC 98000 Monte-Carlo, Principality of Monaco
e-mail: clebohec@centrescientifique.mc

C. Le Bohec · J. D. Whittington · Y. Le Maho
Institut Pluridisciplinaire Hubert Curien, Unité Mixte de Recherche 7178 Centre National de la Recherche Scientifique & Université de Strasbourg,
23 rue Becquerel, 67087 Strasbourg Cedex 02, France
e-mail: yvon.lemaho@iphc.cnrs.fr

J. D. Whittington
Nordic Centre for Research on Marine Ecosystems and Resources under Climate Change (NorMER), Centre for Ecological and Evolutionary Synthesis (CEES), Department of Biology, University of Oslo, P.O. Box 1066 Blindern 0316, Norway
e-mail: jason.whittington@bio.uio.no

C. Verde and G. di Prisco (eds.), *Adaptation and Evolution in Marine Environments, Volume 2*, From Pole to Pole, DOI: 10.1007/978-3-642-27349-0_11,

elaborate network of management and protection treaties that apply to the Antarctic and Sub-Antarctic areas, which can also be difficult to navigate. Monitoring of Arctic ecosystems is logistically easier because of its proximity to more urbanized and industrialized continents, but a longer history of political and cultural claims on marine territories make geopolitical constraints an even larger issue than in the Southern Ocean. Furthermore, the Arctic marine ecosystems are under stronger anthropogenic impacts, e.g. over-exploitation of marine resources and pollution. This can lead to complex and difficult to interpret interactions with climate impacts, and so, in addition to the political complications, the information may be easier to collect, but more difficult to understand. Thus, as a result of these many challenges, direct monitoring of these polar ecosystems would require a huge additional investment in equipment, personnel, and logistic support.

Yet, even if all these constraints could be solved, finding a way to properly monitor marine ecosystems and their associated biodiversity would remain a real challenge. Ecosystems are highly complex and difficult to assess in their totality (Ciannelli et al. 2005), with impacts from climate changes occurring at all trophic levels and with a variety of possible responses (Chambers et al. 2005). Moreover, evidence is mounting that extinctions are altering key processes important to the productivity and sustainability of Earth's ecosystems (Balvanera et al. 2006; Hooper et al. 2005, 2012). Further species loss is expected to accelerate changes in ecosystem processes (Stachowicz et al. 2007; Perrings et al. 2011; Hooper et al. 2012), but it is unclear how these effects, combined with the more direct effects of other forms of environmental change, are driving diversity loss and altering ecosystem function. There is therefore a need for indicators that can provide aggregated information on the 'health of the ecosystem' (Heslenfeld and Enserink 2008).

An alternative solution to assess and monitor these ecosystems is to use selected animals as 'sentinels' because they are sensitive to environmental changes. High trophic level animals typically have large food web interactions and are usually long-lived, both features making them particularly valuable bio-indicators of environmental changes and/or extreme events.

In this context, seabirds have been identified as excellent tools for monitoring changes in many variables that can be used to assess ecosystem health (Furness and Greenwood 1993). Here, we identify which characteristics make them particularly suitable in this role, how seabird populations can rapidly respond to changing environments (including case studies), and lay out an argument and framework for expanding polar seabird studies into larger scale Life Observatories for polar monitoring.

11.2 Seabirds as Bio-Indicators

The value of a bio-indicator is related to the ability of a biological agent to provide large quantities of information from a signaller (the ecosystem) to a receiver (the researcher). This is ideally achieved by the bio-indicator summarizing information from different sources and of different natures into a smaller set of signals (intrinsic to, or carried by, the agent) that can be assessed and understood by scientists. Studies of bio-indicators within the natural environment to assess ecosystems turn them into functional ecological indicators, where the study of characteristics of a particular species can tell us about the state of a variety of ecosystem variables, thus reducing the amount of investment that would be required to otherwise investigate these variables directly and individually.

Seabirds are conspicuous denizens of global marine environments, and have a large representation in both polar regions. Given their relative ease of study, they have frequently been identified as useful ecological indicators of the health and status of marine ecosystems (Montevecchi 1993 in Furness and Greenwood 1993; Furness and Camphuysen 1997; Piatt et al. 2007). Indeed, changes in seabird distribution and abundance, as well as breeding success and timing, growth rates, survival and diet composition, have been closely linked to regional climate variability (e.g. North Atlantic Oscillations and El Niño/La Niña events) and global climate change (Aebischer et al. 1990; Brown 1991; Montevecchi and Myers 1997; Barbraud and Weimerskirch 2006; Grémillet and Boulinier 2009), often due to changes in prey abundance (Cairns 1987; Hamer et al. 1991; Diamond and Devlin 2003).

Seabirds are widely recognized for their important status in marine ecosystems. For example, numerous seabird species have been considered of special interest in marine resource management strategies and in the design of marine protected areas (MPAs; e.g. Pichegru et al. 2010). Qualified as umbrella species, i.e. species whose conservation is expected to confer protection to a large number of naturally co-occurring species (Roberge and Angelstam 2004), seabirds can play an important role in determining the minimum size for conservation areas, identifying sites to be included in reserve networks, and setting minimum standards for the composition, structure, and processes of ecosystems (Roberge and Angelstam 2004). Approaches to seabird MPA site identification consider single- versus multiple-species approaches, mapping of marine biological 'hotspots', and assessment of overlap with risks and threats (Ronconi et al. 2012; Lascelles et al. 2012).

The use of seabirds as ecological indicators, including both pros and cons, has been reviewed by Durant et al. (2009). Identified advantages include high trophic level position, a diverse array of species and food-web interactions, a broad sampling range, and autonomous environmental sampling.

11.2.1 High Trophic Level Position

The ecology of an ecosystem is generally defined as the interactions of organisms with one another and with the environment in which they occur. There are two theoretical ideologies that describe how these ecosystem interactions are controlled. The first, called 'bottom-up control', states that low level processes, such as climate processes, oceanographic conditions, or nutrient and mineral supply to

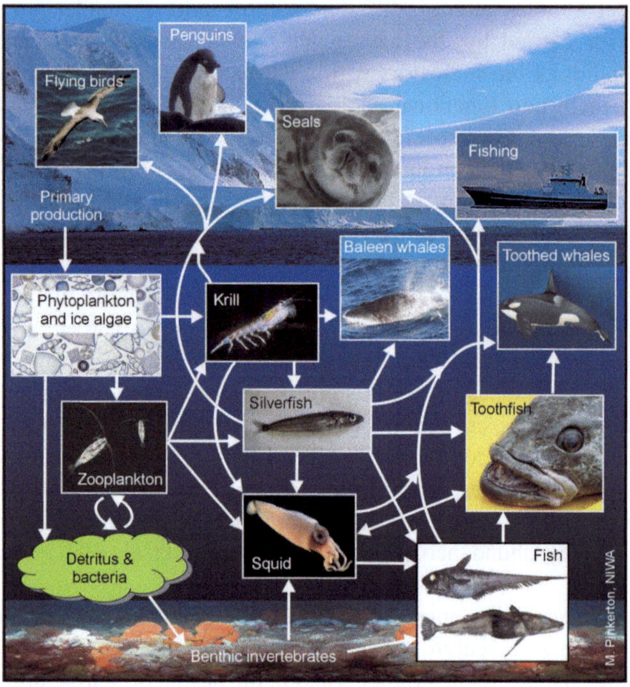

Fig. 11.1 A simplified food-web of the Ross Sea slope and shelf, Antarctica. *Source* M. Pinkerton, New Zealand Institute of Water & Atmospheric Research

phytoplankton determine the dynamics of higher trophic level function, by propagation of this resource availability through the food web. The alternate theory, called 'top-down control', states that consumption of resources by higher trophic levels ultimately controls ecosystem function by a cascade of effects that propagate from the upper trophic levels to progressively lower levels of the food web. However, natural ecosystems typically function as a mixture of these two processes operating at the same time, and focused studies are necessary to determine the relative importance of these processes in different systems. This is supported by recent studies of marine food webs, which reveal changing influence of bottom-up or top-down controls, or a combination of both (Litzow and Ciannelli 2007; Cury et al. 2008).

Seabirds stand at or near the top levels of their trophic chains, especially within the polar regions, where the complex trophic relationships are often short and involve relatively fewer species (see Smith et al. 1999). This is particularly true in the Southern Ocean ecosystems (Fig. 11.1), where the terrestrial component of the food chains are even more limited than in the north polar ecosystems. Although a few species are occasionally preyed upon by other polar marine predators at particular life stages, adult seabirds are generally not the primary prey for higher trophic levels in polar ecosystems. Polar seabirds, such as penguins, can feed at very low trophic levels, thus trophic chains are often very short at these higher latitudes.

There is evidence that seabird populations and reproductive success are strongly dependent on prey availability (Hunt et al. 1990; Monaghan 1996; Montevecchi and Myers 1996; Gaston et al. 2009; Shultz et al. 2009). This is an indication that their population dynamics are driven by bottom-up processes, and that they can be accurately classified as upper trophic level species whose populations fluctuations reflect characteristics of the environments on which they depend.

11.2.2 Diversity of Species and Food Web Interactions

The diversity of seabird species and the breadth and variety of their diets, both within and between species, contribute to the value of these animals as bio-indicators of ecosystem changes (reviewed in Boyd et al. 2006).

Seabirds can be categorized according to their feeding habits as either generalists or specialists. A generalist species has a greater ability to modify its diet to accommodate shifts in prey species resulting from environmental changes. Importantly, more visible effects of environmental changes on seabirds can be observed in specialist feeders, which rely on one or a group of prey species, such as small fish, squid, or plankton. In response to declining food resources, these species are more constrained in their responses, often being forced to either disperse to new locations or suffer population declines (Crawford et al. 2011). Those that eat more generalized diets cope more successfully with climate change than those with more limited environmental and dietary demands (Both et al. 2010). This demonstrates that species with highly specific needs are having greater difficulty adapting to climate change (Ahola et al. 2007) because of their adaptation to specialized niches, and that rapid changes in local environments are likely to favour generalist species (Crick 2004). Because food specialists show the impact of food abundance better than generalists (Pierroti and Annett 1990; Le Maho et al. 1993), their population responses to environmental changes are both more rapid and more visible, especially in the very short trophic chains that exist in polar regions, and these species are more valuable as bio-indicator species.

However, when assessing seabird populations to indirectly monitor marine ecosystem changes, human activities must also be taken into account. Because many seabirds feed on species of fish that are also intensively exploited or indirectly affected by human activities (Crawford et al. 2006), it can sometimes be difficult to determine which factors are driving the observed changes. This concern may be more important for the north polar regions, where proximity to human populations means a greater concentration of fisheries exploitation than in south polar regions. Yet, multi-species studies, which combine both generalist and specialist seabirds, may provide one way of determining the origin of the observed dynamics (Crawford et al. 2008a, b).

11.2.3 Wide Sampling Range

Standard oceanic survey techniques for monitoring invertebrates and fish are generally focused on just a few commercially exploited species, and are limited temporally and spatially.

In contrast, many seabird species are recognized to be able to cover huge distances during foraging or migration trips (e.g. Jouventin and Weimerskirch 1990; Bost and Le Maho 1993; Egevang et al. 2010), and thus have the potential to provide information on ecosystem characteristics for large areas of the oceans. Furthermore, their activity at sea is concentrated on oceanographic features that are of key importance to the functioning of global marine systems, such as oceanic fronts (Bost et al. 2009; Ribic et al. 2011), upwellings (Wynn and Knefelkamp 2004), thermoclines (Ropert-Coudert et al. 2009), or eddies (Cotté et al. 2007).

11.2.4 Autonomous Environmental Samplers

As central-place foragers, seabirds forage at sea but must return to land for specific purposes (breeding, moulting, etc.), making them generally accessible to scientists. Moreover, seabirds exhibit a high degree of philopatry (ca 80 %; Grémillet and Boulinier 2009), i.e. in numerous species, individuals usually return to the same site throughout their lifetimes. Thus, through a long-term monitoring program installation at a single location, researchers are able to perform long-term studies and sampling of individuals.

Polar seabirds are therefore considered autonomous samplers of marine environment and are good ecological indicators of the changes occurring in these areas (Le Maho et al. 1993; Durant et al. 2009). The broad-ranging movements and longevity of seabirds make them valuable tracking environmental changes at spatial and temporal scales that are otherwise difficult to sample (Diamond and Devlin 2003; Huettmann and Diamond 2006). Seabird species are also useful bio-indicators by providing valuable information on unknown pelagic habitats (Springer et al. 1996). Their return to a central location from often remote foraging locations allows researchers to collect biological samples (e.g. blood, feather, tissue, and stomach content) that tell us about these remote ecosystems and their potential changes over the long term. Seabirds finally have the potential to function as indicators of pollutants, particularly because they rapidly bio-accumulate chemicals such as organochlorines (e.g. DDT, PCBs) and organo-metals (e.g. methylmercury) (Chapdelaine et al. 1987; Furness and Camphuysen 1997).

11.3 Understanding Seabird Responses to Environmental Patterns Can Help Us Gauge the Adaptive Capacities to Future Climate Changes

Climate models predict a multi-degree warming of both northern and southern polar regions (IPCC 2007). A research priority is therefore to understand the effect of such changes upon marine organisms, food webs and biogeochemical cycles (Beaugrand et al. 2009; Beaugrand and Kirby 2010). Climate change will also alter the evolutionary forces acting on species' demographic strategies and life-history traits, as these have evolved to cope with the range of environmental fluctuations species have experienced in the past. This is especially important in long-lived species (e.g. Cairns 1992; Stenseth et al. 2002, 2004), such as most seabird species, which are typically the upper trophic level predators in ecosystems. These top-predators are therefore considered to be key indicators of short- and long-term changes in food webs and food availability (Le Maho et al. 1993; Boyd and Murray 2001; Voigt et al. 2003), as they integrate and amplify the effects of climatic forcing on lower levels of food chains (Croxall et al. 2002; Frederiksen et al. 2007). They are also suitable for studying whether the effects of climate change on top predators may be mitigated by phenotypic plasticity (morphological, physiological, behavioural traits) and microevolutionary changes.

Indeed, when faced with new selection pressures, such as those imposed by ongoing climate change, populations can display two general categories of responses (see review in Gienapp et al. 2008). First, they can adjust to the changing conditions by means of phenotypic plasticity (no genetic alterations). Second, they can adapt to the changing conditions by means of genetic changes (evolution).

11.3.1 Phenotypic Flexibility and Plasticity

Climate change affects marine ecosystems worldwide, with broad impacts on distribution and abundance, phenology, productivity of species, species interactions, community organization, and ecosystem functions and services (Parmesan and Yohe 2003; Hoegh-Guldberg and Bruno 2010; Doney et al. 2012). Polar organisms are exposed to strong environmental constraints, and it is important to understand how they have adapted in the past to cope with those challenges in order to assess how current climate changes will impact them in the future.

Phenotypic plasticity, defined as the expression of different phenotypes in response to environmental conditions, can allow organisms to respond rapidly and effectively to environmental changes (Charmantier et al. 2008; Nussey et al. 2005). Individual adjustments, for instance to their behaviour, in response to the environment could enable a population to track a rapidly changing environment very closely, without the immediate necessity of genetic adaptations. The first step in assessing these is therefore to identify traits that play key roles in populations, and then to explore their flexibility and plasticity at the population and species scale. By monitoring how colonies function under environmental constraints, it is possible to gauge global adaptive potential to face future ecosystem changes.

11.3.1.1 Phenology

Changes in climate have been strongly connected to shifts in phenology in many species, across a wide range of taxa and from the individual level to the level of communities (Crick et al. 1997; Parmesan and Yohe 2003; Root et al. 2003; Walther et al. 2002; Jonzen et al. 2007). In birds, seasonal changes in environmental conditions have led to changes in population dynamics (Sæther et al. 2000), timing of migration (Bradley et al. 1999; Forchhammer et al. 2002; Møller et al. 2008), reproduction (Dawson 2008; Visser et al. 2004; Møller 2002), and also foraging patterns (Cotton 2003; Weimerskirch et al. 2012).

Recent climate warming trends are disrupting the phenology of organisms throughout ecosystems, however differences in generation time, reproductive patterns, and species' co-dependence mean that these changes in timing generally occur at different rates for different taxonomic levels (Thackeray et al. 2010). Different aspects of a species' ecological niche may shift at different rates and in different amounts in response to environmental changes, leading to uncertainty in how much a species should be shifting to match the changes in its environment (Visser and Both 2005). This becomes a problem when the timing of the peak energy demand (often breeding) of organisms at higher taxonomic levels (predators) no longer overlaps with the timing of peak resource (prey) availability. This idea is encompassed in the Match-Mismatch hypothesis (Cushing 1970), and the difference in timing within a dependant ecosystem is referred to as a 'trophic mismatch' and has the potential to lead to a cascade of effects throughout an ecosystem.

There is evidence that this 'trophic mismatch' is occurring in the Arctic region, where more complete data are available on predator–prey interactions. In the north Atlantic, Burthe et al. (2012) recently found contrasting phenological responses to climate change (increase of SST) for numerous seabird species (Common guillemot *Uria aalge*, Razorbill *Alca torda*, European shag *Phalacrocorax aristotelis*, black-legged Kittiwake *Rissa tridactyla*, and Atlantic puffin *Fratercula arctica*) from multiple trophic levels of the pelagic food web in the North Sea, suggesting that the system is currently experiencing trophic mismatching. There is evidence this may also be occurring in the north Pacific region, where surveys of seabirds and their prey show earlier seasonal abundance of nekton prey (Batten and Mackas 2009) and a few, but not all, of the avian species that depend on them (Thompson et al. 2012). While some of the difference in seasonal abundances observed in the survey of seabird species by Thompson et al. (2012) reveal population declines, they can also often demonstrate the plasticity of an individual species' response to climate-induced environmental changes.

11.3.1.2 Foraging Strategies

Seabirds are often wide-ranging and highly mobile predators (Bost and Le Maho 1993). This ensures that some seabirds are able to mitigate the effects of environmental changes through shifts in their foraging strategies (see review in Durant et al. 2009).

Provencher et al. (2012) examined changes since the 1970 s in the diet of the Thick-billed murres *Uria lomvia* breeding in the Canadian Arctic. They found a decline in the diversity of the prey foraged by the birds, signalling a potential change in the number of prey species available. They also observed changes in the sea ice-dependent Arctic cod in the birds' diet, and the changing patterns differed according to the latitude. The amount of Arctic cod declined more strongly in the murre diet in the low Arctic *versus* high Arctic, where summer sea ice continues to dominate the seascape in most years. Gradually Arctic cod appeared to be replaced by the sub-Arctic species, capelin, marking an apparent northward expansion of this species over the last 30 years. They also found sea ice conditions averaged over multiple years explained the changes better than single-year indices, indicating that long-term changes in sea ice do impact this prey species.

Other recent work showed a very strong link between summer SST within the foraging areas of little auks *Alle alle*, species composition of local zooplankton communities, and the dietary preferences of the birds, whereby birds had to feed on smaller, less profitable copepod species in warmer water (Karnovsky et al. 2010). However, little auks seem to be able to respond to a wide range of sea surface temperatures thanks to plasticity of their foraging behaviour (Grémillet et al. 2012), strengthening the idea that behavioural plasticity is the primary mechanism by which animals deal with the consequences of climate change (Bradshaw and Holzapfel 2006; Charmantier et al. 2008; Kearney et al. 2009). Another example of behavioral plasticity can be seen in king penguins, which are considered as specialist feeders foraging essentially on myctophid fish whose main biomass is at the Polar Front during the austral summer. As the Polar Front moves further south in warmer years, Possession Island king penguins forage at about 600 km south of their colony compared to about 300 km in colder years (Péron et al. 2012). However, despite this ability to shift their foraging range, this plastic response has a finite range beyond which king penguins will be unable to forage far enough to gain the necessary resources as warming continues.

11.3.1.3 Dispersal

Models assessing ongoing environmental changes suggest that many birds will experience substantial pressure to shift or contract their ranges to accommodate changes in resource availability and also seasonal conditions (La Sorte and Jetz 2010). While resource availability is frequently seen as one of the most influential drivers of these changes (Monaghan 1996; Montevecchi and Myers 1996; Durant et al. 2007), long-term trends of 110 common breeding birds across 20 European countries showed that the sharpest population shifts and local declines were among

Fig. 11.2 Changes in the number of breeding pairs in penguin rookeries near Palmer Station, western Antarctic Peninsula. As the amount of sea-ice declines, ice-dependent Adélie penguins are declining and being replaced by sub-polar Gentoo penguins. *Source* Ducklow et al. (2007); McClintock et al. (2008)

species with lower thermal maxima (Jiguet et al. 2010), demonstrating that physical conditions can also drive such changes in range or contractions.

Seabirds in Antarctic and sub-Antarctic areas have shown large dispersal responses to changes in sea-ice extent over the past century (Ainley et al. 2003; Croxall et al. 2002; Smith et al. 1999). For instance, since 1970, the sea-ice dependent Adélie and emperor penguins (*Pygoscelis adéliae* and *Aptenodytes forsteri*, respectively) have nearly disappeared from their northernmost sites around Antarctica, i.e. in the Antarctic Peninsula. Adélie penguins have declined by 70 % on Anvers Island (Emslie et al. 1998; Fraser et al. 1992), whereas they are thriving at the more southerly Ross Island (Wilson et al. 2001), effectively shifting this species poleward. Emperors have declined from 300 breeding pairs down to just nine in the western Antarctic Peninsula (Gross 2005). In the long-term, sea-ice-dependent birds will suffer a general reduction of habitat as ice shelves contract [e.g. as it has already occurred in the Ross Sea (IPCC 2007)] or collapse [e.g. as did the Larsen Ice Shelves along the Antarctic Peninsula in 2002 (Alley et al. 2005)]. In contrast, sub-Antarctic chinstrap and gentoo penguins (*Pygoscelis antarcticus* and *papua*, respectively) invaded southward along the Antarctic Peninsula between 20 and 50 years ago, with paleontological evidence that gentoo had been previously absent from the Palmer region for at least the past 700 years (Emslie et al. 1998; Fraser et al. 1992; Ducklow et al. 2007; McClintock et al. 2008; Fig. 11.2).

Dispersal responses can also affect migration routes without impacting the presence of populations at historical locations. Populations of a pole–pole migrant, the sooty shearwater, *Puffinus griseus*, have shifted their migration routes by hundreds of kilometers in concert with altered sea surface temperature (SST) in the Pacific (Spear and Ainley 1999). Therefore, their populations at each endpoint of their migration are the same, but the pathway has changed.

11.3.2 Microevolutionary Processes

Environment changes associated with rapid changes in climate are likely to impose strong selection pressures on traits important for fitness across most species. In response to this climate-mediated selection, microevolutionary processes can be important mechanisms that mitigate some of the negative consequences of climate change. Microevolution, which is the changes in allele frequencies that occur over time within a population, results from four general processes: selection, mutation, genetic drift, and gene flow. To better understand these processes, and how they interact with changes in the natural environment and phenotypic plasticity, genetic assessment must to expanded and implemented on a larger scale (see a review of genetic approaches in Taylor and Friesen 2012).

11.3.2.1 Selection

Identification of the genetic basis of traits affecting fitness is a central issue in evolutionary biology (Stearns 1992). There is considerable interest at present in the evolutionary ecology of plasticity due to its relevance to adaptation to human-induced environmental changes, where phenotypes are typically measured as traits across environments in a reaction norm approach (Nussey et al. 2005). Knowledge of the molecular genetic basis of characters, coupled with expression studies, will enable the determination of whether phenotypes in different environments are the same traits, and also potentially the identity of loci ('plasticity genes') that control the expression of these phenotypes. Measuring fitness in natural populations requires dedicated field efforts, which are particularly valuable for genetic studies of fitness by giving access to long-term pedigrees of animals of known relatedness (Ellegren and Sheldon 2008). Estimation of environmental and genetic influences on traits, and their distribution among individuals is vital to gaining unbiased estimates of the selection force on characters.

11.3.2.2 Genetic Drift

Genetic diversity is essential to withstanding ecosystem changes, which sets a lower limit on the effective size of a population needed to maintain evolutionary potential. Franklin and Frankham (1998) suggested that it is possible to quantify this value as the balance between the loss of alleles due to genetic drift and the generation of new alleles through mutation. From this, and the work of others, it has been suggested that the long-term genetic security of a species can be maintained with a target effective population size in the range of 500–5000 individuals (Frankham 1995; Lynch and Lande 1998).

However, it remains unclear how diversity is maintained in philopatric colonial systems typical of polar seabird species, such as in isolated penguin colonies. Yet, the maintenance of genetic variability is crucial to the capacities of populations of seabirds to adapt to environmental changes. Polar seabirds may achieve this by being less philopatric than typically described, or by breeding within such large colonies so that genetic diversity is autonomously maintained and not adversely affected by the effects of genetic drift.

Analyses of genetic markers in European shags, *Phalacrocorax aristotelis*, and the wandering albatross, *Diomedea exulans*, both philopatric colony breeders, showed little or no evidence of genetic structure of different populations (Barlow et al. 2011; Milot et al. 2008). These studies indicated that the open sea, and the high colonial philopatry, are not barriers to effective dispersal, and thus genetic drift is not a critical issue even over very large spatial scales. On the other hand, a study of Cook's petrel, *Pterodroma cookii*, found that a phenological difference caused by different habitat specialization during the non-breeding season can restrict gene flow and lead to genetic differentiation even within populations (Rayner et al. 2011).

11.3.2.3 Genetic Flow

Dispersal is critically linked to the demographic and evolutionary trajectories of populations (Bohonak 1999; Ross 2001). It evolves under the influence of multiple interacting forces that are modified in turn by a feedback effect of dispersal rates (e.g. Gandon and Michalakis 2001). Among these forces, there can be inbreeding avoidance, spatial and temporal heterogeneity in resources, kin competition and selection, competition for partners and resources, and fitness costs incurred by dispersal itself (Clobert et al. 2001). In pelagic seabirds, the apparent paradox between vagility and reluctance for dispersal has raised numerous questions on the benefits of philopatry and coloniality (e.g. Coulson 2002), the barriers to dispersal (Friesen et al. 2007) and the dynamics of colony formation and extinction (Matthiopoulos et al. 2005). These issues are important for our understanding of the persistence of populations in marine environment where foraging areas are often distant from breeding sites. In addition, in the context of natural environmental changes, dispersal might be essential for populations to follow spatial and temporal changes in resource availability. Indeed, gene flow may be a process that could rescue populations from the impacts of climate changes (Roeder et al. 2001; Burg and Croxall 2004).

Although Milot et al. (2008) found evidence for very low dispersion rates among wandering albatross populations, they also found almost no genetic population structure. This indicates that even a very low dispersion rate, e.g. one bird per cohort, can be sufficient to maintain genetic flow between disparate colonies.

11.3.3 Integrating Phenotypic and Microevolutionary Approaches

There are both advantages and disadvantages to pursuing either phenotypic-based, or genetic-based studies of seabird population responses to environmental changes. While dispersal can lead to local extinctions, though often with persistence elsewhere, other phenotypic plasticity mechanisms can avoid this, and behavioural mechanisms are usually reversible. Genetic changes, on the other hand, are fixed or can generally only be reversed after considerable time has passed. However, an integration of both phenotypic plasticity and genetic-based studies can help us understand the true impact of environmental changes on seabird species. Behavioural plasticity may significantly attenuate the predicted climatic effects (Grémillet and Charmantier 2010), undermining the ability of envelope models to realistically mirror future trends (Morin and Thuiller 2009; Grémillet et al. 2012). To counteract this problem, researchers turn to mechanistic models (Fort et al. 2009; Buckley et al. 2010) or coupled individual-based and population models. While there is a use for these models, and they need to be further developed, only an integrated approach where the genetic basis, or underlying patterns, of any phenotypic response can help us understand the real effects of large-scale environmental changes (see Knight et al. 2008).

The king penguin, *Aptenodytes patagonicus*, provides an example of a system that will benefit from the integration of phenotypic and genetic approaches. Fitness measures in adult king penguins were negatively affected by warming events, with a decrease of $\sim$10 % in survival when the sea surface temperature (SST) at their wintering quarters increased by only 0.3 °C (Le Bohec et al. 2008; Fig. 11.3). Consequently, based on IPCC temperature projections, we predicted an extinction of the Crozet population within the next two centuries. However, the fundamental mechanisms driving the dynamics of penguin populations still remain largely unidentified, so it is currently not possible to fully clarify which parameters are sensitive to environmental changes or how to interpret their variability over different temporal and spatial scales (Piatt et al. 2007). It is therefore essential to better understand how they may or may not cope with environmental changes in order to use them appropriately as indicators of ecosystem problems whose importance is increasing in severity along with the exploitation of Southern Ocean resources and global human effects. This is now the focus of a new approach using genetic markers (microsatellites and RAD techniques) to assess the underlying patterns to population structure and life-history traits (Cristofari et al. unpublished).

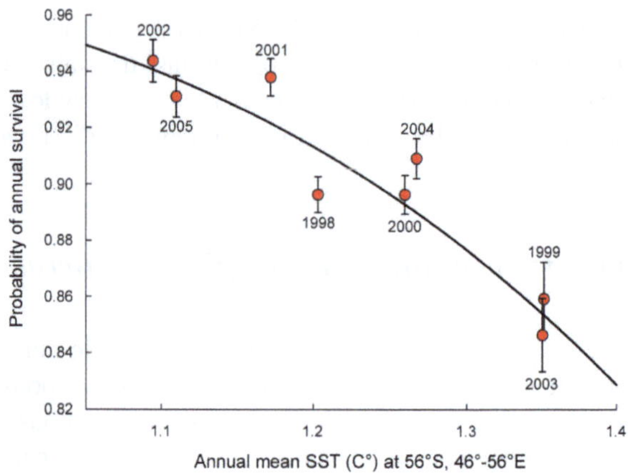

Fig. 11.3 Effect of sea-surface temperature at the Marginal Ice Zone on the survival of adult king penguins breeding in Crozet Archipelago. *Source* Le Bohec et al. 2008

11.4 Polar Life Observatories to Track Changes of Polar Ecosystems

11.4.1 What are Life Observatories?

It is critical to understand how polar ecosystems will respond to changes in climate and other natural and human forcing and the potential for feedbacks. To achieve this enhanced understanding, sustained multi-disciplinary observations are essential. One way to achieve this is through the development and expansion of Life Observatory systems. These sustain and integrate a wide array of scientific platforms within a framework to support existing and emerging technologies for observing living systems and their interaction with the physical environment. They generally include instrumentation, data management and analysis within a single, unified framework.

With the international focus, and the confusing network of regulations and treaties that govern polar regions, Life Observatories could be an essential tool for a more unified scientific approach for policy managers and conservation programs such as, among others, the United Nations Environmental Program (UNEP, http://www.unep.org), the Convention on Biological Diversity (CDB, http://www.cbd.int), the Scientific Committee on Antarctic Research (SCAR, http://www.scar.org), the Commission for the Conservation of Antarctic Marine Living Resources (CCAMLR, http://www.ccamlr.org), or the International Arctic Science Committee (IASC, http://iasc.info). To improve our conservation and sustainable management of natural and man-made resources and biodiversity, Polar Life Observatories would allow us to radically increase our understanding of the response of ocean biology to global environmental changes and the interaction between biological and physical factors. This would be achieved by the integration of existing monitoring schemes into a more organized framework that includes the following components.

11.4.1.1 Demographic Monitoring

Long-term time series and longitudinal data on individuals are fundamental to studies of life-history traits (e.g. age-specific survival or reproductive success; Lebreton et al. 1992), and to understand how variability in these traits influences population dynamics and behavioural strategies. Numerous seabirds live in colonies and are remarkably phylopatric, making them ideal focal species for the development of Polar Life Observatories (see for instance Gendner et al. 2005). However, the collection of such data-sets in natural populations is not simple, requiring the financial and infrastructure support to repeatedly identify individuals over time (see Le Maho et al. 2011).

11.4.1.2 Genetic Assessment and Monitoring

Polar species and ecosystems are notoriously difficult to study, and genetics could contribute to monitoring and conservation in the polar regions. Genetic monitoring programs are fundamental to gauging changing abundance or effective population size, population structure, and migration rates, as well as adaptation to environmental changes or evolution over time and to monitor emerging threats to biodiversity (Schwartz et al. 2007; Laikre et al. 2009; Cook et al. 2012). In that context, a more coordinated program would increase the breadth and diversity of archived samples and stimulate sample exchanges through improved access to polar specimen collections in international repositories (SnowBank; Hart et al. unpublished).

11.4.1.3 Bio-Monitoring Pollution

Pollutants are likely to cause toxic effects in seabird populations that are most evident in embryo or chick development, hatching success, or chick behaviour (Furness and Camphuysen 1997). The fact that seabirds are at or near the top of trophic chains in polar regions imply that they are particularly appropriate as biomonitors of pollutants that amplify in concentration through marine food chains. This is especially a characteristic of pollutants that are lipid-soluble but have low water solubility, such as organo-chlorines (such as DDT or PBC) and organometals (i.e. heavy metal such as mercury). Because they allow non-destructive sampling and permit retrospective study, seabird feathers are particularly convenient for monitoring heavy metal pollution in marine food webs (Monteiro and Furness 1995). However, accurate measurement of long-term trends in pollutant contamination depends on the assumption that seabird diet composition has not changed. This can be assessed by analysis of stable isotopes of N and C from the same feathers used for mercury measurement, a technique that also permits the monitoring of trophic status over time or between regions.

11.4.1.4 Foraging Monitoring

There are many examples of collapses of fish stocks as a result of a combination of excessive human harvesting and environmental changes (Cushing 1988). The complex ecological interactions between fish stocks through both competitive and predator–prey relationships, and also human harvesting pressure, make the changes in stocks even more difficult to predict, further increasing the need for improved broad-scale empirical monitoring of changes in marine resources. Many studies have shown that seabirds are very sensitive to changes in food supply, and therefore have potential as monitors of fish stocks.

Tracking shifts in seabird diets (through direct methodologies: diet composition, number of prey species, and proportions, see the review of Barrett et al. 2007; or indirect ones: stable isotope analysis using feathers, see Jaeger and Cherel 2011) and improving our understanding of the fundamental processes affecting seabird foraging ecology are therefore highly relevant to the effective management and conservation of entire marine ecosystems, including components with direct relevance to human economic and social well-being, such as commercial fish stocks. This effort should start with the identification of a subset of representative seabird species and the selection of sites where diets can be sampled over the long term.

11.4.2 Innovative Technology Development and Ethics

In the last 20 years, the use of bio-logging devices has revolutionized our knowledge of how seabirds can act as monitors of prey stocks. One of the most interesting applications of data-loggers on seabirds is determination of the distribution and availability of prey for which we have little knowledge, such as mesopelagic fish, squid, and krill (Bost et al. 2008). Monitoring the behavioural changes (variation of their foraging ranges and grounds, diving strategies, etc.) of individuals and populations over successive years within a monitoring framework that also gives researchers access to life-history and genetic data has the potential to vastly improve our knowledge of what is occurring within these polar marine systems. However, the data obtained using animal-attached tags must be carefully interpreted because seabirds with attached devices do not behave in a manner identical to unequipped conspecifics (see review in Wilson and Vandenabeele 2012).

For instance, the stress induced by capture and handling (Le Maho et al. 1992) and/or the increased drag devices such as data-loggers to study foraging strategies (Global Position Systems, Accelerometers, etc.) can significantly impair their physiological capacities and reduce their fitness (e.g. Beaulieu et al. 2010). Such disturbances are therefore a major issue, because they introduce scientific biases and raise serious ethical considerations. We therefore need to include studies of animals that are totally undisturbed if we want to avoid scientific bias and trust data reliability.

An example of the severe impact that may result from devices fixed on seabirds and of the resulting scientific bias is provided by penguin flipper bands. Indeed, because the anatomy of penguins' legs does not enable the use of leg bands, flipper bands have been widely used to identify and follow individuals. These have two main advantages. First, flipper bands can be read from a distance with binoculars, therefore avoiding the stress of capture or handling for identification as in most other birds. Second, they make it possible to visually locate individuals within a colony. We have however shown that flipper bands have a major impact on penguins (Gauthier-Clerc et al. 2004; Dugger et al. 2006; Saraux et al. 2011). An analysis of over ten years of data from Radio Frequency IDentification (RFID) antennae buried beneath natural paths that microtagged birds use to pass between land and sea [see technical details in Gendner et al. (2005)] showed the survival of flipper-banded adults was reduced by 16 % and their breeding success by 41 % relative to unbanded birds (Saraux et al. 2011). Since flipper-banded birds have longer foraging trips than non-banded birds, this severe effect appears to result from the increased drag at sea. The impact is even more severe for immature birds: their return rate after three years is ~70 % for unbanded chicks (Gauthier-Clerc et al. 2004) compared to only 6–47 % for banded chicks (Weimerskirch et al. 1992; Brodin et al. 1998). Moreover, different relationships between sea surface temperature and the population growth of banded and non-banded birds have been observed, which makes it problematic to rely on banded penguins to predict the impact of environmental changes (Saraux et al. 2011).

The main limitation of RFID, however, is the short reading range of the tags (about 60 cm), which constrains the use of this technology for the identification of individuals within dense colonies. It is possible to bypass this limitation through the use of delimited passages (entrances/exits) in the animal colonies for the detection of individuals. However, for some species, such as emperor penguins that breed on sea ice, such delimited passages do not exist. Simply multiplying RFID antennae throughout a large colony is not feasible. A second limitation is that because RFID-tags are tiny and implanted under the skin, individuals cannot be visually located from a distance once they are inside the colony. This argument has been used by some scientists to continue the use of flipper bands in penguins in spite of the published deleterious effects. However, a reduction in breeding success is an important trigger for the dispersion of seabirds (Boulinier et al. 2008) and, again, we have shown that lower breeding success is one of the main impacts of flipper banding (Gauthier-Clerc et al. 2004; Saraux et al. 2011). Clearly, in order to avoid any bias, a new methodology had to be developed to improve the assessment of individuals within colonies.

We therefore developed and tested an innovative technological platform: the use of mobile RFID antennae carried by remote controlled robots that can circulate among breeding penguins (Fig. 11.4). In a pilot study in king penguins, we were able to locate and identify microtagged individuals in this way.

Fig. 11.4 Remote controlled robot equipped with two circular RFID antennae to identify and localize tagged penguins into their breeding colony

11.5 Concluding Remarks

There is need for a broad expansion (in terms of key species and geographic coverage) of Polar Life Observatories, and also for international agreement on how these Observatories should be managed (standardization of data collection, benchmarking, metadatabase organization, access rights and management) to make them more effective and to promote further international collaborative work and exchange.

Studies taking an ecosystem approach to marine system dynamics are few. Especially for the polar ecosystems, data time series are often short and quantitative knowledge of the dynamics of interactions between predators, their prey, and the environment remains very limited. Improving the knowledge of polar marine ecology through a framework of research that relies on seabirds as bioindicators has the potential to lead to more accurate ecosystem data for improved marine ecosystem management strategies that satisfy both ecological and economical criteria. The development of Polar Life Observatories, combining international and multidisciplinary projects on polar top-predators into a more organized and integrated framework, will vastly improve our understanding of these globally important ecosystems in a coordinated way and directly meets international priorities regarding environmental protection and management strategies as defined by the Convention on Biological Diversity (CBD), the Intergovernmental Panel on Climate Change (IPCC), and the International Council for Science (ICSU).

We have seen, in recent years, how polar seabirds, such as albatrosses, penguins, or puffins, are very popular with the public, making them potentially valuable biological ambassadors to more effectively engage the public and sound the alarm about the ongoing human-induced global changes. Also, some of the best long-term data series on wild animal populations exist for these species. These two factors further make these species ideal model organisms for Polar Life

Observatories. The improved visibility of how these species are affected by rapid global environmental changes that would come from organizing seabird research into a Polar Life Observatory framework could be the necessary tipping point to convince policy-makers and governments to meet their responsibilities to respond to a major global threat. This comes at an important time as governments, currently facing a challenging human and global agenda for THE next decade, failed in agreeing on strong and courageous decisions at the Rio + 20—United Nations Conference on Sustainable Development.

Acknowledgments We are very grateful to the Institut Polaire Français–Paul-Emile Victor (IPEV, Programme 137 ECOPHY-ANTAVIA) and the Terres Australes et Antarctiques Françaises (TAAF), the Centre National de la Recherche Scientifique (Programme Zone Atelier de Recherches sur l'Environnement Antarctique et Sub-Antarctique), the Agence Nationale de la Recherche (Programme ANR BLANC 1728-01-PICASO) and the Centre Scientifique de Monaco (LEA-647 BioSensib) for all of their support for long-term data collection and their help in developing Polar Life Observatories. Robot development was supported by the TOTAL corporate Foundation. We also wish to thank H. Ducklow, W. Fraser, J. McClintock and M. Pinkerton for figure permissions.

References

Aebischer NJ, Coulson JC, Colebrook JM (1990) Parallel long-term trends across four marine trophic levels and weather. Nature 347:753–755

Ahola MP, Laaksonen T, Eeva T, Lehikoinen E (2007) Climate change can alter competitive relationships between resident and migratory birds. J Anim Ecol 76:1045–1052

Ainley DG, Ballard G, Emslie SD, Fraser WR, Wilson PR, Woehler EJ (2003) Adélie penguins and environmental change. Science 300:429–430

Alley RB, Clark PU, Huybrechts P, Joughin I (2005) Ice-sheet and sea-level changes. Science 310:456–460

Balvanera P, Pfisterer AB, Buchmann N, He JS, Nakashizuka T, Raffaelli D, Schmid B (2006) Quantifying the evidence for biodiversity effects on ecosystem functioning and services. Ecol Lett 9:1146–1156

Barbraud C, Weimerskirch H (2006) Antarctic birds breed later in response to climate change. Proc Natl Acad Sci U S A 103(16):6248–6251

Barlow EJ, Daunt F, Wanless S, Alvarez D, Reid JM, Cavers S (2011) Weak large-scale population genetic structure in a philopatric seabird, the European Shag *Phalacrocorax aristotelis*. Ibis 153:768–778

Barrett RT, Camphuysen CJ, Anker-Nilssen T, Chardine JW, Furness RW, Garthe S, Hüppop O, Leopold MF, Montevecchi WA, Veit RR (2007) Diet studies of seabirds: a review and recommendations. ICES J Mar Sci 64:1675–1691

Batten SD, Mackas DL (2009) Shortened duration of the annual *Neocalanus plumchrus* biomass peak in the Northeast Pacific. Mar Ecol Prog Ser 393:189–198

Beaugrand G, Kirby RR (2010) Climate, plankton and cod. Glob Change Biol 16:1268–1280

Beaugrand G, Luczak C, Edwards M (2009) Rapid biogeographical plankton shifts in the North Atlantic Ocean. Glob Change Biol 15:1790–1803

Beaulieu M, Thierry AM, Handrich Y, Le Maho Y, Massemin-Challet S, Ancel A (2010) Adverse effects of instrumentation in incubating Adélie penguins (*Pygoscelis adeliae*). Polar Biol 33:485–492

Bindoff NL, Willebrand J, Artale V, Cazenave A, Gregory J, Gulev S, Hanawa K, Le Quéré C, Levitus S, Nojiri Y, Shum CK, Talley LD, Unnikrishnan A (2007) Observations: oceanic climate change and sea level. In: Solomon S, Qin D, Manning M, Chen Z, Marquis M, Averyt KB, Tignor M, Miller HL (eds) Climate change 2007: the physical science basis. Contribution of Working Group I to the fourth assessment report of the intergovernmental panel on climate change. Cambridge University Press, Cambridge

Bohonak JA (1999) Dispersal, gene flow, and population structure. Quat Rev Biol 74:21–45

Bost CA, Le Maho Y (1993) Seabirds as bio-indicators of changing marine ecosystems: new perspectives. Acta Oecol Int J Ecol 14:463–470

Bost CA, Jaeger A, Huin W, Koubbi P, Halsey LG, Handrich Y (2008) Monitoring prey availability via data loggers deployed on seabirds: advances and present limitations. In: Tsukamoto K, Kawamura T, Takeuchi T, Beard TDJ, Kaiser MJ (eds) Fisheries for global welfare and environment. The 5th world fisheries congress, Yokohama, pp 121–137

Bost CA, Cotté C, Bailleul F, Cherel Y, Charrassin JB, Guinet C, Ainley DG, Weimerskirch H (2009) The importance of oceanographic fronts to marine birds and mammals of the southern oceans. J Mar Syst 78:363–376

Both C, van Turnhout CAM, Bijlsma RG, Siepel H, van Strien AJ, Foppen RPB (2010) Avian population consequences of climate change are most severe for long-distance migrants in seasonal habitats. Proc R Soc Lond Ser B 277:1259–1266

Boulinier T, McCoy KD, Yoccoz NG, Gasparini J, Tveraa T (2008) Public information affects breeding dispersal in a colonial bird: kittiwakes cue on neighbours. Biol Lett 4:538–540

Boyd IL, Murray AWA (2001) Monitoring a marine ecosystem using responses of upper trophic level predators. J Anim Ecol 70:747–760

Boyd IL, Wanless S, Camphuysen CJ (2006) Top predators in marine ecosystems: their role in monitoring and management. Cambridge University Press, Cambridge

Bradley NL, Leopold AC, Ross J, Huffaker W (1999) Phenological changes reflect climate change in Wisconsin. Proc Natl Acad Sci U S A 96(17):9701–9704

Bradshaw WE, Holzapfel CM (2006) Evolutionary response to rapid climate change. Science 312:1477–1478

Brodin A, Olsson O, Clark CW (1998) Modelling the breeding cycle of long-lived birds: why do king penguins try to breed late. Auk 115:767–771

Brown RGB (1991) Marine birds and climatic warming in the northwest Atlantic. In: Montevecchi WA, Gaston AJ (eds) Studies of high latitude seabirds 1: behavioural, energetic and oceanographic aspects of seabird feeding ecology. Canadian wildlife service occasional paper 68, Ottawa

Buckley LB, Urban MC, Angilletta MJ, Crozier LG, Rissler LJ, Sears MW (2010) Can mechanism inform species' distribution models? Ecol Lett 13:1041–1054

Burg TM, Croxall JP (2004) Global population structure and taxonomy of the wandering albatross species complex. Mol Ecol 13:2345–2355

Burthe S, Daunt F, Butler A, Elston DA, Frederiksen M, Johns D, Newell M, Thackeray SJ, Wanless S (2012) Phenological trends and trophic mismatch across multiple levels of a North Sea pelagic food web. Mar Ecol Prog Ser 454:119–133

Cairns DK (1987) Seabirds as indicators of marine food supplies. Biol Ocean 5:261–271

Cairns DK (1992) Population regulation of seabird colonies. In: Power DM (ed) Current ornithology. Dordrecht, Kluwer Academic/Plenum, pp 37–62

Chambers LE, Hughes L, Weston MA (2005) Climate change and its impact on Australia's avifauna. Emu 105:1–20

Chapdelaine G, Laporte P, Nettleship DN (1987) Population, productivity and DDT contamination of Northern Gannets at Bonaventure Island, Quebec 1967–1984. Can J Zool 65:2922–2926

Charmantier A, McCleery RH, Cole LR, Perrins C, Kruuk LEB, Sheldon BC (2008) Adaptive phenotypic plasticity in response to climate change in a wild bird population. Science 320:800–803

Ciannelli L, Hjermann D, Lehodey P, Ottersen G, Duffy-Anderson J, Stenseth NC (2005) Climate forcing, food web structure and community dynamics in pelagic marine ecosystems. In: Belgrano A (ed) Aquatic food webs: an ecosystem approach. Oxford University Press, Oxford, pp 143–169

Clobert J, Danchin E, Dhondt AA, Nichols JD (2001) Dispersal. Oxford University Press, Oxford

Cook JA, Brochmann C, Fedorov V, Talbot SL, Taylor EB, Väinölä R, Hoberg EP, Kholodova M, Magnusson KP (2012) Genetic perspectives on Arctic biodiversity. In Arctic biodiversity assessment. Conservation of Arctic fauna and flora Committee, Copenhagen (in press)

Cotté C, Park YH, Guinet C, Bost CA (2007) Movements of foraging king penguins through marine mesoscale eddies. Proc Roy Soc Lond Ser B 274:2385–2391

Cotton PA (2003) Avian migration phenology and global climate change. Proc Natl Acad Sci U S A 100:12219–12222

Coulson JC (2002) Colonial breeding in seabirds. In: Schreiber EA and Burger J (eds) Biology of marine birds 2002. CRC Press, London, pp 87–114

Crawford RJM, Goya E, Roux JP, Zavalaga CB (2006) Comparison of assemblages and some life-history traits of seabirds in the Humboldt and Benguela systems. Afr J Mar Sci 28:553–560

Crawford RJM, Sabarros PS, Fairweather T, Underhill LG, Wolfaardt AC (2008a) Implications for seabirds off South Africa of a long-term change in the distribution of sardine. Afr J Mar Sci 30:177–184

Crawford RJM, Tree AJ, Whittington PA, Visagie J, Upfold L, Roxburg KJ, Martin AP, Dyer BM (2008b) Recent distributional changes of seabirds in South Africa: is climate having an impact? Afr J Mar Sci 30:189–193

Crawford RJM, Altwegg R, Barham BJ, Barham PJ, Durant JM, Dyer BM, Geldenhuys D (2011) Collapse of South Africa's penguins in the early 21st century. Afr J Mar Sci 33:139–156

Crick HQP (2004) The impact of climate change on birds. Ibis 146:48–56

Crick HQP, Dudley C, Glue DE, Thomson DL (1997) UK birds are laying eggs earlier. Nature 388:526

Croxall JP, Trathan PN, Murphy EJ (2002) Environmental change and Antarctic seabird populations. Science 297:1510–1514

Cury PM, Shin YJ, Planque B, Durant JM, Fromentin JM, Kramer-Schadt S, Stenseth NC, Travers M, Grimm V (2008) Ecosystem oceanography for global change in fisheries. Trends Ecol Evol 23:338–346

Cushing DH (1970) Marine ecology and fisheries. Cambridge University Press, Cambridge

Cushing DH (1988) The provident sea. Cambridge University Press, Cambridge

Dawson A (2008) Control of the annual cycle in birds: endocrine constraints and plasticity in response to ecological variability. Phil Trans R Soc B Biol Sci 363:1621–1633

Diamond AW, Devlin CM (2003) Seabirds as indicators of changes in marine ecosystems: Ecological monitoring on Machias Seal Island. Environ Monit Assess 88:153–181

Doney SC, Ruckelshaus M, Duffy JE, Barry JP, Chan F, English CA, Galindo HM, Grebmeier JM, Hollowed AB, Knowlton N, Polovina J, Rabalais NN, Sydeman WJ, Talley LD (2012) Climate change impacts on marine ecosystems. Annu Rev Mar Sci 4:11–37

Dugger KM, Ballard G, Ainley DG, Barton KJ (2006) Effects of flipper bands on foraging behaviour and survival of Adélie penguins (*Pygoscelis adeliae*). Auk 123:858–869

Ducklow HW, Baker K, Martinson DG, Quetin LB, Ross RM, Smith RC, Stammerjohn SE, Vernet M, Fraser WR (2007) Marine pelagic ecosystems: the West Antarctic Peninsula. Phil Trans R Soc B 362:67–94.

Durant JM, Hjermann DØ, Ottersen G, Stenseth NC (2007) Climate and the match or mismatch between predator requirements and resource availability. Clim Res 33:271–283

Durant JM, Hjermann DØ, Frederiksen M, Charrassin JB, Le Maho Y, Sabarros PS, Crawford RJM, Stenseth NC (2009) Pros and cons of using seabirds as ecological indicators. Clim Res 39:115–129

Egevang C, Stenhouse IJ, Phillips RA, Petersen A, Fox JW, Silk JRD (2010) Tracking of Arctic terns *Sterna paradisaea* reveals longest animal migration. Proc Natl Acad Sci U S A 107:2078–2081

Ellegren H, Sheldon BC (2008) Genetic basis of fitness differences in natural populations. Nature 452:169–175

Emslie SD, Fraser W, Smith RC, Walker W (1998) Abandoned penguin colonies and environmental change in the Palmer Station area, Anvers Island, Antarctic Peninsula. Antarct Sci 10:257–268

Forchhammer MC, Post E, Stenseth NC (2002) North Atlantic oscillation timing of long- and short-distance migration. J Anim Ecol 71:1002–1014

Fort J, Porter WP, Grémillet D (2009) Thermodynamic modeling predicts energetic bottleneck for seabirds wintering in the northwest Atlantic. J Exp Biol 212:2483–2490

Frankham R (1995) Effective population size/adult population size ratios in wildlife: a review. Genet Res 66:95–107

Franklin IR, Frankham R (1998) How large must populations be to retain evolutionary potential? Anim Conserv 1:69–70

Fraser WR, Trivelpiece WZ, Ainley DC, Trivelpiece SG (1992) Increases in Antarctic penguin populations: reduced competition with whales or a loss of sea ice due to environmental warming? Polar Biol 11:525–531

Frederiksen M, Mavor RA, Wanless S (2007) Seabirds as environmental indicators: the advantages of combining data sets. Mar Ecol Prog Ser 352:205–211

Friesen VL, Burg TM, McCoy KD (2007) Mechanisms of population differentiation in seabirds. Mol Ecol 16:1765–1785

Furness RW, Camphuysen K (1997) Seabirds as monitors of the marine environment. ICES J Mar Sci 54:726–737

Furness RW, Greenwood JJD (1993) Birds as monitors of environmental change. Chapman & Hall, London

Gandon S, Michalakis Y (2001) Multiple causes of the evolution of dispersal. In: Clobert J, Danchin E, Dhondt AA, Nichols JD (eds) Dispersal. Oxford University Press, Oxford, pp 155–167

Gaston AJ, Gilchrist HG, Mallory ML, Smith PA (2009) Changes in seasonal events, peak food availability, and consequent breeding adjustment in a marine bird: a case of progressive mismatching. Condor 111:111–119

Gauthier-Clerc M, Gendner JP, Ribic CA, Fraser WR, Woehler EJ, Descamps S, Gilly C, Le Bohec C, Le Maho Y (2004) Long-term effects of flipper bands on penguins. Proc R Soc Lond B 271:423–426

Gendner JP, Gauthier-Clerc M, Le Bohec C, Descamps S, Le Maho Y (2005) A new application for transponders in studying of penguins. J Field Ornithol 76:138–142

Gienapp P, Teplitsky C, Alho J, Mills J, Merila J (2008) Climate change and evolution: disentangling environmental and genetic responses. Mol Ecol 17:167–178

Graversen RG, Mauritsen T, Tjernström M, Källen E, Svensson G (2008) Vertical structure of recent Arctic warming. Nature 451:53–56

Grémillet D, Boulinier T (2009) Spatial ecology and conservation of seabirds facing global climate change: a review. Mar Ecol Prog Ser 391:121–137

Grémillet D, Charmantier A (2010) Shifts in phenotypic plasticity constrain the value of seabirds as ecological indicators of marine ecosystems. Ecol Appl 20:1498–1503

Grémillet D, Welcker J, Karnovsky NJ, Walkusz W, Hall ME, Fort J, Brown ZW, Speakman JR, Harding AMA (2012) Little auks buffer the impact of current Arctic climate change. Mar Ecol Prog Ser 454:197–206

Gross L (2005) As the Antarctic ice pack recedes, a fragile ecosystem hangs in the balance. PLoS Biol 3(4):e127

Hamer KC, Furness RW, Caldow RWG (1991) The effects of changes in food availability on the breeding ecology of great skuas, *Catharacta skua*, in Shetland. J Zool 223:175–188

Hansen J, Sato M, Ruedy R, Lo K, Lea DW, Medina-Elizade M (2006) Global temperature change. Proc Natl Acad Sci U S A 103:14288

Heslenfeld P, Enserink EL (2008) OSPAR ecological quality objectives: the utility of health indicators for the North Sea. ICES J Mar Sci 65:1392–1397

Hoegh-Guldberg O, Bruno JF (2010) The impact of climate change on the world's marine ecosystems. Science 328:1523

Hooper DU, Chapin FS, Ewel JJ, Hector A, Inchausti P, Lavore S, Lawton JH, Lodge DM, Loreau M, Naeem S, Schmid B, Setälä H, Symstad AJ, Vandermeer J, Wardle DA (2005) Effects of biodiversity on ecosystem functioning: a consensus of current knowledge. Ecol Monogr 75:3–35

Hooper DU, Adair EC, Cardinale BJ, Byrnes JEK, Hungate BA, Matulich KL, Gonzalez A, Duffy JE, Gamfeldt L, O'Connor MI (2012) A global synthesis reveals biodiversity loss as a major driver of ecosystem change. Nature. doi:10.1038/nature11118

Huettmann F, Diamond AW (2006) Large-scale effects on the spatial distribution of seabirds in the Northwest Atlantic. Landsc Ecol 21:1089–1108

Hunt GL, Piatt JF, Erikstadt KE (1990) How do foraging seabirds sample their environment?. Acta XX Congressus Internationalis Ornithologici, Christchurch, pp 2272–2280

IPCC (2007) Climate Change 2007: synthesis report. contribution of Working Groups I, II and III to the fourth assessment report of the intergovernmental panel on climate change (eds Core Writing Team, Pachauri RK, Reisinger A)

Jaeger A, Cherel Y (2011) Isotopic investigation of contemporary and historic changes in penguin trophic niches and carrying capacity of the Southern Indian Ocean. PLoS ONE 6(2):e16484

Jiguet F, Devictor V, Ottvall R, van Turnhout C, van der Jeugd H, Lindström Å (2010) Bird population trends are linearly affected by climate change along species thermal ranges. Proc R Soc Lond B 277:3601–3608

Jonzen N, Ergon T, Linden A, Stenseth NC (2007) Introduction to CR Special 17: bird migration and climate. Clim Res 35:1–3

Jouventin P, Weimerskirch H (1990) Satellite tracking of wandering albatrosses. Nature 343:746–748

Karnovsky N, Harding AMA, Walkusz W, Kwamniewski S, Goszczko I, Wiktor J Jr, Routti H, Bailey A, McFadden L, Brown Z, Beaugrand G, Grémillet D (2010) Foraging distributions of little auks (*Alle alle*) across the Greenland Sea: implications of present and future climate change. Mar Ecol Prog Ser 415:283–293

Kearney M, Shine R, Porter WP (2009) The potential for behavioral thermoregulation to buffer "coldblooded'' animals against climate warming. Proc Natl Acad Sci U S A 106:3835–3840

Knight TM, Barfield M, Holt RD (2008) Evolutionary dynamics as a component of stage-structured matrix models: an example using *Trillium grandiflorum*. Am Nat 172(3):375–392

La Sorte FA, Jetz W (2010) Avian distributions under climate change: towards improved projections. J Exp Biol 213:862–869

Laikre L, Nilsson TR, Primmer CR, Ryman N, Allendorf FW (2009) Importance of genetics in the interpretation of favourable conservation status. Cons Biol 23:1378–1381

Lascelles BG, Langham GM, Ronconi RA, Reid JB (2012) From hotspots to site protection: identifying marine protected areas for seabirds around the globe. Biol Cons. doi:10.1016/j.biocon.2011.12.008

Le Bohec C, Durant JM, Gauthier-Clerc M, Stenseth NC, Park YH, Pradel R, Grémillet D, Gendner JP, Le Maho Y (2008) King Penguin population threatened by Southern Ocean warming. Proc Natl Acad Sci U S A 105:2493–2497

Le Maho Y, Karmann H, Briot D, Handrich Y, Robin JP, Mioskowski E, Cherel Y, Farni J (1992) Stress in birds due to routine handling and a technique to avoid it. Am J Phys 263:775–781

Le Maho Y, Gendner JP, Challet E, Bost CA, Gilles J, Verdon C, Plumeré C, Robin JP, Handrich Y (1993) Undisturbed breeding penguins as indicators of changes in marine resources. Mar Ecol Prog Ser 95:1–6

Le Maho Y, Saraux C, Durant J, Viblanc VA, Gauthier-Clerc M, Yoccoz N, Stenseth NC, Le Bohec C (2011) An ethical issue on biodiversity science: the monitoring of penguins with flipper-bands. Comp Rend Biol 334:378–384

Lebreton J, Burnham K, Clobert J, Anderson D (1992) Modeling survival and testing biological hypotheses using marked animals: a unified approach with case studies. Ecol Monogr 62:67–118

Litzow MA, Ciannelli L (2007) Oscillating trophic control induces community reorganization in a marine ecosystem. Ecol Lett 10:1124–1134

Lynch M, Lande R (1998) The critical effective size for a genetically secure population. Anim Conserv 1:70–72

Matthiopoulos J, Harwood J, Thomas L (2005) Metapopulation consequences of site fidelity for colonially breeding mammals and birds. J Anim Ecol 74:716–727

McClintock J, Ducklow H, Fraser W (2008) Ecological Responses to climate change on the Antarctic Peninsula. Am Sci 96:302–310

Milot E, Weimerskirsch H, Bernatchez L (2008) The seabird paradox: dispersal, genetic structure and population dynamics in a highly mobile, but philopatric albatross species. Mol Ecol 17:1658–1673

Møller AP (2002) North Atlantic Oscillation (NAO) effects of climate on the relative importance of first and second clutches in a migratory passerine bird. J Anim Ecol 71:201–210

Møller AP, Rubolini D, Lehikoinen E (2008) Populations of migratory bird species that did not show a phenological response to climate change are declining. Proc Natl Acad Sci U S A 105:16195–16200

Monaghan P (1996) Relevance of the behaviour of seabirds to the conservation of marine environments. Oikos 77:227–237

Monteiro LR, Furness RW (1995) Seabirds as monitors of mercury in the marine environment. Water Air Soil Pollut 80:851–870

Montevecchi WA (1993) Birds as indicators of change in marine prey stocks. In: Furness RW, Greenwood JJD (eds) Birds as monitors of environmental change. Chapman and Hall, London, pp 217–266

Montevecchi WA, Myers RA (1996) Dietary changes of seabirds indicate shifts in pelagic food webs. Sarsia 80:313–322

Montevecchi WA, Myers RA (1997) Centurial and decadal oceanographic influences on changes in northern gannet populations and diets in the north-west Atlantic: implications for climate change. ICES J Mar Sci 54:608–614

Morin X, Thuiller W (2009) Comparing niche- and process-based models to reduce prediction uncertainty in species range shifts under climate change. Ecology 90:1301–1313

Nussey DH, Clutton-Brock T, Elston DA, Albon SD, Kruuk LEB (2005) Phenotypic plasticity in a maternal trait in red deer. J Anim Ecol 74:387–396

Parmesan C, Yohe G (2003) A globally coherent fingerprint of climate change impacts across natural systems. Nature 421:37–42

Parry ML et al (2007) Climate change 2007: impacts, adaptation and vulnerability. Summary for Policymakers (a report of Working Group II of the intergovernmental panel on climate change) and technical summary (a report accepted by Working Group II of the IPCC but not yet approved in detail: part of the Working Group II contribution to the fourth assessment report of the intergovernmental panel on climate change)

Péron C, Weimerskirch H, Bost CA (2012) Projected poleward shift of king penguins' (*Aptenodytes patagonicus*) foraging range at the Crozet Islands, southern Indian Ocean. Proc R Soc Lond B 279:2515–2523

Perrings C, Naeem S, Ahrestani FS, Bunker DE, Burkill P, Canziani G, Elmqvist T, Fuhrman JA, Jaksic FM, Kawabata Z, Kinzig A, Mace GM, Mooney H, Prieur-Richard AH, Tschirhart J, Weisser W (2011) Ecosystem services, targets, and indicators for the conservation and sustainable use of biodiversity. Front Ecol Environ 9:512–520

Piatt JP, Sydeman WJ, Wiese F (2007) Introduction: a modern role for seabirds as indicators. Mar Ecol Prog Ser 352:199–204

Pichegru L, Grémillet D, Crawford RJM, Ryan PG (2010) Marine no-take zone rapidly benefits endangered penguin. Biol Lett 6(4):498–501

Pierroti R, Annett CA (1990) Diet and reproductive output in seabirds. BioScience 40:568–574

Provencher JF, Gaston AJ, O'Hara PD, Gilchrist HG (2012) Seabird diet indicates changing Arctic marine communities in eastern Canada. Mar Ecol Prog Ser 454:171–182

Rayner MJ, Hauber ME, Steeves TE, Lawrence HA, Thompson DR, Sagar PM, Bury SJ, Landers TJ, Phillips RA, Ranjard L, Shaffer SA (2011) Contemporary and historical separation of transequatorial migration between genetically distinct seabird populations. Nat Commun 2:332

Ribic CA, Ainley DG, Ford RG, Fraser WR, Tynan CT, Woehler E (2011) Water masses, ocean fronts, and the structure of Antarctic seabird communities: putting the eastern Bellingshausen Sea in perspective, Deep-Sea Res 2. Top Stud Oceanogr 58:1695–1709

Roberge JM, Angelstam P (2004) Usefulness of the umbrella species concept as a conservation tool. Conserv Biol 18:76–85

Roeder AD, Marshall RK, Mitchelson AJ, Visagathilagar T, Ritchie PA, Love DR, Pakai TJ, McPartlan HC, Murray ND, Robinson NA (2001) Gene flow on the ice: genetic differentiation among Adélie penguin colonies around Antarctica. Mol Ecol 10:1645–1656

Ronconi RA, Lascelles BG, Langham GM, Reid JB, Oro D (2012) The role of seabirds in marine protected area identification, delineation, and monitoring: introduction and synthesis. Biol Conserv. doi:10.1016/j.biocon.2012.02.016

Root TL, Price JT, Hall KR, Schneider SH, Rosenzweig C, Pounds JA (2003) Fingerprints of global warming on wild animals and plants. Nature 421:57–60

Ropert-Coudert Y, Kato A, Chiaradia A (2009) Impact of small-scale environmental perturbations on local marine food resources: a case study of a predator, the little penguin. Proc R Soc Lond Ser B 276:4105–4109

Ross KG (2001) Molecular ecology of social behaviour: analyses of breeding systems and genetic structure. Mol Ecol 10:265–284

Sæther BE, Tufto J, Engen S, Jerstad K, Røstad OW, Skåtan JE (2000) Population dynamical consequences of climate change for a small temperate songbird. Science 287:854–856

Saraux C, Le Bohec C, Durant JM, Viblanc VA, Gauthier-Clerc M, Beaune D, Park YH, Yoccoz NG, Stenseth NC, Le Maho Y (2011) Reliability of flipper-banded penguins as indicators of climate change. Nature 469:203–206

Schwartz MK, Luikart G, Waples RS (2007) Genetic monitoring as a promising tool for conservation and management. Trends Ecol Evol 22:25–33

Serreze MC, Holland MM, Stroeve J (2007) Perspectives on the Arctic's shrinking sea ice cover. Science 315:1533–1536

Shultz MT, Piatt JF, Harding AMA, Kettle AB, Van Pelt TI (2009) Timing of breeding and reproductive performance in murres and kittiwakes reflect mismatched seasonal prey dynamics. Mar Ecol Prog Ser 393:247–258

Smith RC, Ainley D, Kaber K, Domack E, Emslie S, Fraser B, Kennett J, Leventer A, Mosley-Thompson E, Stammerjohn S, Vernet M (1999) Marine ecosystem sensitivity to historical climate change in the Antarctic Peninsula. BioScience 49:393–404

Spear LB, Ainley DG (1999) Migration routes of sooty shearwaters in the Pacific Ocean. Condor 101:205–218

Springer AM, Piatt JF, van Vliet G (1996) Sea birds as proxies of marine habitats and food webs in the western Aleutian arc. Fish Oceanogr 5(1):45–55

Stachowicz JJ, Bruno JF, Duffy JE (2007) Understanding the effects of marine biodiversity on communities and ecosystems. Annu Rev Ecol Evol Syst 38:739–766

Stearns SC (1992) The evolution of life histories. Oxford University Press, Oxford

Stenseth NC, Mysterud A, Ottersen G, Hurrell JW, Chan KS, Lima M (2002) Ecological effects of climate fluctuations. Science 297:1292–1296

Stenseth NC, Ottersen G, Hurell JW, Belgrano A (2004) Marine ecosystems and climate variation. The North Atlantic: a comparative perspective. Oxford University Press, Oxford

Taylor SA, Friesen VL (2012) Use of molecular genetics for understanding seabird evolution, ecology and conservation. Mar Ecol Prog Ser 451:285–304

Thackeray SJ, Sparks TH, Frederiksen M, Burthe S (2010) Trophic level asynchrony in rates of phenological change for marine, freshwater and terrestrial environments. Glob Change Biol 16:3304–3313

Thompson SA, Sydeman WJ, Santora JA, Morgan KH, Crawford W, Burrows MT (2012) Phenology of pelagic seabird abundance relative to marine climate change in the Alaska Gyre. Mar Ecol Prog Ser 454:159–170

Visser ME, Both C (2005) Shifts in phenology due to global climate change: the need for a yardstick. Proc R Soc Lond Ser B 272:2561–2569

Visser ME, Both C, Lambrechts MM (2004) Global climate change leads to mistimed avian reproduction. Adv Ecol Res 35:89–110

Voigt W, Perner J, Davis AJ, Eggers T, Schumacher J, Bährmann R, Fabian B, Heinrich W, Köhler G, Lichter D, Marstaller R, Sander FW (2003) Trophic levels are differentially sensitive to climate. Ecology 84:2444–2453

Walther GR, Post E, Convey P, Menzel A, Parmesan C, Beebee TJC, Fromentin JM, Hoegh-Guldberg O, Bairlein F (2002) Ecological responses to recent climate change. Nature 416:389–395

Weimerskirch H, Stahl JC, Jouventin P (1992) The breeding biology and population dynamics of King Penguin *Aptenodytes patagonicus* on the Crozet Islands. Ibis 134:107–117

Weimerskirch H, Louzao M, de Grissac S, Delord K (2012) Changes in wind pattern alter albatross distribution and life-history traits. Science 335:211–214

Wilson RP, Vandenabeele SP (2012) Technological innovation in archival tags used in seabird research. Mar Ecol Prog Ser 451:245–262

Wilson PR, Ainley DG, Nur N, Jacobs SS, Barton KJ, Ballard G, Comiso JC (2001) Adélie penguin population change in the pacific sector of Antarctica: relation to sea-ice extent and the Antarctic Circumpolar Current. Mar Ecol Prog Ser 213:301–309

Wynn RB, Knefelkamp B (2004) Seabird distribution and oceanic upwelling off northwest Africa. British Birds 97:323–335

Conclusions

This book adds to assembling the outstanding contribution of Marine Biology to IPY. The quality of research, the frequent international collaborations and the remarkable boost in pursuing multidisciplinarity are values that also characterise Volume 2 of "Adaptation and Evolution in Marine Environments—The Impacts of Global Change on Biodiversity". These values help in having the contribution of Marine Biology be considered as an invaluable component of the general scientific outcome of IPY 2007–2009.

The responses of cold-adapted polar organisms provide information to analyse the effect of changes in general, and foresee their impact at lower latitudes. In this general scenario, all research reported in the eleven chapters is under the umbrella of the international, multi- and cross-disciplinary programme "Evolution and Biodiversity in the Antarctic—The Response of Life to Change" (EBA). Launched by the Scientific Committee for Antarctic Research (SCAR) in 2004, EBA covers most of Antarctic biological research in the marine, terrestrial and freshwater realms. EBA's relevance to Global Change underscores the importance of this programme in environmental research during IPY. EBA addresses the impacts of change on Antarctic biodiversity, evolutionary adaptations and community dynamics, with the ambitious aim to extrapolate the implications and seek forecasts concerning the whole planet.

The programme envisages links with the Arctic. The latter is undergoing rapid climate change, with progressive and fast decrease of sea and land ice, only matched—for the time being—by what is happening in the Antarctic Peninsula. The similarities and, conversely, the differences that the northern and southern polar environments are revealing as consequences of current global warming suggest that, in the context of future polar research, initiatives that envisage bipolar activities and investigations will become more and more common, especially considering the key role of Antarctica in driving the climate of the whole planet, on one hand, and on the other the growing political and socio-economical importance of the Arctic. The official polar international institutions, namely

C. Verde and G. di Prisco (eds.), *Adaptation and Evolution in Marine Environments, Volume 2*, From Pole to Pole, DOI: 10.1007/978-3-642-27349-0,

SCAR and IASC (International Arctic Science Committee, www.iasc.info), are well aware of the importance of this trend, and have been customarily holding joint meetings since several years. An increasing number of national institutions is following this trend. The research projects reflect this factor, in that they increasingly include collaborative bipolar activities. At least five of the eleven chapters of this Volume underscore the importance of bipolar research and Arctic/Antarctic comparisons in studying adaptations to the impacts of current global climate change. This aspect is also a hallmark of IPY.

Climate change and its effects on biological systems, evolution and biodiversity in a changing environment is a major multi-disciplinary theme; as such, it has been the primary target of EBA's sponsored research. EBA's wide umbrella has fulfilled its outstanding coordination role for many years, providing fertile ground for organisation and multi-national efforts, but now needs to be replaced by more focussed projects. EBA is approaching 2013, the end of its planned lifespan. Since 2009, the EBA community has been planning the course EBA's philosophy will take in future polar science, Now is the time for achieving progress and proceed further. The Antarctic biology community is proposing two programmes on distinct but complementary aspects of polar biology, across marine, freshwater and terrestrial environments: "State of the Antarctic Ecosystem (AntEco)", and "Antarctic Thresholds—Ecosystem Resilience and Adaptation (AnT-ERA)". This is the legacy of EBA, and the key to keep understanding and protecting biodiversity. The cooperative and cross-disciplinary structure, in particular for evolutionary and biodiversity information, is a long-term legacy and must be retained. Antarctic research is very expensive: it must be excellent, relevant, multinational and well planned.

This development will inherit the message and values of the mother programme, and keep feeding inputs into IPY 2007–2009, just as EBA did, to conquer its deserved place in the history of international polar science.

Cinzia Verde
Guido di Prisco

Perspectives and Implications

The SCAR (www.scar.org) international, multidisciplinary biology programme "Evolution and Biodiversity in the Antarctic—The Response of Life to Change" (EBA) sponsors the chapters of this Volume 2 of "Adaptation and Evolution in Marine Environments—The Impacts of Global Change on Biodiversity" of the Series "From Pole to Pole: Polar Environmental Research during the International Polar Year 2007–2009". The EBA and IPY activities were conceived in parallel, and EBA was chosen by the IPY Science Plan as a Lead IPY Project. Assembling almost one hundred teams and covering most of Antarctic biological research in the marine, terrestrial and freshwater realms, the EBA community is now submitting proposals for more focussed developments.

EBA continues to contribute relevant science to the Antarctic Treaty System, highlighting the need to understand the impact of change on Antarctic ecosystems, relating the biology to the northern polar regions and liaising with physics, climatology, earth sciences and history. In pursuit of its mission to protect biodiversity, studies in both the field and the laboratory help to predict how organisms and communities will respond to current and future environmental change. Taking advantages of new technologies and especially of the development of molecular approaches, EBA is a continuing reminder of all of the best qualities of IPY.

The contributions, organised as three parts of this volume 2 (Part I: Biodiversity Evolution and Data Management; Part II: Evolution: A Molecular Perspective; Part III: Monitoring and Management.), are all in the framework of EBA. They are interconnected and complementary. Studies often merge into wider programmes, e.g. CAML, the part of CoML that deals with the Southern Ocean (SO). The importance of CAML has been referred to in 8 of the 11 chapters of Volume 1.

Further essential information on the accomplishments of CAML is summarised in the two chapters of Part I: Biodiversity Evolution and Data Management (history, organisation, targets, main expeditions, main results, workshops, coordination with SCAR-MarBIN, DNA barcoding, legacy). After five years,

C. Verde and G. di Prisco (eds.), *Adaptation and Evolution in Marine Environments, Volume 2*, From Pole to Pole, DOI: 10.1007/978-3-642-27349-0,

the legacy of CAML, funded by the Alfred P. Sloan Foundation, includes inventories, biodiversity databases, and extensive use of genomic techniques (Barcode of Life), that will allow us to answer important questions on marine genetic diversity and distribution of species and their links with areas north of the Polar Front. Thanks to a strong interaction with the sister project, Arctic Ocean Diversity (ArcOD), which addressed the census of Arctic marine biodiversity, comparisons can be drawn between differences in ecological structure and dynamics of of the Arctic Ocean and the SO. CAML revealed many new species. Sampling sites from which biodiversity data were obtained will be re-sampled in the future to track changes in marine biodiversity. Specimens to be used for research for many years to come will be collected. Data sets will be freely accessed (http://data.biodiversity.aq). The Southern Ocean Observing System (SOOS) will coordinate multi-disciplinary pan-Antarctic long-term monitoring which will help to assess the impacts of global change on these SO ecosystems.

IPY generated massive amounts of data, which need to be interrelated to understand complex problems such as environmental change and its impact on Antarctic biodiversity. The SCAR Marine Biodiversity Information Network (SCAR-MarBIN, www.scarmarbin.be), and the new Antarctic Biodiversity Information Facility (ANTABIF, www.biodiversity.aq) will be especially important in this with active data management to ensure that ANTABIF will maintain universal open access to biodiversity data.

The SO and the Antarctic continent pose extreme survival challenges for organisms, from bacteria to vertebrates. Considering the strong environmental constraints that polar organisms had to face to successfully cope with progressive cooling over millions of years, evolutionary adaptation has been, and will continue to be, a major theme of research in IPY. Climate change is calling for more and more work on the consequences that even a slight modification of the climate may entail for cold-adapted organisms, whose physiology has previously succeeded in adapting to allow species to escape extinction. It will be essential to increase our understanding of how polar organisms have adapted to cope with past challenges, to what extent adaptations may be upset by current changes, and—most important—whether it will be possible to minimise future threats of extinction, at our latitudes. Whilst terrestrial species are adapted to very variable conditions, marine species face more severe problems, because the large thermal capacity of water means that the most stable thermal environments are aquatic. Although warming will occur in the atmosphere to a much greater extent, the impacts of even a small increase in temperature in a marine environment are likely to affect organisms and their biodiversity much more severely. Therefore, studies on marine organisms seem likely to provide the most important insights into adaptation.

Part II (Evolution: A Molecular Perspective) comprises five chapters. Polar science is taking advantage of the development of molecular biology, that now has the potential to revolutionise evolutionary biology and ecology, and provides the tools to explore the function of individual genes and to use DNA barcode sequences to maximise taxonomic and geographic coverage. Changes in polar regions may lead to extinctions, and a reference baseline of sequences is under way, also thanks to CAML; it will be possible to focus on gaps, set priorities for the most important taxa and avoid duplicated collections, and make use of molecular phylogeny in drawing evolutionary trees. Thanks to "omic" technologies, new biochemical pathways, evolutionary adaptation and tolerance/ resistance to extreme conditions can now be investigated, gaining insights into how low temperatures may affect the physiology of vertebrates and invertebrates, in particular of microorganisms, thereby shedding light on microbial adaptations to cold.

Chapter 3 deals with evolution in the SO in a global context. In the SO, with the exception of studies on vertebrates and commercially valuable species, molecular ecology and phylogenetics are in their infancy. Yet they are transforming our understanding of connectivity within the ocean itself, and between the SO and other oceans, including the Arctic. At least half of the findings stem from DNA barcoding, in most cases with the help of CAML. Molecular ecology should focus on additional genes and markers (nuclear markers, including microsatellite markers). To date, microsatellites have been developed for only a few invertebrates: krill, some isopods, and three octopus species. Future molecular studies should also include haplotype networks and look for network patterns that can be predicted under different refugial scenarios at glacial maxima. Investigations should also look for congruent patterns between nuclear and mitochondrial markers and seek evidence in molecular signatures for recent population expansions. They may also be able to confirm that historical seaways acted as a conduit for gene flow.

Species bipolarity has raised the intriguing question whether co-specific Antarctic and Arctic populations evolved independently, since separation between the Arctic and Antarctic regions or genetic continuity is ensured by trans-tropical gene flow. Classical approaches (morphology, analysis of genetic variation) have limitations. Ciliates are ideal organisms to analyse the breeding structure of microbial populations and obtain data which satisfy the interbreeding criterion on which the Darwinian concept of species is founded. Living laboratory material can readily be available in unlimited amounts. Chapter 4 describes how ciliates govern gene exchanges through sexual conjugation. Breeding analyses of Antarctic, Fuegian and Arctic populations of a species show that they are genetically interconnected by gene flow and form a unique interbreeding species. These strains share the same gene pool, indicating that bipolar populations may maintain genetic continuity in spite of ecological discontinuity and ensure a pole-to-pole gene flow by the cold currents that cross the equatorial depths. Nuclear and mitochondrial ribosomal gene sequences and single nucleotide polymorphisms (genetic markers of the evolutionary history) need to be elucidated. Polymorphisms may reflect natural hybridisation between Arctic and Antarctic populations, and may also be able to interact *via* signalling cross-reactive pheromones (synthesised to promote mating and growth), whose structures secure wide-range dispersal in any environment.

Such molecular approaches will be invaluable in evolutionary and ecological studies. Gene flow, genetic drift, selection, and other factors, affect the evolution of biodiversity. The ACC ought to homogenise the structure of populations of the SO, and strong connectivity is indeed found for some species, with genotypes being shared across the full range. However, species-specific life-history traits influence the patterns of most taxa such that distinct populations can be identified.

The next two Chapters deal with the physiological and physico-chemical role of temperature, a fundamental driver in shaping biotic and abiotic factors.

Chapter 5 discusses the concept of oxygen and capacity limited thermal tolerance (OCLTT), looking at brachyuran and anomuran crabs. These live at the border of the polar oceans and have settled in "warmer" water bodies, excluded from permanent life at the coldest temperatures. A cause/effect understanding of thermal limitation and adaptation at various levels of biological organisation is crucial in elaboration of how climate has shaped the functional properties of the fauna. Crustaceans are a group where such cause/effect understanding is still in its infancy. They may display features resulting from excess oxygen availability rather than cold-induced oxygen limitation. Further studies ought to explore whether sub-polar species show a trend to uncouple oxygen shortage from capacity limitation at low threshold temperatures. This approach suggests that the patterns of oxygen and capacity limited thermal tolerance are linked with life-history consequences typically seen in permanent cold. Future research needs to address these interrelationships, and also to consider how climate challenges not only involve temperature changes but also the effect of additional stressors, e.g. ocean acidification. These aspects are largely unexplored. A relevant question is whether high CO_2 concentrations exert negative effects on physiological processes modified during cold adaptation. OCLTT reaches the ecosystem level, where biotic interactions shape ecosystem structure and functioning. Interacting species coexist where thermal niches overlap. Each species will be affected by changing conditions, with consequences on coexistence and competitiveness. Therefore, studies must also address the consequences of thermal stress for species interactions. Comparative research should look at the changes in species-specific sensitivities and performance of interacting species with the goal of identifying the mechanisms causing relevant shifts in interactions. Temperature is the prime candidate for building a matrix on which to understand effects of other stressors (hypoxia, CO_2, etc).

The rate of growth of a microorganism is related to the rate of the metabolic reactions, catalysed by enzymes that require proper folding and stability. Chapter 6 stresses that thermodynamic analysis is of primary importance, since activation entropy and energy make reactions possible. In thermophiles, high temperatures favour reaction rates, but also induce unfolding/misfolding. Proteins have slightly modified their structure to resist high temperatures, while folding is assisted by chaperones. In psychrophiles, the main problem is to secure metabolic fluxes. This problem has been solved through evolution of enzymes displaying lower activation energy and lower thermal dependence of the activity. Limitation of the tools permitting folding at low temperature is correlated with reduction of physico-chemical constrains that can counteract folding and also with the energetic care to limit protein synthesis. The high number of enzymes with a key role in all these aspects underscores the need for thermodynamic characterisation. Comparison between phylogenetically related bacteria in freezing and non-freezing habitats may help to understand whether extreme environments require adaptations at species level, or the action of a few genes is sufficient for defining the preference for a given environment.

Part III assembles five chapters. Monitoring and Management are an extremely important field of activities, because they yield tools that are essential to advance knowledge in virtually all other fields of biological science.

In Chapter 7 we learn about the benefit of satellite technology in investigating seals thriving in both polar environments. Many species are pelagic except for breeding and moulting, unlike those staying in close proximity to the coast or the ice edge, where they can be observed throughout the year. Satellite-linked dive recorders permit us to learn about the whereabouts and behaviour of ecologically and economically important pelagic species, as well as about seasonal distribution and diet composition. Satellite-linked dive recorders can elucidate location and diving behaviour throughout the year in the Arctic and Antarctic, determining haul-out patterns, useful in converting aerial sightings into population numbers. Besides implications in ecology and population dynamics, such investigations have socio-economical importance especially in the Arctic, where some species impact on the economy of coastal communities. The North Atlantic and the Barents Sea-White Sea hold some important fish stocks, so the ecological and economical implications of millions of seals will benefit from further investigation.

Chapters 8 and 9 complement each other and deal with monitoring and management issues related to environmental processes, biodiversity, and environmental assessment in Admiralty Bay, King George Island (KGI). The sensitivity of Admiralty Bay is high, due to its location close to the Antarctic Peninsula, and the high risk of human impacts besides climate change. Close cooperation between operators and programmes will add value to KGI studies by providing a regional, continental and global context for interpretations and comparisons. Standardised techniques and measurements, common sets of variables, and open access to data will be essential for cross-comparison of data sets from diverse locations and scientific teams, not only at KGI but also more broadly in Antarctica. Because of the lifescience, geoscience, and physical science studies conducted at KGI, multidisciplinary integration is possible in a way that cannot be achieved at most other locations. These outcomes will benefit all national programmes in KGI. Mutually beneficial cooperation and partnerships can be realised if duplication of efforts is minimised, infrastructure and logistics are shared to reduce costs and impacts, and standard techniques and sets of variables are agreed between national scientists collaborating with each other and with SCAR. The realisation of these opportunities can only happen if there is a desire for partnerships and if bi- and multilateral agreements to advance common goals can be negotiated.

Chapter 10 discusses anthropogenic impacts on sub-Antarctic and Antarctic islands (including KGI) and the adjacent environments. Marine pollution, diversity changes caused by introduction of non-indigenous species, and global environmental changes are the main consequences of human activities. Some isolated islands remain pristine, but are threatened by human visitation. Management of marine ecosystems requires the ability to distinguish the effects of human impacts from those of climate change. Marine reserves (target areas and ecosystem reserves) are needed to monitor the impacts of global changes. Multidisciplinary studies, identification of impact origins, and long-term monitoring are needed to assess the effects of human activities on environments and biodiversity, as management tools, especially when taking biodiversity hotspots into account.

Chapter 11 conjugates behaviour, ecology and technology, in the framework of climate change. There is a need for a broad expansion (in terms of key species and geographic coverage) of Polar Life Observatories (PLOs), and also for international agreement on how they should be managed (standardisation of data collection, benchmarking, metadatabase organisation, access rights) to make them effective and promote international collaborative work. In fact, studies taking an ecosystem approach to marine dynamics are few, data time series are often short and quantitative knowledge of the dynamics of interactions between predators, prey and environment is limited. Seabirds (albatrosses, penguins, puffins) sensitive to changes are indicated as monitoring sentinels, excellent to assess ecosystem health, since we know which stocks of marine organisms they feed on. Some of the best data series existing are on birds, making them ideal models for PLOs. How they are affected by environmental changes could be the necessary tipping point to convince policy-makers and governments to respond to a major global threat. PLOs meet priorities regarding environmental protection defined by the Convention on Biological Diversity (CBD), the Intergovernmental Panel on Climate Change (IPCC), and the International Council for Science (ICSU). Bio-logging devices act as monitors of prey stocks (mesopelagic fish, squid, krill, etc). Single penguins cannot be visually localised inside the colony, thus Radio Frequency IDentification (RFID) antennae are carried by remotely controlled robots that can circulate among penguins.

Polar regions experience greater rates of climate change than elsewhere. Ecosystems are adapted to extreme environments, and may become vulnerable. Research will continue to increase our knowledge of the Antarctic marine fauna of the continental shelf, the slopes and the deep sea. One of the most urgent and challenging tasks for the next decade will be to incorporate thinking from the physiological/biochemical viewpoint at the molecular level into evolutionary biology.

This view appears firmly embedded in the minds of the authors of the chapters of this Volume 2 of "Adaptation and Evolution in Marine Environments—The Impacts of Global Change on Biodiversity". What appears very clearly is the maintenance of the momentum of IPY 2007–2009 through the scientific relevance of the research performed by the authors, and the planning of future ventures, aimed at identifying impacts and threats of global climatic processes.

The concept of a "sustainable world" is essentially linked to the concept of "biodiversity". Biodiversity makes Earth what Earth is today: it makes life possible, and human life depends on the continuing variety of other life forms. It is one of the most crucial values of life on Earth, which calls for the best of our efforts to preserve it both for its own sake and as a primary support for our own future survival.

Cinzia Verde
Guido di Prisco

MIX
Papier aus verantwortungsvollen Quellen
Paper from responsible sources
FSC® C105338

If you have any concerns about our products,
you can contact us on
ProductSafety@springernature.com

In case Publisher is established outside the EU,
the EU authorized representative is:
Springer Nature Customer Service Center GmbH
Europaplatz 3, 69115 Heidelberg, Germany

Printed by Libri Plureos GmbH
in Hamburg, Germany